Interior Point Techniques in Optimization

Applied Optimization

Volume 6

Series Editors:

Panos M. Pardalos
University of Florida, U.S.A.

Donald Hearn
University of Florida, U.S.A.

The titles published in this series are listed at the end of this volume.

Interior Point Techniques in Optimization

Complementarity, Sensitivity and Algorithms

by

Benjamin Jansen
Centre for Quantitative Methods CQM B.V.,
Eindhoven, The Netherlands

SPRINGER-SCIENCE+BUSINESS MEDIA, B.V.

A C.I.P. Catalogue record for this book is available from the Library of Congress.

ISBN 978-1-4419-4772-7 ISBN 978-1-4757-5561-9 (eBook)
DOI 10.1007/978-1-4757-5561-9

Printed on acid-free paper

CONTENTS

LIST OF FIGURES

Chapter 8

Chapter 9

Chapter 10

Chapter 11

Appendix A

LIST OF TABLES

PREFACE

The area of *interior point methods* has been one of the liveliest in mathematical programming in the last decade. In 1984 the mathematician Karmarkar introduced a new algorithm for linear programming which he showed to run in polynomial time, and claimed to outperform the commonly used simplex method on large practical problems. This caused a tremendous amount of research on interior point methods, involving people from all over the world. Important outcomes of the work done until now are (among others):

- very efficient techniques for linear programming have been developed that really outperform the simplex method on large problems;
- the new interior point methods are closely related to the barrier methods dating from the 1960s;
- stable and fast interior point algorithms have been developed for many nonlinear programming problems;
- a link between nonlinear programming and combinatorial optimization has been given practical relevance and importance via the development of efficient methods for *semidefinite programming.*

The literature on interior point methods to date mainly consists of papers in international journals and conference proceedings. Some books on mathematical programming contain a chapter on these methods (unfortunately often describing some technical aspects, not the underlying interpretation), the book by Fiacco and McCormick from 1968 is still a relevant reference, there are a few books dedicated to the area and recently a survey book edited by T. Terlaky appeared [48].

In this book we will deal with several aspects of interior point methods in a broad spectrum of mathematical programming: from linear programming to nonlinear programming to combinatorial optimization. The words in the subtitle of this book, *complementarity, sensitivity* and *algorithms*, will be the leading threads running through the book. Although there is a definite structure in the ordering of the chapters, they can be read independently and used in the classroom. The book contains an extensive up-to-date bibliography of the field.

In Chapter 1 we give the historical background of the field and describe the contents of this book in some detail. In Chapters 2 and 3 the theory and sensitivity analysis in linear programming will be developed from an interior point of view. Using so-called strictly complementary solutions we show how to circumvent traps to be encountered when applying standard sensitivity analysis with the simplex method. In Chapter 4 this is extended to sensitivity analysis for convex quadratic programming. In Chapters 5 and 6 we describe a class of interior point methods that is shown to work in polynomial time for linear as well as certain (nonconvex) nonlinear programming problems. Chapter 7 gives computational results with these methods for some optimization problems arising in statistics. Chapters 8 and 9 offer a framework for the interpretation and analysis of many interior point methods for linear and nonlinear programming, thereby summarizing hundreds of pages in the literature into a relevant few. In Chapter 10 we discuss two applications of semidefinite programming and their relationship with interior point methods, while Chapter 11 concerns the applicability of interior point methods within decomposition techniques.

The research performed for this book would not have been possible without the help of many people, whom I kindly acknowledge. I thank Kees Roos and Tamás Terlaky, with whom I had the pleasure to work for several years. As excellent co-authors and fellow-workers I would like to acknowledge Jean-Philippe Vial, Yinyu Ye, Akiko Yoshise, and Olaf Flippo, Koos Ballintijn of SHELL (KSLA, Amsterdam), Jurjen de Jong, Arjan Berkelaar, and Etienne de Klerk. I'm grateful for the hospitality of the members of the department SSOR at Delft University of Technology where I stayed for four years, especially mentioning my office-mates Dick den Hertog, Jan van Noortwijk and Bernd Kraan. I acknowledge the Dutch Organization for Scientific Research (NWO) for their financial support during that period. John Martindale and Arlene Apone of Kluwer Publishers greatly helped in getting this book into print.

Last, but not least, I am especially grateful to my parents and Inez, who luckily are not familiar with the ins and outs of this book but are familiar with many other interesting things in life.

Eindhoven, August 1996
Benjamin Jansen

1

INTRODUCTION

1.1 HISTORICAL BACKGROUND

Major breakthroughs in mathematical programming are often related to linear programming (LP). First of all, the introduction of the simplex algorithm by Dantzig in 1947 [37] had both a theoretical and practical impact in the field, maybe even initiated it. As direct consequences we mention the development of LP and its extensions [39], network problems and algorithms [56], nonlinear programming (NLP) [146], decomposition schemes [41], complementarity theory [33, 148], stochastic programming [38], cutting plane methods for large integer programs [40], etc. The reader is referred to [149] for an overview of the early history of mathematical programming. When in 1979 Khacijan [130] showed that the ellipsoid algorithm applied to the LP problem runs in polynomial time, this was not only of importance for the complexity theory of LP, it also had important implications for the complexity theory of many combinatorial optimization problems, as shown by Grötschel et al. [92]. Unfortunately, the good complexity didn't lead to good computational efficiency in practice, causing the method to become merely a theoretical tool. However, no single development since the introduction of the simplex method has influenced the field of mathematical programming to such an extent as did the 1984 paper by Karmarkar [126] which had (and still has) a great impact on both the theory *and* practice of mathematical programming. Describing a new polynomial time algorithm (called *projective scaling algorithm*) for LP with better complexity than the ellipsoid method and claiming it to be extremely efficient in practice, Karmarkar triggered a tremendous amount of research on what is now commonly called *interior point methods.* Hundreds of researchers all over the world went into the subject, over 2000 papers were written (see Kranich [144] for a bibliography). For an overview of the developments in the theory of interior point methods for LP the reader is referred to surveys by Gonzaga [87] and Den Hertog [100]; the computational state–of–the–art is described in Lustig et al. [157].

The massive interest in interior point methods is even more remarkable if one takes into account that interior point techniques were extensively investigated in the 1960s (see Fiacco and McCormick [53]) and beginning 1970s as part of *sequential unconstrained minimization techniques.* One of the important techniques is the logarithmic barrier method, introduced by Frisch [62] in 1955. Also, the *affine scaling algorithm* proposed by Barnes [16] and Vanderbei et al. [238] as a simplified version of Karmarkar's method appeared to be just a rediscovery of a method developed by Dikin [42] in 1967.

Several reasons can be given why interior point methods were out of sight since the early 1970s, but regained so much interest in the mathematical programming society after Karmarkar's work. The first is of theoretical nature. Interior point techniques were originally developed to solve NLP problems with inequality constraints. For LP the simplex method performed reasonably well, and there was no incentive to investigate the theoretical properties of the interior methods when applied to LP, as theoretical complexity of the algorithm was not regarded to be an issue in the 1960s. In fact it was only around 1970 that complexity theory was developed, mainly in the field of combinatorial optimization (see Karp [127] and Garey and Johnson [67]), and for convex optimization by Judin and Nemirovskii [124]. It was shown by Klee and Minty [131] that certain variants of the simplex method need, in the worst case, an exponential number of arithmetic operations. Since then, the search for a polynomial method being efficient in practice was alive, without considering the possibility that existing methods, when sufficiently adjusted, could satisfy these requirements. Shortly after the publication of Karmarkar's paper Gill et al. [71] showed that Karmarkar's projective algorithm was closely related to the logarithmic barrier method. Following this connection theoretical work on interior point methods soon led to the introduction of the analytic center by Sonnevend [218] and analysis of the *central path* in a primal–dual setting by Megiddo [168] which are the central themes in both theoretical work as well as in practical implementations of interior point techniques. In 1987 Roos and Vial [210] derived a very elegant and simple complexity proof of the basic logarithmic barrier method, showing a new property of an essentially old method. Renegar [206] derived the complexity of a method using analytic centers which can be traced back to Huard [108]. Anstreicher [8] analyzed SUMT [189], an old implementation of an interior point method and showed it to be polynomial.

A second reason for the revival and popularity of interior point methods comes from the computational side. Hardware and software (particularly for sparse linear algebra) have been improved so much in the last decade that the computationally expensive task in any interior point method (viz., solving a sparse linear system) can be performed efficiently and with great accuracy. Particularly, new preprocessing techniques and research on sparse Cholesky factorization with various ordering heuristics have contributed to the success of interior point methods, see e.g., Lustig et al. [157]. Recently, solving sparse indefinite systems has become popular, see e.g., Fourer and Mehrotra [58] and Gondzio and Terlaky [83]. The use of new theoretical insights on the intrinsic nature of interior point techniques when applied to the LP problem has enhanced the implementation of interior point methods in such a way that very efficient codes now exist and are available at both commercial and academic level; we mention CPLEX [35], OSL [196], LOQO [237], IPMOS [247], LIPSOL [255]. To be honest we have to mention that the improvements in the simplex codes have been equally flabbergasting since the early 1980s. Bixby claims[1] an improvement in computation time with a factor 1000000, where a factor 1000 is due to more sophisticated methods and the other factor 1000 to evolving computer hardware.

A third reason that interior point techniques were out of sight for fifteen years can be found in the difference between applying the logarithmic barrier method to an LP problem as compared to an NLP problem. First, as was shown by Lootsma [152] and Murray [188] the Hessian of the logarithmic barrier function (with which a system needs to be solved in each iteration) becomes increasingly ill–conditioned when the iterates approach an optimal solution. This was experienced in SUMT [189]. However, in the linear case this behavior is less apparent and typically doesn't influence the efficiency and effectiveness of interior point methods in practice (see e.g., Stewart [221], Wright [245] and Vavasis [239]). For certain NLP problems Wright [245] extends the results in [152] and [188], and more importantly, suggests a way to get around the indicated difficulties in practice. Secondly, the LP environment proposed a natural opportunity to develop *primal–dual* methods (i.e., methods generating primal and dual solutions in *each* iteration), initially by Monteiro and Adler [180] and Kojima et al. [139]. Comparing early computational results with pure primal or dual methods and primal–dual implementations (compare e.g., [164] and [156]) shows a big decrease in computational effort with the use of the latter. Using infeasible iterates as proposed by Lustig [155] among others, improved the codes even more, as was the case with Mehrotra's proposal [170] to use a

[1] Talk at "Workshop Optimization in Production and Transportation", November 10, 1994, Scheveningen, The Netherlands.

predictor–corrector scheme. Incidentally, after the success of infeasible primal–dual predictor–corrector methods for LP it has, with some success, been tried to transfer the use of primal–dual techniques to NLP, see e.g., Vial [240] and Yamashita [249].

Apart from its impact on LP, Karmarkar's algorithm is also ultimately responsible for several other trends in optimization. We mention the revival (c.q., rehabilitation) of decomposition schemes (see e.g., Goffin and Vial [74, 77], Bahn et al. [13], Den Hertog et al. [102]), the revival of Newton's method in NLP combined with a beautiful analysis of certain interior point methods for NLP by Nesterov and Nemirovskii [193], the use of semidefinite programming (SDP), e.g., in control theory, linear algebra and combinatorial optimization (see Boyd et al. [28, 236], Alizadeh [5]), the development of efficient practical algorithms for NLP problems (see Yamashita [249], Vial [240], Shanno [215], Andersen and Ye [7], Breitfeld and Shanno [30] and Ben–Tal and Roth [19]) and reconsidering sensitivity analysis (Adler and Monteiro [2], Jansen et al. [111], Greenberg [91]). Some of these developments were, to a certain extent, pointed to by Karmarkar (see [126, pp. 394-396]), others are still surprising.

1.2 SCOPE OF THE BOOK

In this book interior point methodology is used in various areas of mathematical programming, viz., linear, quadratic, nonlinear, combinatorial and semidefinite optimization. It concerns results on theoretical aspects as well as on algorithms, and unifies existing literature in interior point methods. The leading thread running through this book is *complementarity* which we view as a basic concept in mathematical programming and in interior point methods in specific. Fundamentally, complementarity is a certificate for optimality. For nonoptimal solutions the (error in) complementarity is a measure for the distance of the solution to optimality; for optimal solutions the complementarity is zero. Among all optimal solutions the *strictly complementary solutions* are of special interest. While much of the literature on interior point techniques (specifically those for NLP) concentrates on the central path and considers methods in (small) neighborhoods of the central path, we will not make such a restriction.

In this book we will first use (strict) complementarity to derive results in the theory (Chapter 2) for LP. This concerns well–known results for which we provide alternative proofs, as well as some new results. Here we also introduce the *v–space* of the LP problem which gives a characterization of nonoptimal solutions. In Chapters 3 respectively 4 we use strictly respectively *maximal complementary solutions* in the sensitivity analysis for linear and convex quadratic programming (CQP). The v–space is used in the analysis of a class of primal–dual algorithms for (non)linear complementarity problems (Chapters 5-7) and in developing a unifying, called *target–following*, framework for analyzing interior point methods (Chapters 8 and 9). Finally, we use complementarity in two other subjects: in Chapter 10 we analyze two applications of semidefinite programming and we consider interior point based decomposition methods (Chapter 11). Each chapter is concluded with a section containing a summary, conclusions and directions for further research. We will now explain the contents of this book in some more detail.

1.2.1 Theory and sensitivity (Chapters 2-4)

In all classical textbooks on LP, developing the dual problem, weak and strong duality and the complementary slackness theorem are contained in one of the first chapters. While for optimal solutions the complementarity is zero, this is not true for feasible but nonoptimal solutions. A less well–known property of the LP problem is the existence of a strictly complementary solution first shown by Goldman and Tucker [80] in 1956. A strictly complementary solution is an optimal solution with a useful special property. While historically these theoretical results are taught using *a basis* (which is intimately related to the simplex method) a natural question is whether they can also be obtained using ideas from interior point techniques. For the theory of LP this was first done by Güler et al. [96]. In Chapter 2 of this book we derive these results much more easily using a skew–symmetric reformulation of the LP problem. We give proofs for duality and the existence of a strictly complementary solution in the general LP problem, using only analytical arguments. The idea of using a skew–symmetric reformulation is due to Tucker [234] and Dantzig [39]. Recently, Ye et al. [254] put new life into the skew–symmetric formulation by showing that an interior point method applied to it can easily handle infeasible starting points and provide a certificate for infeasibility. We also introduce the so–called v–*space*, or *the space of complementarity products* of the LP problem, which will play an important role in the chapters to follow. The v–space gives a description of the set of strictly feasible solutions, since any point in the v–space can be shown to be in one–to–one correspondence with such a strictly feasible solution (Kojima et al. [136]). The v–space provides us a tool for developing new interior point algorithms as well as for describing and analyzing many interior point methods in a uniform and coherent way.

For a long time it was generally believed that *sensitivity analysis* based on interior point solutions were impossible, since in principle no optimal basis is obtained; instead, interior point methods compute a strictly complementary solution (see Güler and Ye [97]). Megiddo [169] developed a strongly polynomial algorithm to compute an optimal basis from a strictly complementary solution, which was implemented by Bixby and Saltzmann [24], among others. Normally (in textbooks and in software) sensitivity analysis is performed using one optimal solution; however, this may give ambiguous results in case of degeneracy, see e.g., Gal [65]. In Chapter 3 we propose a unifying analysis of sensitivity based on *optimal sets*, stressing the important connection with the *optimal value function* of a perturbed LP problem. The importance of a strictly complementary solution here is that it provides the *optimal partition* of the problem which can be used to describe the optimal set. Within this framework we compare three ways of describing optimal sets: using bases, using the optimal partition and using the optimal value. Since the optimal partition is provided by interior point solutions we then derive a way of performing sensitivity analysis in LP on the basis of interior point methods. These three approaches have been applied to a practical LP model used at and provided by a major oil company.

In Chapter 4 we investigate to what extent the approach to sensitivity analysis using optimal partitions can be generalized to CQP. A major difference is that in CQP no strictly complementary solution need to exist. Instead, we use *maximal complementary solutions* (see Güler and Ye [97]) which define the *tripartition* of the problem. We characterize the optimal value function using tripartitions. An application we consider is Markowitz's mean–variance model for portfolio analysis [163].

1.2.2 Primal–dual affine scaling algorithms (Chapters 5-7)

In Chapter 5 we introduce and analyze a class of algorithms, called *primal–dual Dikin-affine scaling algorithms.* The derivation and the analysis are almost exclusively done within the v–space introduced in Chapter 2. Dikin's original primal affine scaling algorithm [42] in each iteration only generates a feasible solution of the primal LP problem, and it is generally believed not to have a polynomial complexity bound. In 1987 Monteiro et al. [182] introduced an algorithm that could be interpreted as a primal–dual variant of the affine scaling algorithm, and proved its polynomial complexity. Specifically, their algorithm requires at most $\mathcal{O}(n(\ln(1/\epsilon))^2)$ iterations, where n is the number of variables and ϵ the required accuracy in the duality gap. We show that there is a more natural generalization of the affine scaling idea to the primal–dual setting, which leads to an algorithm having better complexity than the one in [182]: it requires at most $\mathcal{O}(n \ln(1/\epsilon))$ iterations, with the same amount of work per iteration as in [182]. The search–direction in the algorithm is derived by minimizing the complementarity over an ellipsoid in the primal–dual feasible space. We show that in the v–space the resulting direction is exactly Dikin's direction.

The algorithm being originally derived for LP in Jansen et al. [113] can be transferred to *linear and nonlinear complementarity problems.* The class of linear complementarity problems (LCPs) contains LP and CQP as special cases; the class of (monotone) nonlinear complementarity problems ((M)NCPs) contains for instance convex programming problems. It is an analysis for the LCP that we present in this book. Due to second order effects in the computation of the search–direction we investigate whether the use of *corrector steps* can improve the complexity of the algorithm. Correctors have shown great potential in implementations of interior point methods for LP, see e.g., Mehrotra [170]. We prove that correctors improve the theoretical complexity of the algorithm to (asymptotically) the best known bound for LP.

The scaling used in the algorithm can be modified, which leads to a family of algorithms including both the classical primal–dual affine scaling algorithm of [182] and the primal–dual Dikin–affine algorithm as special cases. In Chapter 6 we analyze this family for NCPs. The analysis is more complicated than in the linear case since it requires a smoothness condition on the nonlinear mapping involved. We introduce such a condition and relate it to conditions used in the literature, namely the *scaled Lipschitz condition* [258], the *relative Lipschitz condition* [118] and the *self–concordance condition* [193]. An advantage of our condition over the others is that it can also be applied to nonmonotone mappings. Moreover, we show that our condition is suitable for analyzing primal–dual algorithms working in a large neighborhood, while the others are typically used for small neighborhood methods.

To investigate the computational behavior of the family of algorithms we apply it to several NLP problems arising in statistics (Chapter 7). Specifically, we consider nonlinear statistical regression problems, entropy optimization problems and maximum likelihood estimation. From our experiments we may conclude that the primal–dual Dikin–affine algorithm performs best within our family, is stable and requires little tuning. So the primal–dual Dikin–affine scaling method not only theoretically outperforms the algorithm of Monteiro et al. [182], it also does so in practice.

1.2.3 Target–following framework (Chapters 8-9)

A major application of the v–space or space of complementarity products is a unifying framework for the analysis of primal–dual interior point methods for LP and dual methods for nonlinear problems, which we develop in Chapters 8 and 9; we call it the *target-following* approach. Here we make use of the one–to–one correspondence between points in the v–space and strictly feasible primal–dual solutions. For LP we first prove general results on the behavior of the complementarity, feasibility and distance to the central path after a step, which we then specialize to different methods. The methods and their analysis are described in the v–space, which appears to be very convenient.

Our approach leads to simple and uniform complexity proofs for various methods that were previously analyzed in several separate papers in the literature. Furthermore, the analysis in the v–space suggests some new methods, for which an evenly simple analysis is provided. These methods use the Dikin–affine step derived in Chapter 5 and have the important property of combining approaching optimality as well as the central path simultaneously. So we analyze central path–following methods (e.g., [139, 180]), weighted path–following methods [45, 104], variants of the primal–dual Dikin–affine scaling algorithm [113], a variant of the cone–affine scaling method [224], Freund's shifted barrier method [59], and computing analytic or weighted centers [10, 104, 174, 176]. We first consider *short–step algorithms* which have the best known complexity bound for LP. We extend the results to the more practical *long–step methods* which unfortunately have a worse theoretical complexity (cf., Den Hertog [100]). One of the outcomes of this work is a negative influence upon the complexity when a method does not closely follow the central path.

In Chapter 9 we derive a similar target–following methodology for NLP problems as well as for variational inequalities, thereby analyzing several methods for which the analysis has not yet been transferred from LP to NLP. Here we use the self–concordance condition introduced by Nesterov and Nemirovskii [193]. A major conceptual difference with their approach is that in our applications the self–concordance parameters involved change from one iteration to another, which requires some new machinery to be developed. In the nonlinear case we restrict ourselves to the usual logarithmic barriers (cf., Den Hertog [100]) and show that the target–following concept is closely related to the use of weighted barrier functions.

In the study of (monotone) variational inequalities we extend some of our results to more general spaces, not restricting to logarithmic barriers. Here we take a new approach by considering a *self-concordant barrier operator* which can be viewed as a mapping having similar properties as the gradient of a barrier function. This has the advantage that we may analyze target-following methods for problems defined on cones other than the nonnegative orthant.

1.2.4 Semidefinite programming and decomposition (Chapters 10-11)

For some time, it has been attempted to use interior point methods in combinatorial optimization. Mitchell and Todd [173] and Borchers and Mitchell [26] implemented a branch-and-cut respectively branch-and-bound method using interior point methods for solving the LP subproblems. However, the results are often comparable or worse than using simplex based solvers. One of the reasons for this is that a good warm-start strategy is still missing in interior point methods, see e.g., Andersen and Ye [6]. The application of interior point methods in combinatorial optimization, however, comes from a different side, namely from nonlinear relaxations to combinatorial problems. Such a relaxation scheme was developed by Lovász and Schrijver [154], see also Lovász [153] and Grötschel et al. [92]. Often, these approximations appear to be problems defined over the cone of positive semidefinite matrices, so-called *semidefinite programming* (SDP) problems. Nesterov and Nemirovskii [193] analyzed an interior point method for SDP, being theoretically efficient; practical primal-dual versions were developed by Boyd and Vandenberghe [236] among others. It is to be attributed to Alizadeh [5] to have brought the possible applications of SDP and interior point methods in combinatorial optimization to the foreground. Goemans and Williamson [72] showed that a solution of the semidefinite relaxation of certain combinatorial problems (as MAXCUT and MAXSAT) can be rounded to a provably good solution, with a better worst-case bound than for previous algorithms. This initiated an extensive new field of research, see e.g., Laurent and Poljak [147], Poljak et al. [200], Helmberg et al. [99].

In Chapter 10 we consider a relaxation of the problem of minimizing a general quadratic form over ellipsoids, which can be used in approximating quadratic $0-1$-problems. We show that our relaxation is essentially equivalent to other relaxations in Shor [217] and Boyd and Vandenberghe [236]. The main difference is that our relaxation is nonlinear, while the others are linear in a higher dimension. Other applications of SDP can be found in, e.g., control and system theory and linear algebra. In the latter area we analyze a method for computing the smallest eigenvalue of a symmetric matrix. Although the method itself is closely related to well-known Newton–Raphson type methods our interior point methodology enables us to derive strong theoretical results as polynomiality and quadratic convergence, without making assumptions on the multiplicity of the smallest eigenvalue.

Finally, in Chapter 11 we present a result on the use of interior point methods in decomposition schemes. Dantzig–Wolfe [41] and Benders decomposition [22] were supposed to be slow for solving large linear problems with a specific structure. However, as noted by Magnanti et al. [158, 160] for Benders decomposition an enhancement can drastically reduce computation time. Paying the price of solving one extra LP problem per iteration a *Pareto-optimal cut* can be computed, being in a certain sense the best cut that can locally be obtained. We show that modern *interior cutting plane methods* (e.g., Goffin and Vial [77], Bahn et al. [13]) generate such a Pareto–optimal cut for free. We also give a condition guaranteeing a Pareto–optimal cut to be obtained for free in the classical setting, where this condition appears to be weakest if a strictly complementary solution to the subproblem is computed.

1.3 NOTATIONAL PRELIMINARIES AND ABBREVIATIONS

To simplify formulas we allow ourselves some notational abuse with regard to manipulation of vectors. Specifically, if $x, s \in \mathbb{R}^n$ then xs, x/s, $\sqrt{x}$ and x^α for $\alpha \in \mathbb{R}$ denote the vectors obtained from componentwise operations. For instance,

$$w = xs \quad \text{means} \quad w \in \mathbb{R}^n,\ w_i = x_i s_i.$$

Alternatively, in convenient situations we adopt the approach taken in earlier literature on interior point methods: if $x \in \mathbb{R}^n$, then X is the diagonal matrix with the elements of x on its diagonal; then $w = Xs$ has the same meaning as $w = xs$.

With respect to norms, $\langle \cdot, \cdot \rangle$ represents a general inner product. Given $x, y \in \mathbb{R}^n$ and a positive semidefinite matrix H we define

$$\|x\|_H := \sqrt{x^T H x}, \quad \text{and} \quad H[x, y] := x^T H y.$$

For a trilinear form H in $\mathbb{R}^{n\times n\times n}$ and $x, y, z \in \mathbb{R}^n$ we define

$$H[x,y,z] := \sum_{i,j,k=1}^{n} x_i y_j z_k H_{jk}^{(i)},$$

where $H^{(i)}$ are $n \times n$ matrices. Table 1.1 summarizes the notation.

For $x, s \in \mathbb{R}^n$, $\alpha \in \mathbb{R}$:		For $X \in \mathbb{R}^{n\times n}$:	
X	$X = \mathrm{Diag}\,(x_1, \ldots, x_n)$	$X \succeq 0$	X positive semidefinite
$w = xs$	$w_i = x_i s_i, \quad i = 1, \ldots, n$	$X \succ 0$	X positive definite
$w = Xs$	$w_i = x_i s_i, \quad i = 1, \ldots, n$	$\mathrm{tr}(X)$	the trace of X
$u = x^\alpha$	$u_i = x_i^\alpha, \quad i = 1, \ldots, n$	$\lVert X \rVert$	Frobenius norm, $\sqrt{\sum_{i,j} X_{ij}^2}$
$v = \sqrt{xs}$	$v_i = \sqrt{x_i s_i}, \quad i = 1, \ldots, n$	$\lVert X \rVert_\infty$	$\max_{i,j} \lvert X_{ij} \rvert$
$d^2 = x/s$	$d_i^2 = x_i/s_i, \quad i = 1, \ldots, n$		
$x^{(\alpha)}$	x at parameter value α		
$\sigma(x)$	$\{\, i \,:\, x_i > 0\}$		
$\min(x)$	$\min_{1\le i\le n} x_i$		
$\max(x)$	$\max_{1\le i\le n} x_i$		
$\omega(x)$	$\min(x)/\max(x)$		
$x \geq 0$	$x_i \geq 0, \quad i = 1, \ldots, n$		
$x > 0$	$x_i > 0, \quad i = 1, \ldots, n$		
e	all–one vector		
$e^{(i)}$	ith unit vector		
$\mathbb{R}_+^n$	$\{\, x \in \mathbb{R}^n \,:\, x \geq 0 \,\}$		
$\mathbb{R}_{++}^n$	$\{\, x \in \mathbb{R}^n \,:\, x > 0 \,\}$		
$\mathcal{S}^n$	$\{\, X \in \mathbb{R}^{n\times n} \,:\, X = X^T, X \succeq 0 \,\}$		

Table 1.1 Notation used in the book.

We use the following abbreviations:

LP	linear programming
CQP	convex quadratic programming
NLP	nonlinear programming
LCP	linear complementarity problem
(M)NCP	(monotone) nonlinear complementarity problem
SDP	semidefinite programming

2

THE THEORY OF LINEAR PROGRAMMING

2.1 COMPLEMENTARITY AND V–SPACE

In textbooks and papers that consider the theory of linear programming (LP) various techniques are used to prove strong duality and the existence of a strictly complementary solution (Goldman–Tucker's Theorem [80]). Among others, Balinski and Tucker [14] and Dantzig [39] basically use the simplex method , Farkas Lemma is used by Schrijver [214] and Stocr and Witzgall [222], mathematical induction by Goldman and Tucker [80] and Tucker [234], while Von Neumann and Morgenstern [195] and Rockafellar [207] apply a separation theorem for convex sets. Recently, Güler et al. [96] presented a complete duality theory for LP based on the concepts of interior point methods, making the field of interior point methods for LP self–supporting. Their proofs of the well–known results use almost only analytical arguments.

A fundamental concept in LP is self–duality. By a self–dual LP problem we mean one which equals its own dual program. From the early days of LP symmetric self–dual structures and algorithms have been recognized for their importance. Tucker [234] considered a skew–symmetric self–dual system. The existence of a strictly complementary solution for this system was proved by induction. Using this construction, Goldman and Tucker [80] proved the existence of a strictly complementary solution for general LP problems. Dantzig [39] presented a self–dual parametric algorithm for LP. He also discussed the symmetric form of the primal and the dual LP problems. Later, Terlaky [227] constructed a simple self–dual pivot algorithm, called criss–cross. Recently, Ye et al. [254] introduced the notion of self–duality in the field of interior point methods. They formulated a new self–dual problem, to which a standard interior point method can be applied to derive the best known complexity bound for an infeasible start interior point method. The approach is also computationally efficient and very effective in discovering primal and/or dual infeasibility (Xu et al. [247]).

The purpose of this chapter is to give a very easy proof of strong duality and the existence of a strictly complementary solution in LP by combining the methodology in [96] with a self–dual formulation like the one in [254]. We first introduce some notation and state the two results mentioned above. Let $c, x \in \mathbb{R}^n$, $b \in \mathbb{R}^m$ and A an $m \times n$ matrix. The primal LP problem in standard form is given by

$$(\mathrm{P}) \quad \min_x \left\{ c^T x \; : \; Ax = b, \; x \geq 0 \right\}.$$

The associated dual problem is

$$(\mathrm{D}) \quad \max_{y,s} \left\{ b^T y \; : \; A^T y + s = c, \; s \geq 0 \right\}.$$

The sets of feasible solutions of (P) and (D) are denoted by $\mathcal{P}$ and $\mathcal{D}$ respectively. Problem (P) is called feasible if the set $\mathcal{P}$ is nonempty; if $\mathcal{P}$ is empty then (P) is infeasible; if there is a sequence of feasible solutions for which the objective value goes to minus infinity then (P) is said to be unbounded; analogous statements hold for (D). We assume throughout that A has full row rank. This implies that y follows from a given feasible $s \geq 0$ in a unique way, and we may identify a feasible solution of (D) just by s. The first theorem is the main result in the theory of LP.

Theorem 2.1.1 (Strong duality) *For (P) and (D) one of the following alternatives holds:*
(i) (P) and (D) are feasible and there exist $x^* \in \mathcal{P}$ and $(y^*, s^*) \in \mathcal{D}$ such that $c^T x^* = b^T y^*$;
(ii) (P) is infeasible and (D) is unbounded;
(iii) (D) is infeasible and (P) is unbounded;
(iv) Both (P) and (D) are infeasible.

An alternative way of writing the optimality condition in Theorem 2.1.1(*i*) is the *complementary slackness condition*

$$x_i^* s_i^* = 0, \quad i = 1, \ldots, n.$$

Because of the nonnegativity condition on x^* and s^* this is also equivalent to $(x^*)^T s^* = 0$. Note that for arbitrary complementary solutions we might have $x_i^* = s_i^* = 0$. In the analysis of interior point methods *strict complementarity* is a central theme; it is involved in theoretical analyses, in sensitivity analysis as well as in the development and analysis of polynomial time interior point methods.

Theorem 2.1.2 (Strict complementarity) *If (P) and (D) are feasible then there exist $x^* \in \mathcal{P}$ and $(y^*, s^*) \in \mathcal{D}$ such that $(x^*)^T s^* = 0$ and $x_i^* + s_i^* > 0$, $i = 1, \ldots, n$. The solution (x^*, s^*) is called strictly complementary.*

The strict complementarity condition implies that for each index i exactly one of x_i^* and s_i^* is zero, while the other is positive. This result was first shown in 1956 by Goldman and Tucker [80]. For convenience we define the *support* of a vector $x \in \mathbb{R}^n$ as follows

$$\sigma(x) := \{\, i \;:\; x_i > 0 \,\}.$$

Then a feasible solution (x, s) is optimal if and only if $\sigma(x) \cap \sigma(s) = \emptyset$, and strictly complementary if and only if it is optimal and $\sigma(x) \cup \sigma(s) = \{1, \ldots, n\}$. Using a strictly complementary solution the *optimal partition* of the LP problem is defined as follows.

Definition 2.1.3 (Optimal partition) *Let (x^*, s^*) be a strictly complementary solution of (P) and (D). The partition $\pi = (B, N)$ of the index set $\{1, \ldots, n\}$ defined by $B = \sigma(x^*)$ and $N = \sigma(s^*)$ is called the optimal partition of (P) and (D).*

Combining the conditions for feasibility and optimality, optimal solutions for (P) and (D) are characterized by solutions of the nonlinear system of equations

$$\begin{aligned} Ax &= b, \quad x \geq 0, \\ A^T y + s &= c, \quad s \geq 0, \\ x_i s_i &= 0, \quad i = 1, \ldots, n. \end{aligned}$$

All methods for LP use this system. Dantzig's simplex method [37] is an iterative method that relaxes one of the sets of inequality constraints $s \geq 0$ (primal simplex) or $x \geq 0$ (dual simplex) for intermediate iterates. Interior point methods are characterized by the fact that they keep one or both sets of inequality constraints strictly satisfied during the process. Instead the set of nonlinear complementarity constraints is relaxed; in infeasible start methods this is combined with a relaxation of the linear equality constraints. In (feasible) primal–dual interior point methods each iterate satisfies the system

$$\begin{aligned} Ax &= b, \quad x > 0, \\ A^T y + s &= c, \quad s > 0, \\ x_i s_i &= w_i, \quad i = 1, \ldots, n. \end{aligned} \tag{2.1}$$

for some positive vector $w \in \mathbb{R}^n_{++}$. Observe that for (x, s) satisfying the system the (error in) complementarity is given by

$$x^T s = e^T w.$$

In most interior point methods system (2.1) with $w_i = \mu > 0$ for all i plays a special role, since its solutions for varying μ characterize the *central path* of the problem. The central path in this primal–dual setting was introduced and investigated by Megiddo [168], Bayer and Lagarias [17] and Sonnevend [218]. For decreasing values of μ the central path leads to a strictly complementary solution of the problem (cf. Theorem 2.2.9), hence it is used as a guideline to optimality in *path-following methods.* We introduce the following definition.

Definition 2.1.4 (Positive primal–dual pair) *Let x be feasible in (P) and (y, s) in (D) such that $x > 0$ and $s > 0$; then we call (x, s) a positive primal-dual pair.*

The following theorem establishes a one–to–one correspondence between positive primal–dual pairs (x, s) and positive vectors in $\mathbb{R}^n$. The theorem was proved by McLinden [166], Kojima et al. [136], see also Güler et al. [96].

Theorem 2.1.5 *Let there exist at least one positive primal–dual pair for (P) and (D). Then for each $w \in \mathbb{R}^n_{++}$ there exists a unique positive primal-dual pair (x, s) such that $x_i s_i = w_i, \quad i = 1, \ldots, n$.*

We now define the *v–space* of a given LP problem as the space of (the square roots of) the complementary products of positive primal–dual pairs:

$$\mathcal{V} = \{\, v \in \mathbb{R}^n \; : \; v_i = \sqrt{x_i s_i}, \; Ax = b, \; A^T y + s = c, \; x > 0, \; s > 0 \,\}.$$

Note that if $v = \sqrt{xs}$ then $\|v\|^2 = x^T s$, so in the v–space the points with constant norm represent all positive primal–dual pairs with a fixed duality gap. Observe that all optimal pairs (x, s) correspond to the vector $v = 0$. In

Chapter 5 we will use the v–space to derive a new primal–dual (affine scaling) interior point method. In Chapter 8 we will extensively use it in the analysis and development of algorithms.

To derive the duality results with an interior point approach we proceed as follows. In Section 2.2 we first discuss a very special, almost trivial skew–symmetric self–dual LP problem. Using the logarithmic barrier approach, the concept of the central path and the special structure of the problem, it is quite simple to prove the results for this problem. In Section 2.3 these results are applied to derive the strong duality and the existence of a strictly complementary solution for general LPs. Theorem 2.1.5 will follow as a byproduct of the results in this chapter.

2.2 DUALITY THEORY FOR SKEW–SYMMETRIC SELF–DUAL LP-S

We define a specific skew–symmetric self–dual LP problem in the following form

$$\text{(SP)} \quad \min_x \left\{ a^T x \,:\, Cx \geq -a,\ x \geq 0 \right\},$$

where C is an $n \times n$ skew–symmetric matrix (i.e., $C^T = -C$) and $a, x \in \mathbb{R}^n$. We require $a \geq 0$. Observe that for each $x \in \mathbb{R}^n$ it holds

$$x^T C x = 0. \tag{2.2}$$

The associated dual program is given by

$$\text{(SD)} \quad \max_y \left\{ -a^T y \,:\, C^T y \leq a,\ y \geq 0 \right\},$$

with $y \in \mathbb{R}^n$. Obviously the skew–symmetry of C implies that the primal and dual feasible sets are identical. The strong duality for these problems is easy.

Lemma 2.2.1 *(SP) and (SD) are feasible and for both the zero vector is an optimal solution.*

Proof: Since $a \geq 0$ the zero vector is primal and dual feasible. For each primal feasible x it holds

$$0 = x^T C x \geq -a^T x$$

by (2.2), so $a^T x \geq 0$; analogously $a^T y \geq 0$ for each dual feasible y. Hence the zero vector is an optimal solution for (SP) and also for (SD). □

Corollary 2.2.2 *Let x be feasible for (SP) and define $s = Cx + a$. Then x is optimal if and only if $x^T s = 0$.*

Proof: Using (2.2) it holds

$$a^T x = s^T x - x^T C^T x = s^T x. \tag{2.3}$$

The statement follows from Lemma 2.2.1. □

Observe that (SP) is trivial from a computational point of view since an optimal solution is readily available. However, the problem is interesting from a theoretical point of view. To complete the duality theory of (SP) we need to prove the existence of a strictly complementary solution. Since (SP) and (SD) are identical it suffices to work with the primal problem (SP). The feasible region of (SP) will be denoted by

$$\mathcal{SP} := \{ (x, s) \ : \ Cx - s = -a, \ x \geq 0, \ s \geq 0 \}.$$

The set of positive vectors in $\mathcal{SP}$ is denoted as $\mathcal{SP}^0$:

$$\mathcal{SP}^0 := \{ (x, s) \ : \ Cx - s = -a, \ x > 0, \ s > 0 \}.$$

The set of optimal solutions of (SP) is denoted by $\mathcal{SP}^*$. From Corollary 2.2.2 we have

$$\mathcal{SP}^* = \{ (x, s) \ : \ Cx - s = -a, \ x^T s = 0, \ x \geq 0, \ s \geq 0 \}.$$

We will need the following well-known result from elementary convex analysis, see e.g. Rockafellar [207].

Lemma 2.2.3 *Let $f : D \to \mathbb{R}$ be a convex differentiable function, where $D \subseteq \mathbb{R}^n$ is an open convex set. Then $x \in D$ minimizes f over D if and only if $\nabla f(x) = 0$.*

We also use the following straightforward lemma from calculus.

Lemma 2.2.4 *Let $\mu \in \mathbb{R}_{++}$ and $p \in \mathbb{R}^n_{++}$ be given. The function $h(x) := p^T x - \mu \sum_{i=1}^n \ln x_i$, where $x \in \mathbb{R}^n_{++}$, has a unique minimizer.*

Proof: We introduce the following notation: $h(x) = \sum_{i=1}^{n} \tilde{h}_i(x_i)$, where $\tilde{h}_i(x_i) := p_i x_i - \mu \ln x_i$. Let

$$h_i(x_i) := \tilde{h}_i(x_i) - \mu + \mu \ln \mu - \mu \ln p_i = \mu \left(\frac{p_i x_i}{\mu} - \ln \frac{p_i x_i}{\mu} - 1 \right).$$

The functions $h_i(x_i)$ are strictly convex and nonnegative on their domain $(0, \infty)$; furthermore $h_i(x_i) \to \infty$ as $x_i \to 0$ or $x_i \to \infty$. Hence all level sets of the functions $h_i(x_i)$ are bounded, and bounded away from zero. Consider a nonempty τ-level set $\mathcal{L} := \{ x : h(x) \leq \tau \}$ of $h(x)$. Note that $\mathcal{L}$ is nonempty if we take $\tau := h(x^{(0)})$ for some $x^{(0)} > 0$. For $x \in \mathcal{L}$ and for each i, we have

$$\begin{aligned} h_i(x_i) &\leq \sum_{i=1}^{n} h_i(x_i) = \sum_{i=1}^{n} (\tilde{h}_i(x_i) - \mu + \mu \ln \mu - \mu \ln p_i) \\ &= h(x) - n\mu(1 - \ln \mu) - \mu \sum_{i=1}^{n} \ln p_i \leq \tau - n\mu(1 - \ln \mu) - \mu \sum_{i=1}^{n} \ln p_i. \end{aligned}$$

So $\mathcal{L}$ is a subset of the Cartesian product of level sets of the functions h_i, and we conclude that the level set $\mathcal{L}$ is bounded. Since $h(x)$ is continuous, it has a minimizer in $\mathcal{L}$. The uniqueness of the minimizer follows from the strict convexity of $h(x)$. □

For $\mu > 0$ we define the function $\tilde{f}_\mu : \mathbb{R}^n_{++} \times \mathbb{R}^n_{++} \to \mathbb{R}$ by

$$\tilde{f}_\mu(x, s) := a^T x - \mu \left(\sum_{i=1}^{n} \ln x_i + \sum_{i=1}^{n} \ln s_i \right),$$

and $f_\mu : \mathbb{R}^n_{++} \to \mathbb{R}$ by

$$f_\mu(x) := a^T x - \mu \left(\sum_{i=1}^{n} \ln x_i + \sum_{i=1}^{n} \ln(c_{i\cdot} x + a_i) \right), \tag{2.4}$$

where $c_{i\cdot}$ denotes the ith row of C. Note that $f_\mu(x) = \tilde{f}_\mu(x, s)$ for $(x, s) \in \mathcal{SP}^0$. The function f_μ is the *logarithmic barrier function* for (SP) with barrier parameter μ. Due to (2.3) the term $a^T x$ can equally well be replaced by $x^T s$, which shows that $\tilde{f}_\mu(x, s)$ is symmetric in x and s on $\mathcal{SP}$.

Lemma 2.2.5 *Let $\mu > 0$. The following two statements are equivalent:*
(i) The function $f_\mu(x)$ has a (unique) minimizer;

(ii) There exist $x, s \in \mathbb{R}^n$ such that

$$\begin{array}{rcll} Cx - s & = & -a, & , x \geq 0,\ s \geq 0, \\ xs & = & \mu e. & \end{array} \tag{2.5}$$

Further, if one of the statements holds then x minimizes f_μ if and only if x and s satisfy (2.5).

Proof: First note that whenever (x, s) solves (2.5), then both x and s are positive, due to the second equation. So the nonnegativity conditions for x and s in (2.5) can be replaced by requiring that x and s are positive. One easily checks that $f_\mu(x)$ is strictly convex, and hence it has at most one minimizer. Since the domain of f_μ is open, Lemma 2.2.3 applies and it follows that f_μ has x as a minimizer if and only if $\nabla f_\mu(x) = 0$, i.e.,

$$a - \mu X^{-1}e - \mu C^T S^{-1} e = 0, \tag{2.6}$$

where $X = \mathrm{Diag}\,(x)$ and $S = \mathrm{Diag}\,(s)$. Using $s = Cx + a$ and $C^T = -C$, we can write (2.6) as

$$\mu X^{-1}e - s = C(\mu S^{-1}e - x).$$

Rearranging terms we obtain

$$0 = (C - X^{-1}S)S^{-1}(\mu e - Xs).$$

Since C is skew–symmetric and the matrices $X^{-1}S$ and S^{-1} are positive definite and diagonal, the last equation holds if and only if $Xs = \mu e$. This proves the lemma. □

Assume that $\mathcal{SP}^0$ is nonempty and let $(x^{(0)}, s^{(0)}) \in \mathcal{SP}^0$. By (2.2) we have for any $(x, s) \in \mathcal{SP}$

$$(x - x^{(0)})^T(s - s^{(0)}) = (x - x^{(0)})^T C(x - x^{(0)}) = 0. \tag{2.7}$$

Property (2.7) is known as the orthogonality property and often used in pivoting algorithms, see, e.g., Terlaky and Zhang [230]. Equivalently it holds,

$$(s^{(0)})^T x + (x^{(0)})^T s = x^T s + (x^{(0)})^T(s^{(0)}) = a^T x + a^T x^{(0)},$$

which gives

$$a^T x = (s^{(0)})^T x + (x^{(0)})^T s - a^T x^{(0)}. \tag{2.8}$$

Defining the function $g_\mu : \mathbb{R}^n_{++} \times \mathbb{R}^n_{++} \to \mathbb{R}$ by

$$g_\mu(x, s) := (s^{(0)})^T x + (x^{(0)})^T s - \mu \left(\sum_{i=1}^n \ln x_i + \sum_{i=1}^n \ln s_i \right),$$

we have for any $(x, s) \in \mathcal{SP}^0$

$$g_\mu(x, s) := f_\mu(x) + a^T x^{(0)},$$

so $g_\mu(x, s)$ and $f_\mu(x)$ differ by a constant on $\mathcal{SP}^0$. We now prove the following theorem.

Theorem 2.2.6 *Let $\mu > 0$. The following statements are equivalent:*
(i) The set $\mathcal{SP}^0$ is nonempty;
(ii) The function $f_\mu(x)$ defined in (2.4) has a (unique) minimizer;
(iii) The system (2.5) has a (unique) solution.

Proof: The equivalence of *(ii)* and *(iii)* is contained in Lemma 2.2.5. Earlier we noted the obvious fact that *(iii)* implies *(i)*. So it suffices to show that *(i)* implies *(ii)*. Assuming *(i)*, let $(x^{(0)}, s^{(0)}) \in \mathcal{SP}^0$. Due to (2.8) minimizing $f_\mu(x)$ over $\mathbb{R}^n_+$ is equivalent to minimizing $g_\mu(x, s)$ over $\mathcal{SP}^0$. So it suffices to show that g_μ has a minimizer in $\mathcal{SP}^0$. Note that g_μ is defined on the intersection of $\mathbb{R}^{2n}_{++}$ and an affine space. By the proof of Lemma 2.2.4 the level sets of g_μ are bounded, hence g_μ has a (unique) minimizer. This completes the proof. □

Observe that Theorem 2.2.6*(i)*, *(iii)* constitute Theorem 2.1.5 for (SP). In the remainder of this chapter, we make the basic assumption that statement *(i)* of Theorem 2.2.6 holds.

Assumption 2.2.7 *$\mathcal{SP}$ contains a positive vector $(x^{(0)}, s^{(0)})$, i.e., the set $\mathcal{SP}^0$ is nonempty.*

For each positive μ we denote the minimizer of $f_\mu(x)$ as $x(\mu)$, and define $s(\mu) := Cx(\mu) + a$. The set $\{\, x(\mu) \,:\, \mu > 0 \,\}$ is called the *central path* of (SP). We now prove that any section $(0 < \mu \le \bar{\mu})$ of the central path is bounded.

Lemma 2.2.8 *Let $\bar{\mu} > 0$. The set $\{\, (x(\mu), s(\mu)) \,:\, 0 < \mu \le \bar{\mu} \,\}$ is bounded.*

Proof: Let $(x^{(0)}, s^{(0)}) \in \mathcal{SP}^0$. Using the orthogonality property (2.7) and the fact that (2.5) holds with $x(\mu)$ we get for any i, $1 \le i \le n$,

$$\begin{aligned} s_i^{(0)} x_i(\mu) &\le (s^{(0)})^T x(\mu) + (x^{(0)})^T s(\mu) = x(\mu)^T s(\mu) + (x^{(0)})^T s^{(0)} \\ &= n\mu + (x^{(0)})^T s^{(0)} \le n\bar{\mu} + (x^{(0)})^T s^{(0)}. \end{aligned}$$

This shows that $x_i(\mu) \leq (n\overline{\mu} + (x^{(0)})^T s^{(0)})/s_i^{(0)}$. So the set $\{ x(\mu) : 0 < \mu \leq \overline{\mu} \}$ is bounded. The proof for $\{ s(\mu) : 0 < \mu \leq \overline{\mu} \}$ is similar. □

We proceed by showing the existence of a strictly complementary solution $(x^*, s^*) \in \mathcal{SP}$ under Assumption 2.2.7, that is, a solution satisfying $(x^*)^T s^* = 0$ and $x^* + s^* > 0$.

Theorem 2.2.9 *If Assumption 2.2.7 holds, then there exist $(x^*, s^*) \in \mathcal{SP}^*$ such that $x^* + s^* > 0$.*

Proof: Let $\{\mu_k\}_{k=1}^{\infty}$ be a positive sequence such that $\mu_k \to 0$ if $k \to \infty$. By Lemma 2.2.8 the set $\{ (x(\mu_k), s(\mu_k)) \}$ is bounded, hence it contains a subsequence converging to a point (x^*, s^*). Since $(x^*, s^*) \in \mathcal{SP}$ and $x(\mu_k)^T s(\mu_k) = n\mu_k \to 0$, we conclude $(x^*)^T s^* = 0$, so (x^*, s^*) is an optimal solution. We show that (x^*, s^*) is strictly complementary. By (2.7)

$$(x(\mu_k) - x^*)^T (s(\mu_k) - s^*) = 0.$$

Rearranging terms and noting that $x(\mu_k)^T s(\mu_k) = n\mu_k$ and $(x^*)^T s^* = 0$, we arrive at

$$\sum_{i \in \sigma(x^*)} x_i^* s_i(\mu_k) + \sum_{i \in \sigma(s^*)} x_i(\mu_k) s_i^* = n\mu_k.$$

Dividing both sides by μ_k and recalling $x_i(\mu_k) s_i(\mu_k) = \mu_k$, we obtain

$$\sum_{i \in \sigma(x^*)} \frac{x_i^*}{x_i(\mu_k)} + \sum_{i \in \sigma(s^*)} \frac{s_i^*}{s_i(\mu_k)} = n.$$

Letting $k \to \infty$, we see that the first sum becomes equal to the number of nonzero coordinates in x^*. Similarly, the second sum becomes equal to the number of nonzero coordinates in s^*. We conclude that (x^*, s^*) is strictly complementary. □

Observe that the proof of Theorem 2.2.9 shows that the central path has a subsequence converging to an optimal solution. This suffices for proving the existence of a strictly complementary solution. However, it can be shown that the central path is an analytic curve and converges itself. Since this will be used in the following chapters, we also prove it here. The limiting behavior of the central path as $\mu \to 0$ has been an important subject in the research on interior point methods since long. In the book by Fiacco and McCormick [53] the convergence of the path to an optimal solution is investigated for general convex programming problems. McLinden [166] considered the limiting behavior of the path for monotone complementarity problems and introduced the idea for the proof–technique of Theorem 2.2.9, which was later adapted by Güler and Ye [97]. Megiddo [168] extensively investigated the properties of the central path, which motivated Monteiro and Adler [180] and Kojima et al. [139] for research on primal–dual algorithms.

Lemma 2.2.10 *If Assumption 2.2.7 holds then the central path converges to a unique primal–dual feasible pair.*

Proof: The proof very much resembles the one of Theorem 2.2.9. Let $\bar{x}$ be optimal in (SP) and $(\tilde{y}, \tilde{s} = C\tilde{y}+a)$ in (SD), and let (x^*, s^*) be the accumulation point of the central path as defined in Theorem 2.2.9. It easily follows that

$$\sum_{i\in\sigma(x^*)} \frac{\bar{x}_i}{x_i^*} + \sum_{i\in\sigma(s^*)} \frac{\tilde{s}_i}{s_i^*} = n.$$

Using the arithmetic–geometric mean inequality we obtain

$$\left(\prod_{i\in\sigma(x^*)} \frac{\bar{x}_i}{x_i^*} \prod_{i\in\sigma(s^*)} \frac{\tilde{s}_i}{s_i^*}\right)^{1/n} \le \frac{1}{n}\left(\sum_{i\in\sigma(x^*)} \frac{\bar{x}_i}{x_i^*} + \sum_{i\in\sigma(s^*)} \frac{\tilde{s}_i}{s_i^*}\right) = 1.$$

Applying the inequality with $\tilde{s} = s^*$ gives

$$\prod_{i\in\sigma(x^*)} \bar{x}_i \le \prod_{i\in\sigma(x^*)} x_i^*,$$

and with $\bar{x} = x^*$ it gives

$$\prod_{i\in\sigma(s^*)} \tilde{s}_i \le \prod_{i\in\sigma(s^*)} s_i^*.$$

This implies that x^* maximizes the product $\prod_{i\in\sigma(x^*)} x_i$ and s^* the product $\prod_{i\in\sigma(s^*)} s_i$ over the optimal set. Hence the central path of (SP) has a unique limit point. □

The proof of the lemma shows that the limitpoint of the central path solves an optimization problem over the optimal set. Actually, we proved that the limitpoint is the *analytic center* of the optimal set.

Definition 2.2.11 (Analytic center) *Let $D \subset \mathbb{R}^n$ be a bounded convex set. The analytic center of D is the unique minimizer of*

$$\min_x \left\{ -\sum_{i=1}^n \ln x_i \ : \ x \in D \right\}.$$

The analytic center was introduced by Sonnevend [218] and plays an important role in interior point methods. We will further encounter it in the next chapters. We remark that the central path is the set of analytic centers of the level–sets of the LP problem.

2.3 DUALITY THEORY FOR GENERAL LP-S

The results of the previous section can easily be applied to prove the strong duality theorem of LP. In this way we present a new proof of this classical result. We also obtain Goldman–Tucker's Theorem for the general case. In this section we consider the LP problem in symmetric form instead of in the standard form (P). Obviously, this can be done without loss of generality since every LP problem can be rewritten from one of these forms to the other, without increasing the number of variables and constraints. So let the primal be given by

$$(\overline{\mathrm{P}}) \quad \min_x \left\{ c^T x \ : \ Ax \geq b, \ x \geq 0 \right\},$$

where A is an $m \times n$ matrix, $c, x \in \mathbb{R}^n$, and $b \in \mathbb{R}^m$. The associated dual problem $(\overline{\mathrm{D}})$ is

$$(\overline{\mathrm{D}}) \quad \max_y \left\{ b^T y \ : \ A^T y \leq c, \ y \geq 0 \right\}.$$

Expressed in this form, a pair (x^*, y^*) is strictly complementary if x^* is feasible in $(\overline{\mathrm{P}})$, y^* is feasible in $(\overline{\mathrm{D}})$ and moreover

$$\begin{aligned}(Ax^* - b)^T y^* = (c - A^T y^*)^T x^* = 0,\\ y^* + (Ax^* - b) > 0,\\ x^* + (c - A^T y^*) > 0.\end{aligned}$$

We formulate a skew–symmetric self–dual LP problem, that incorporates the information contained in $(\overline{\mathrm{P}})$ and $(\overline{\mathrm{D}})$. A similar embedding of the primal and dual problem in a skew–symmetric self–dual problem was considered in

[80, 254]. Let $x^{(0)}, r^{(0)} \in \mathbb{R}^n_{++}$, $y^{(0)}, u^{(0)} \in \mathbb{R}^m_{++}$ and $\vartheta_0, \tau_0, \mu_0, \nu_0 \in \mathbb{R}_{++}$ be arbitrary. Further, we define $\overline{c} \in \mathbb{R}^n$, $\overline{b} \in \mathbb{R}^m$ and $\alpha, \beta \in \mathbb{R}$ as follows:

$$\begin{aligned}
\overline{b} &= (\tau_0 b - Ax^{(0)} + r^{(0)})/\vartheta_0, \\
\overline{c} &= (\tau_0 c - A^T y^{(0)} - u^{(0)})/\vartheta_0, \\
\alpha &= (c^T x^{(0)} - b^T y^{(0)} + \mu_0)/\vartheta_0, \\
\beta &= \alpha\tau_0 + \overline{b}^T y^{(0)} - \overline{c}^T x^{(0)} + \nu_0 \\
&= ((y^{(0)})^T r^{(0)} + (x^{(0)})^T u^{(0)} + \tau_0\mu_0)/\vartheta_0 + \nu_0.
\end{aligned}$$

It is worthwhile to note that if $x^{(0)}$ is strictly feasible for $(\overline{\mathrm{P}})$ and $r^{(0)} := Ax^{(0)} - b$, then we have $\overline{b} = 0$ by setting $\vartheta_0 = \tau_0 = 1$. If $y^{(0)}$ is strictly feasible for $(\overline{\mathrm{D}})$ and $u^{(0)} := c - A^T y^{(0)}$, then $\overline{c} = 0$ if $\vartheta_0 = \tau_0 = 1$. So, $\overline{b}$ and $\overline{c}$ measure the infeasibility of the given vectors $x^{(0)}, r^{(0)}, y^{(0)}$ and $u^{(0)}$. We define the problem

$$(\overline{\mathrm{SP}}) \qquad \begin{array}{lrcrcrcrl}
\min\limits_{y,x,\vartheta,\tau} & & & & & \beta\vartheta & & & \\
\text{s.t.} & & & Ax & + & \overline{b}\vartheta & - & b\tau & \geq 0, \\
& -A^T y & & & - & \overline{c}\vartheta & + & c\tau & \geq 0, \\
& -\overline{b}^T y & + & \overline{c}^T x & & & - & \alpha\tau & \geq -\beta, \\
& b^T y & - & c^T x & + & \alpha\vartheta & & & \geq 0, \\
& y \geq 0, & & x \geq 0, & & \vartheta \geq 0, & & \tau \geq 0. &
\end{array}$$

The selection of the parameters implies that the positive solution $x = x^{(0)}$, $y = y^{(0)}$, $\vartheta = \vartheta_0$, $\tau = \tau_0$ is feasible for $(\overline{\mathrm{SP}})$, hence Assumption 2.2.7 holds. Also, the coefficients in the objective function are nonnegative. So the results of the previous section apply to this problem, and we can derive the following theorem.

Theorem 2.3.1 *For $(\overline{P})$ and $(\overline{D})$ one of the following alternatives holds:*
(i) $(\overline{P})$ and $(\overline{D})$ are feasible and there exists a strictly complementary solution $(\overline{x}^*, \overline{y}^*)$;
(ii) $(\overline{P})$ is infeasible and $(\overline{D})$ is unbounded;
(iii) $(\overline{D})$ is infeasible and $(\overline{P})$ is unbounded;
(iv) Both $(\overline{P})$ and $(\overline{D})$ are infeasible.

Proof: Problem $(\overline{\mathrm{SP}})$ is skew–symmetric and self–dual, the objective has nonnegative coefficients and Assumption 2.2.7 holds. Hence Theorem 2.2.9 guarantees the existence of a strictly complementary solution $(x^*, y^*, \vartheta^*, \tau^*)$. By Lemma 2.2.1 we also know that $\vartheta^* = 0$, since $\beta \geq \nu_0 > 0$. Two possibilities may occur. If $\tau^* > 0$ then $\overline{x}^* := x^*/\tau^*$ and $\overline{y}^* := y^*/\tau^*$ are feasible in $(\overline{\mathrm{P}})$ and

$(\overline{\mathrm{D}})$ respectively, and they constitute a strictly complementary pair. So case (i) holds. On the other hand, if $\tau^* = 0$ then it follows that $Ax^* \geq 0$, $x^* \geq 0$, $A^T y^* \leq 0$, $y^* \geq 0$ and $b^T y^* - c^T x^* > 0$. If $b^T y^* > 0$ then $(\overline{\mathrm{P}})$ is infeasible, since by assuming $\overline{x}$ to be primal feasible one has $0 \geq \overline{x}^T A^T y^* \geq b^T y^*$, which is a contradiction. Also, it follows immediately that *if* $(\overline{\mathrm{D}})$ is feasible then it is unbounded in this case. If $c^T x^* < 0$ then $(\overline{\mathrm{D}})$ is infeasible, since by assuming $\overline{y}$ to be dual feasible we have $0 \leq \overline{y}^T A x^* \leq c^T x^*$, which is a contradiction; also, $(\overline{\mathrm{P}})$ is unbounded if it is feasible. If $b^T y^* > 0$ and $c^T x^* < 0$ then both $(\overline{\mathrm{P}})$ and $(\overline{\mathrm{D}})$ are infeasible, which can be seen in just the same way. □

The proof reveals that the construction $(\overline{\mathrm{SP}})$ cannot always determine which of the alternatives in the theorem actually applies. It is an open question whether a variant of this approach can be found that does not solve an additional feasibility problem, nor uses a 'big M'–parameter, and still identifies exactly which of the four holds for a given pair of LP problems. Now we only have the following corollary.

Corollary 2.3.2 *Let* $(x^*, y^*, \vartheta^*, \tau^*)$ *be a strictly complementary solution of* $(\overline{SP})$*. If* $\tau^* > 0$ *then (i) of Theorem 2.3.1 applies; if* $\tau^* = 0$ *then one of (ii), (iii) or (iv) holds.*

2.4 CONCLUDING REMARKS

In this chapter we considered the theory of LP using an interior point approach. Although most of the theorems in the theory of LP are well–known and established we gave alternative proofs using a skew–symmetric self–dual reformulation of the general LP problem and only analytical arguments. It appears that deriving essential ingredients of interior point methods, as the central path, is closely connected to proving the existence of a strictly complementary solution in LP.

To solve arbitrary LP problems the skew–symmetric self–dual formulation can be used, provided it is solved by an interior point method. If the selected initial solution $(x^{(0)}, r^{(0)}, y^{(0)}, u^{(0)}, \vartheta_0, \tau_0, \mu_0, \nu_0)$ is taken to be the vector of ones, then a point on the central path is obtained for free. This solution can be used to start a path–following algorithm, yielding a strictly complementary solution. As shown by Xu et al. [247] such an approach not only is very efficient but also enables the application of Theorem 2.3.1.

3

SENSITIVITY ANALYSIS IN LINEAR PROGRAMMING

3.1 INTRODUCTION

The merits of LP are nowadays well–established and it is widely accepted as a useful tool in Operations Research and Management Science. In many companies this way of modeling is used to solve various kinds of practical problems. Applications include transportation problems, production planning, investment decision problems, blending problems, location and allocation problems, among others. Often use is made of some standard code, most of which use a version of Dantzig's simplex method as solution procedure (for a recent survey we refer to [216]).

Many LP packages do not only solve the problem at hand, but provide additional information on the solution, in particular information on the sensitivity of the solution to certain changes in the data. This is referred to as ***sensitivity analysis*** or ***postoptimal analysis***. This information can be of tremendous importance in practice, where parameter values may be estimates, where questions of type "What if..." are frequently encountered, and where implementation of a specific solution may be difficult. Sensitivity analysis serves as a tool for obtaining information about the bottlenecks and degrees of freedom in the problem. Unfortunately, interpreting this information and estimating its value is often difficult in practice; misuse is common, which may lead to expensive mistakes (see e.g., Rubin and Wagner [212]). In the literature there are several references where (often partially) the correct interpretation of sensitivity results is stressed. We mention Gal [64, 65], Ward and Wendell [242], Rubin and Wagner [212], Greenberg [90], among others.

The purpose of this chapter is manyfold. Our first objective is to convince the reader of a correct way of considering and applying sensitivity analysis in LP. The important observation is that knowledge of the *set of optimal solutions* is needed, instead of knowing just one optimal solution. Second, we show that, contrary to a popular belief, sensitivity analysis with interior point methods is possible and even natural, using the optimal partition of the LP problem. Research in this area was triggered by Adler and Monteiro [2] and Jansen et al. [111] (see also Mehrotra and Monteiro [171]). Greenberg [91] gives some examples where the interior approach has important practical impact. Third, we unify various viewpoints on sensitivity analysis, namely approaches using optimal bases ('simplex approach'), optimal partitions ('interior approach'), and the optimal value ('value approach'). This unification lingers on the fact that these are three approaches to characterize the optimal set. Finally, we present some computational results obtained with these approaches to sensitivity analysis applied to an LP model of oil distribution and sales developed by a major oil company.

3.2 ILLUSTRATIVE EXAMPLE

We first give an example showing that different LP packages may give different sensitivity information on the same problem. We focus on the output that most commercial packages give, viz.

- the optimal value,
- an optimal solution,
- an optimal dual solution,
- a range for each of the coefficients in the objective function and the right–hand side.

Example 3.2.1 Consider the (unbalanced) transportation problem with three suppliers and three markets. Each supplier can serve each of the markets at a transportation cost of 1 per unit. The capacity of the suppliers is equal to 2, 6 and 5 units respectively. The markets each require at least 3 units. We formulate this problem as an LP problem with variables

x_{ij} : the amount of units transported from supplier i to market j,

s_i : excess supply at supplier i,
d_j : shortage demand at market j.

Then we solve

$$\begin{aligned}
\min_{x,s,d} \quad & \sum_{i=1}^{3}\sum_{j=1}^{3} x_{ij} \\
\text{s.t.} \quad & x_{11}+x_{12}+x_{13}+s_1 = 2, && (3.1)\\
& x_{21}+x_{22}+x_{23}+s_2 = 6, && (3.2)\\
& x_{31}+x_{32}+x_{33}+s_3 = 5, && (3.3)\\
& x_{11}+x_{21}+x_{31}-d_1 = 3, && (3.4)\\
& x_{12}+x_{22}+x_{32}-d_2 = 3, && (3.5)\\
& x_{13}+x_{23}+x_{33}-d_3 = 3, && (3.6)\\
& x_{ij},\ s_i,\ d_j \geq 0 \quad i,j = 1,2,3.
\end{aligned}$$

The cost structure implies that any solution that exactly transports the total required demand of 9 units is optimal. The results of five commercially available LP packages are given in Tables 3.1 and 3.2.

Package	Optimal primal solution								
	x_{11}	x_{12}	x_{13}	x_{21}	x_{22}	x_{23}	x_{31}	x_{32}	x_{33}
CPLEX	0	2	0	2	1	3	1	0	0
LINDO	2	0	0	0	0	2	1	3	1
OSL	0	2	0	2	1	3	1	0	0
PC-PROG	0	0	0	0	3	1	3	0	2
XMP	0	0	2	3	3	0	0	0	1

Package	Optimal dual solution					
	(3.1)	(3.2)	(3.3)	(3.4)	(3.5)	(3.6)
CPLEX	0	0	0	1	1	1
LINDO	0	0	0	1	1	1
OSL	0	0	0	1	1	1
PC-PROG	0	0	0	1	1	1
XMP	0	0	0	1	1	1

Table 3.1 Optimal primal and dual solution in Example 3.2.1.

Even in this simple example no two packages yield the same result: four different optimal solutions are given and all packages return different ranges for the coefficients in the objective function. Note that CPLEX and OSL give the same optimal solution, but that the ranges for the objective coefficients are not

Package	COST–ranges								
	x_{11}	x_{12}	x_{13}	x_{21}	x_{22}	x_{23}	x_{31}	x_{32}	x_{33}
CPLEX	[1,∞)	(-∞,1]	[1,∞)	[1,1]	[1,1]	[0,1]	[1,1]	[1,∞)	[1,∞)
LINDO	(−∞,1]	[1,∞)	[1,∞)	[1,∞)	[1,∞)	[1,1]	[1,1]	[0,1]	[1,1]
OSL	[1,∞)	[1,1]	[1,∞)	[1,1]	[1,1]	[1,1]	[1,1]	[1,∞)	[1,∞)
PC–PROG	[1,∞)	[1,∞)	[1,∞)	[1,∞)	[0,1]	[1,1]	[0,1]	[1,∞)	[1,1]
XMP			(-∞,1]	[1,1]	[0,1]		[1,1]		[1,1]

Package	RHS–ranges					
	(3.1)	(3.2)	(3.3)	(3.4)	(3.5)	(3.6)
CPLEX	[0,3]	[4,7]	[1,∞)	[2,7]	[2,5]	[2,5]
LINDO	[1,3]	[2,∞)	[4,7]	[2,4]	[1,4]	[1,7]
OSL	[0,3]	[4,7]	[1,∞)	[2,7]	[2,5]	[2,5]
PC–PROG	[0,∞)	[4,∞)	[3,6]	[2,5]	[0,5]	[2,5]
XMP	[0,3]	[3,6]	[1,∞)	[3,7]	[3,6]	[2,7]

Table 3.2 **Ranges in Example** 3.2.1.

all the same. Even though the dual solutions are equal in all packages, the ranges for the right–hand side coefficients are not. ◇

Although the differences are completely explainable, showing the above tables to users of LP often gives reactions of unbelief and mistrust. Hence the interpretation of the output should be carefully considered. We feel, that in many textbooks on LP this is not properly done. Particularly, the relation between solutions and ranges is often not stressed enough. A correct interpretation, as we show in this chapter, is as follows.

1. The primal and dual values given are optimal (basic) solutions to the primal and dual problem; they are not necessarily unique.

2. The primal solution gives the rate of change in the objective value if the corresponding coefficient in the objective is changed within the range provided for this coefficient; the primal solution remains optimal in this range.

3. The dual solution gives the rate of change in the objective value value if the corresponding coefficient in the right–hand side is changed within the range provided for this coefficient; the dual solution remains optimal in this range.

It is important to mention that the ranges where the primal or dual solutions remain optimal can be larger than the ones provided by the package. This can be seen from Table 3.2 for the dual solution, since all the packages return the same optimal solution. Taking the RHS–coefficient for constraint (3.5), we can conclude that the dual solution remains optimal at least in the interval $[0, 6]$ (which is the union of the ranges provided by the packages).

3.3 OPTIMAL VALUE FUNCTIONS, OPTIMAL SETS AND OPTIMAL PARTITIONS

We consider the primal and dual LP problems (P) and (D) from Section 2.1:

$$\begin{array}{ll} \text{(P)} & \min_{x} \left\{ c^T x \ : \ Ax = b, \ x \geq 0 \right\} \\ \text{(D)} & \max_{y,s} \left\{ b^T y \ : \ A^T y + s = c, \ s \geq 0 \right\}. \end{array}$$

The sets of feasible solutions are denoted by $\mathcal{P}$ and $\mathcal{D}$, the sets of optimal solutions are given by $\mathcal{P}^*$ and $\mathcal{D}^*$, respectively. Let the index sets B and N be defined as and

$$\begin{aligned} B &:= \{ i \ : \ x_i > 0 \text{ for some } x \in \mathcal{P}^* \}, \\ N &:= \{ i \ : \ s_i > 0 \text{ for some } (y, s) \in \mathcal{D}^* \}. \end{aligned}$$

Combining the Duality Theorem 2.1.1 with the existence of a strictly complementary solution (Theorem 2.1.2) it is easy to see that B and N form the optimal partition (Definition 2.1.3) of the LP problem, so $\pi = (B, N)$. Using the optimal partition we may rewrite the primal and dual optimal sets as

$$\begin{aligned} \mathcal{P}^* &= \{ x \ : \ Ax = b, \ x_B \geq 0, \ x_N = 0 \}, \\ \mathcal{D}^* &= \{ (y, s) \ : \ A^T y + s = c, \ s_B = 0, \ s_N \geq 0 \}. \end{aligned}$$

Since we assume A to have full rank we can identify any feasible $s \geq 0$ with a unique y such that $A^T y + s = c$, and vice versa; hence we sometimes just use $y \in \mathcal{D}^*$ or $s \in \mathcal{D}^*$ instead of $(y, s) \in \mathcal{D}^*$.

We study the pair of LP problems (P) and (D) as b and c change; the matrix A will be constant throughout. Therefore, we index the problems as (P(b, c)) and (D(b, c)). We denote the optimal value function by $z(b, c)$. We will call the pair (b, c) a *feasible pair* if both $(P(b,c))$ and $(D(b,c))$ are feasible. If $(P(b,c))$ is unbounded then we define $z(b,c) := -\infty$; if $(D(b,c))$ is unbounded then we define $z(b,c) := \infty$. If both $(P(b,c))$ and $(D(b,c))$ are infeasible then $z(b,c)$ is undefined. We are specifically interested in the behavior of the optimal value function as one parameter changes. Although this is a severe restriction, it is both common from a theoretical and a computational point of view, since the multi–parameter case is very hard (see e.g. Ward and Wendell [242] for a practical approximative approach). So, let Δb and Δc be given perturbation vectors and define

$$b(\beta) := b + \beta\Delta b, \qquad f(\beta) := z(b(\beta), c),$$
$$c(\gamma) := c + \gamma\Delta c, \qquad g(\gamma) := z(b, c(\gamma)).$$

In the next lemma we prove a well–known elementary fact on the optimal value function.

Lemma 3.3.1 *The optimal value function $f(\beta)$ is convex and piecewise linear in β, while $g(\gamma)$ is concave and piecewise linear in γ.*

Proof: By definition

$$f(\beta) = \max_y \left\{ b(\beta)^T y \ : \ y \in \mathcal{D} \right\}.$$

If $f(\beta)$ has a finite value, the optimal value is attained at the analytic center of one the faces of $\mathcal{D}$ (Lemma 2.2.10). Since the number of faces is finite it holds

$$f(\beta) = \max_y \{ b(\beta)^T y \ : \ y \in \mathcal{S} \},$$

where $\mathcal{S}$ is the finite subset of $\mathcal{D}$ consisting of the analytic centers of its faces. For each $y \in S$ we have

$$b(\beta)^T y = b^T y + \beta \Delta b^T y$$

which is linear in β. So $f(\beta)$ is the maximum of a finite set of linear functions, which implies the first statement. The second can be shown similarly. □

The proof of the lemma is an 'interior point variation' of a well–known proof using for $\mathcal{S}$ the vertices of $\mathcal{D}$. The intervals for β (or γ) on which the optimal value function $f(\beta)$ (or $g(\gamma)$) is linear are called *linearity intervals*. The points where the slope of the optimal value function changes are called *breakpoints*. We give four typical questions a user might ask once an LP problem has been solved for a certain value of, say, β:

Question 1 What is the rate of change the optimal value is affected with by a change in β?

Question 2 In what interval may β be varied such that this rate of change is constant?

Question 3 In what interval may β be varied such that the optimal solution of (D) obtained from our solution procedure remains optimal?

Question 4 What happens to the optimal solution of (P) obtained from our solution procedure?

Questions 1 and 2 clearly have an intimate connection with the optimal value function. It will need some analysis to show that the same is true for Questions 3 and 4. The answer to Question 1 is that the derivative (slope) of the optimal value function is the rate at which the optimal value changes. This rate of change is called the *shadow price* (in case of varying objective we speak of *shadow cost*). However, if β is a breakpoint then we distinguish between increasing and decreasing β, since the rate of change is different in these cases. Moreover, the shadow price is constant on a linear piece of the optimal value function. Hence the answer to Question 2 must be a linearity interval. One of the reasons that Questions 3 and 4 are more involved is that the answer depends on the type of solution that is computed by the solution procedure used.

The next two lemmas show that the set of optimal solutions for $(D(b(\beta), c))$ (being denoted by $\mathcal{D}^*_\beta$) is constant on a linearity interval of $f(\beta)$ and changes in its breakpoints. Similar results can be obtained for variations in c and are therefore omitted.

Lemma 3.3.2 *If $f(\beta)$ is linear on the interval $[\beta_1, \beta_2]$ then the optimal set $\mathcal{D}^*_\beta$ is constant on (β_1, β_2).*

Proof: Let $\overline{\beta} \in (\beta_1, \beta_2)$ and $\overline{y} \in \mathcal{D}^*_{\overline{\beta}}$ be arbitrary. By definition

$$f(\overline{\beta}) = b^T\overline{y} + \overline{\beta}\Delta b^T\overline{y}.$$

Since $\overline{y}$ is feasible in $(D(b(\beta), c))$ for all β it holds

$$b(\beta_1)^T\overline{y} = b^T\overline{y} + \beta_1\Delta b^T\overline{y} \le f(\beta_1), \quad \text{and} \quad b(\beta_2)^T\overline{y} = b^T\overline{y} + \beta_2\Delta b^T\overline{y} \le f(\beta_2).$$

Using the linearity of $f(\beta)$ on $[\beta_1, \beta_2]$ yields

$$\Delta b^T \overline{y} \leq \frac{f(\beta_2) - f(\overline{\beta})}{\beta_2 - \overline{\beta}} = \frac{f(\overline{\beta}) - f(\beta_1)}{\overline{\beta} - \beta_1} \leq \Delta b^T \overline{y}.$$

So the above inequalities are equalities and we obtain $f'(\overline{\beta}) = \Delta b^T \overline{y}$, which in turn implies

$$f(\tilde{\beta}) = b^T \overline{y} + \tilde{\beta} \Delta b^T \overline{y} = b(\tilde{\beta})^T \overline{y}, \qquad \forall\, \tilde{\beta} \in [\beta_1, \beta_2].$$

Hence $\overline{y} \in \mathcal{D}^*_{\tilde{\beta}}$ for all $\beta \in [\beta_1, \beta_2]$. From this we conclude that the sets $\mathcal{D}^*_{\beta}$ are constant for $\beta \in (\beta_1, \beta_2)$. □

Corollary 3.3.3 *Let $f(\beta)$ be linear on the interval $[\beta_1, \beta_2]$ and denote $\overline{\mathcal{D}}^* := \mathcal{D}^*_{\beta}$ for arbitrary $\beta \in (\beta_1, \beta_2)$. Then $\overline{\mathcal{D}}^* \subseteq \mathcal{D}^*_{\beta_1}$ and $\overline{\mathcal{D}}^* \subseteq \mathcal{D}^*_{\beta_2}$.*

Observe that the proof of Lemma 3.3.2 reveals that $\Delta b^T y$ has the same value for all $y \in \mathcal{D}^*_{\beta}$ for all $\beta \in (\beta_1, \beta_2)$. We next deal with the converse implication.

Lemma 3.3.4 *Let β_1 and β_2 be such that $\mathcal{D}^*_{\beta_1} = \mathcal{D}^*_{\beta_2} =: \overline{\mathcal{D}}^*$. Then $\mathcal{D}^*_{\beta} = \overline{\mathcal{D}}^*$ for $\beta \in [\beta_1, \beta_2]$ and $f(\beta)$ is linear on this interval.*

Proof: Let $\overline{y} \in \overline{\mathcal{D}}^*$ be arbitrary. Then

$$f(\beta_1) = b(\beta_1)^T \overline{y}, \quad \text{and} \quad f(\beta_2) = b(\beta_2)^T \overline{y}.$$

Consider the linear function $h(\beta) := b(\beta)^T \overline{y}$. Note that $h(\beta_1) = f(\beta_1)$ and $h(\beta_2) = f(\beta_2)$. Since f is convex it holds $f(\beta) \leq h(\beta)$ for $\beta \in [\beta_1, \beta_2]$. On the other hand, since $\overline{y}$ is feasible for all β we have

$$f(\beta) \geq b(\beta)^T \overline{y} = h(\beta).$$

Hence $f(\beta)$ is linear on $[\beta_1, \beta_2]$ and $\overline{y} \in \mathcal{D}^*_{\beta}$ for all $\beta \in [\beta_1, \beta_2]$. So $\overline{\mathcal{D}}^*$ is a subset of the optimal set on (β_1, β_2). From Corollary 3.3.3 we know that the reverse also holds, hence for all $\beta \in (\beta_1, \beta_2)$ the optimal set equals $\overline{\mathcal{D}}^*$. □

As we have seen in the proof of Lemma 3.3.2 the quantity $\Delta b^T y$ is the same for all $y \in \mathcal{D}^*_{\beta}$ for β in a linearity interval. The next lemma shows that this property distinguishes a linearity interval from a breakpoint. Gauvin [68] was one of the first[1] to show this result and to emphasize the need to discriminate between *left and right shadow prices*, i.e., between decreasing and increasing the parameter.

Lemma 3.3.5 *Let $f'_-(\beta)$ and $f'_+(\beta)$ be the left and right derivative of $f(\cdot)$ in β. Then*

$$\begin{aligned} f'_-(\beta) &= \min_y \left\{ \Delta b^T y \; : \; y \in \mathcal{D}^*_\beta \right\} \\ f'_+(\beta) &= \max_y \left\{ \Delta b^T y \; : \; y \in \mathcal{D}^*_\beta \right\}. \end{aligned}$$

Proof: We give the proof for $f'_+(\beta)$; the one for $f'_-(\beta)$ is similar. Let $\overline{\beta}$ be in the linearity interval just to the right of β and let $\overline{y} \in \mathcal{D}^*_{\overline{\beta}}$. Then

$$f(\overline{\beta}) = b(\overline{\beta})^T \overline{y} \geq (b + \overline{\beta}\Delta b)^T y, \quad \forall y \in \mathcal{D}^*_\beta.$$

Since $\overline{y} \in \mathcal{D}^*_\beta$ by Corollary 3.3.3 we also have $(b+\beta\Delta b)^T y = (b+\beta\Delta b)^T \overline{y}$, $\forall y \in \mathcal{D}^*_\beta$. Hence

$$\Delta b^T y \leq \Delta b^T \overline{y}, \quad \forall y \in \mathcal{D}^*_\beta.$$

Since $\overline{y} \in \mathcal{D}^*_\beta$ and $f'_+(\beta) = f'(\overline{\beta}) = \Delta b^T \overline{y}$ the result follows. □

Next we show how a linearity interval can be computed.

Lemma 3.3.6 *Let β_1, β_2 be two consecutive breakpoints of the optimal value function $f(\beta)$. Let $\overline{\beta} \in (\beta_1, \beta_2)$ and define $\overline{\mathcal{D}}^* := \mathcal{D}^*_{\overline{\beta}}$. Then*

$$\begin{aligned} \beta_1 &= \min_{\beta, x} \left\{ \beta \; : \; Ax - \beta\Delta b = b, \; x \geq 0, \; x^T s = 0 \;\; \forall s \in \overline{\mathcal{D}}^* \right\}, \\ \beta_2 &= \max_{\beta, x} \left\{ \beta \; : \; Ax - \beta\Delta b = b, \; x \geq 0, \; x^T s = 0 \;\; \forall s \in \overline{\mathcal{D}}^* \right\}. \end{aligned}$$

Proof: We only give the proof for β_1; the one for β_2 is similar. Lemma 3.3.2 shows that $\overline{\mathcal{D}}^*$ is the optimal set for all $\beta \in (\beta_1, \beta_2)$. Observe that the minimization problem is convex; let (β^*, x^*) be an optimal solution. Obviously, x^* is also optimal in $(\mathrm{P}(b(\beta^*), c))$ with optimal value $(b+\beta^*\Delta b)^T y$ for arbitrary $y \in \overline{\mathcal{D}}^*$. Hence $\beta^* \geq \beta_1$. On the other hand, let $x^{(1)}$ be optimal in $(\mathrm{P}(b(\beta_1), c))$.

[1] Personal communication 1992; Gauvin's paper is not mentioned in the historical survey by Gal [65].

By Corollary 3.3.3 it holds $(x^{(1)})^T s = 0, \ \forall s \in \overline{\mathcal{D}}^*$. Hence the pair $(\beta_1, x^{(1)})$ is feasible in the minimization problem and we have $\beta^* \leq \beta_1$. This completes the proof. □

Summarizing the results sofar we conclude that correct shadow prices and linearity intervals are obtained with the use of optimal sets. While usually just one optimal solution is used in sensitivity analysis, we next give three approaches based on the use of optimal sets, motivated by three different but equivalent ways of describing the optimal set. The first uses (optimal) bases, the second optimal partitions and the third optimal values.

3.4 USING OPTIMAL BASES

Using the simplex method for solving an LP problem gives an optimal basic solution. A basis $\mathcal{B}$ of A is a set of m indices, such that the submatrix $A_\mathcal{B}$ of A is nonsingular. The corresponding variables are the basic variables. The indices of the remaining nonbasic variables are in $\mathcal{N}$. For basis $\mathcal{B}$, the associated primal basic solution x is given by

$$x = \begin{pmatrix} x_\mathcal{B} \\ x_\mathcal{N} \end{pmatrix} := \begin{pmatrix} A_\mathcal{B}^{-1} b \\ 0 \end{pmatrix},$$

and the dual basic solution by

$$y = A_\mathcal{B}^{-T} c_\mathcal{B}, \quad s = \begin{pmatrix} s_\mathcal{B} \\ s_\mathcal{N} \end{pmatrix} := \begin{pmatrix} 0 \\ c_\mathcal{N} - A_\mathcal{N}^T y \end{pmatrix}.$$

If $x_\mathcal{B} \geq 0$ then $\mathcal{B}$ is a primal feasible basis; if $s_\mathcal{N} \geq 0$ then $\mathcal{B}$ is dual feasible. We call a basis *optimal* if it is both primal and dual feasible; a basis is called *primal optimal* if the associated primal basic solution is optimal for (P); analogously, a basis is called *dual optimal* if the associated dual basic solution is optimal for (D). Note that a primal (dual) optimal basis need not be dual (primal) feasible. A basis $\mathcal{B}$ is called *primal degenerate* if the associated primal solution x has $x_i = 0$ for some $i \in \mathcal{B}$. Analogously, a basis $\mathcal{B}$ is called *dual degenerate* if the associated dual solution s has $s_i = 0$ for some $i \in \mathcal{N}$.

3.4.1 Shadow prices and shadow costs

An important aspect of postoptimal analysis is the determination of shadow prices (shadow costs). As follows from Lemma 3.3.5 the left and right shadow prices (costs) can be obtained from solving auxiliary LP problems. Let $\Delta b := e^{(i)}$, where $e^{(i)}$ is the ith unit vector. Let us denote the shadow prices by p_i^- and p_i^+. Then

$$p_i^- = \min_k \{ y_i^{(k)} \}, \quad p_i^+ = \max_k \{ y_i^{(k)} \}, \tag{3.7}$$

where $y^{(k)}$, $k = 1, \ldots, K$, are the optimal dual basic solutions. This result has been derived in [4, 11, 68, 90, 132]. We illustrate the notion of left and right shadow prices with the following example, where the dual variables are not necessarily equal to a shadow price.

Example 3.4.1 Consider the primal–dual pair of LP problems:

$$\min_x \{ -2x_2 + 2x_3 + 4x_4 + 5x_5 + 6x_6 \ : \ -x_1 - 2x_2 + x_4 + x_5 = 1 \ ,$$
$$-x_2 - x_3 - x_4 + x_6 = -1, \ x \geq 0 \} ,$$
$$\max_y \{ y_1 - y_2 \ : \ -y_1 \leq 0, -2y_1 - y_2 \leq -2, \ -y_2 \leq 2, \ y_1 - y_2 \leq 4 \ ,$$
$$y_1 \leq 5, \ y_2 \leq 6 \} .$$

The optimal dual basic solutions are $y^{(1)} = (2, -2)^T$, $y^{(2)} = (5, 1)^T$. Using (3.7) the left and right shadow prices are

$$p_1^- = 2, \ p_1^+ = 5; \ \ p_2^- = -2, \ p_2^+ = 1.$$

The optimal objective value is 4. From $y^{(1)}$ it could erroneously be concluded that a unit increase in b_1 (from 1 to 2) would yield a value 6, whereas the correct value is 9; also $y^{(2)}$ suggests that decreasing b_2 from -1 to -2 gives an optimal value 3 instead of 6. ◇

The theory and the example show that in case of multiple optimal dual basic solutions (primal degeneracy) one has to distinguish between the rate of change as a consequence of decreasing and increasing the parameter β. In this case, the widespread belief that *the* shadow price is given by *the* dual value is not valid. Rubin and Wagner [212] indicate the traps and give a number of tips for correct interpretation of results of the dual problem in practice. Analogously, shadow costs are not uniquely defined in a breakpoint of the optimal value function $g(\gamma)$ (cf. Greenberg [90]). This leads to the introduction of left and right shadow costs for which similar results can be derived. The validity of a shadow price (cost) can be checked by computing the range where it is correct, which is the next subject.

3.4.2 Linearity intervals

The classical approach to sensitivity analysis is to pose the question in what interval the objective coefficient c_j (or right–hand side b_i) can vary such that a given (computed) optimal basis $\mathcal{B}$ remains optimal. In fact, one is interested in the range where the shadow price (cost) is valid. Let us consider the case of varying primal objective, and assume that $\Delta c = e^{(j)}$. Hence we are interested in the problem (P($b, c(\gamma)$)) and its dual. Let us denote by $T_{\mathcal{B}}$ the interval for γ for which $\mathcal{B}$ is an optimal basis. Then

$$T_{\mathcal{B}} = \{ \gamma \; : \; \{ (x,y,s) \; : \; Ax = b, \; x_{\mathcal{B}} \geq 0, \; x_{\mathcal{N}} = 0, \\ A^T y + s = c + \gamma \Delta c, \; s_{\mathcal{B}} = 0, \; s_{\mathcal{N}} \geq 0 \} \neq \phi \}.$$

It is well known that $T_{\mathcal{B}}$ is an interval which can be computed at low cost by twice computing m ratios and comparing them. The results in Tables 3.1 and 3.2 were produced as outlined above. The reason that this approach gives so different answers is explained by the degeneracy apparent in the problem, whence the optimal basis might not be unique and/or the optimal primal or dual solution might not be unique. Recall from Section 3.3 that optimal sets should be used, which in the context of the simplex method implies (by definition) that *primal optimal bases* are required. Let x^* be the optimal basic solution for the original problem and denote the set of primal optimal bases associated with x^* by $S(x^*)$. Ward and Wendell [242] introduce the *optimal coefficient set* of an optimal solution x^* of (P(b, c)) as

$$T(x^*) := \{ \gamma \; : \; x^* \text{ is an optimal solution of } (\text{P}(b, c(\gamma))) \}.$$

A similar definition is given by Mehrotra and Monteiro [171]. Let us also define

$$R(x^*) := \{ \gamma \; : \; g(\gamma) = g(0) + \gamma x_j^* \}.$$

Since x^* is optimal in (P(b, c)), $R(x^*)$ is either a linearity interval of $g(\gamma)$ with slope x_j^*, or the set $\{0\}$; in the latter case $\gamma = 0$ is a breakpoint of $g(\gamma)$. The following lemma contains the main result of this paragraph.

Lemma 3.4.2 *(i) If x^* is an optimal solution of (P(b, c)) then $T(x^*) = R(x^*)$; (ii) If x^* is an optimal basic solution of (P(b, c)) then $T(x^*) = \bigcup_{\mathcal{B} \in S(x^*)} T_{\mathcal{B}}$.*

Proof: (i) For $\gamma \in T(x^*)$ it holds

$$g(\gamma) = (c + \gamma e^{(j)})^T x^* = c^T x^* + \gamma x_j^* = g(0) + \gamma x_j^*,$$

so $\gamma \in R(x^*)$. If $\gamma \in R(x^*)$ then

$$g(\gamma) = g(0) + \gamma x_j^* = c^T x^* + \gamma x_j^* = (c + \gamma e^{(j)})^T x^*,$$

which shows that x^* is optimal in $(P(b, c(\gamma)))$.
(ii) If $\gamma \in \bigcup_{\mathcal{B} \in S(x^*)} T_{\mathcal{B}}$ then clearly x^* is optimal in $(P(b, c(\gamma)))$, so $\gamma \in T(x^*)$. Conversely, if $\gamma \in T(x^*)$ there is a basis $\mathcal{B}$ which is optimal in $(P(b, c(\gamma)))$ and associated with x^*; so $\gamma \in T_{\mathcal{B}}$. Since $\mathcal{B}$ is primal feasible for $(P(b, c(\gamma)))$ it is primal feasible for $(P(b, c))$. Hence $\mathcal{B} \in S(x^*)$ by the definition of a primal optimal basis. □

A few remarks are in order. Item *(ii)* of the lemma was shown by Ward and Wendell [242, Th. 17], probably being the first to stress the use of primal optimal bases. Note that the basis $\mathcal{B}$ used in its proof is primal feasible for $(P(b, c))$ but not necessarily dual feasible. From Lemma 3.4.2 we may conclude that either the optimal basic solution is only optimal in the breakpoint, or it corresponds to a linearity interval of the optimal value function in the sense that for each value of the parameter in this interval this solution is an optimal solution of the corresponding problem. If $\gamma = 0$ is a breakpoint of $g(\gamma)$ then obviously there exist multiple optimal basic solutions of $(P(b, c))$. The following lemma implies that whenever the intersection of optimal coefficient sets corresponding to different optimal basic solutions is nontrivial, then the sets coincide.

Lemma 3.4.3 *Let x^* and $\overline{x}^*$ be optimal basic solutions of $(P(b, c))$ and let $T(x^*) \cap T(\overline{x}^*) \neq \{0\}$. Then $T(x^*) = T(\overline{x}^*)$.*

Proof: By assumption, there exists $\overline{\gamma} \neq 0$ such that

$$\begin{aligned} g(\overline{\gamma}) &= g(0) + \overline{\gamma}\, x_j^* = c^T x^* + \overline{\gamma}\, x_j^* \\ g(\overline{\gamma}) &= g(0) + \overline{\gamma}\, \overline{x}_j^* = c^T \overline{x}^* + \overline{\gamma}\, \overline{x}_j^*. \end{aligned}$$

From $c^T x^* = c^T \overline{x}^*$ we may conclude $x_j^* = \overline{x}_j^*$. From this the result immediately follows. □

To the best of our knowledge, all commercial LP packages offering the opportunity of performing sensitivity analysis take the approach using one optimal basis, independently of whether degeneracy is present or not; also this approach is standard in textbooks often without referring to degeneracy problems. Earlier attempts have been made to circumvent the shortcomings of this classical approach, e.g., [51, 64, 65, 90, 132]. They suggest to compute the interval for

γ where at least one of the optimal bases associated with x^* remains optimal. Obviously the overall *critical region* given by such an approach is the union of intervals, each being one where an optimal basis remains optimal. This requires more computational effort, since (possibly) all optimal bases have to be generated. Evans and Baker [51] suggest to solve a sequence of LP problems to find this interval. Knolmayer [132] proposes an algorithm which does not need to generate all optimal bases associated with x^*; however, the statement of his algorithm is not clear nor complete. Gal [66] provides a parametric algorithm inspired by [159] that does not necessarily need all optimal bases associated with x^*; however, this approach still does not generate the complete linearity interval as desired. The following example illustrates the difference between using one optimal basis, optimal bases and primal optimal bases for the computation of intervals in sensitivity analysis.

Example 3.4.4 Consider the pair of primal–dual LP problems with parameter γ:

$$\min_x \{ -2x_2 + (1+\gamma)x_3 + 4x_4 + 5x_5 + 6x_6 \ : \ -x_1 - 2x_2 + x_4 + x_5 = 0 \ ,$$
$$-x_2 - x_3 - x_4 + x_6 = -1, \ x \geq 0 \}$$
$$\max_y \{ -y_2 \ : \ -y_1 \leq 0, -2y_1 - y_2 \leq -2, \ -y_2 \leq 1+\gamma, \ y_1 - y_2 \leq 4 \ ,$$
$$y_1 \leq 5, \ y_2 \leq 6 \} .$$

In Figure 3.1 the dual feasible region is depicted. Solving these problems for the initial value $\gamma_0 = 0$, the unique optimal primal solution is $x_3^* = 1, \ x_i^* = 0, \ i \neq 3$; the optimal dual basic solutions are $y^{(1)} = (1.5, -1)^T$ and $y^{(2)} = (3, -1)^T$. The set of optimal bases associated with x^* is

$$F(x^*) = \{ \{2,3\}, \ \{3,4\} \ \}.$$

Figure 3.1 shows that the bases in $F(x^*)$ are optimal in the following intervals for γ:

$$\mathcal{B}_1 = \{2,3\} \quad \longrightarrow \quad \gamma \in [-3, 1],$$
$$\mathcal{B}_2 = \{3,4\} \quad \longrightarrow \quad \gamma \in [-2, 1].$$

The set of primal optimal bases associated with x^* is given by

$$S(x^*) = \{ \{1,3\}, \ \{2,3\}, \ \{3,4\}, \ \{3,5\} \ \}.$$

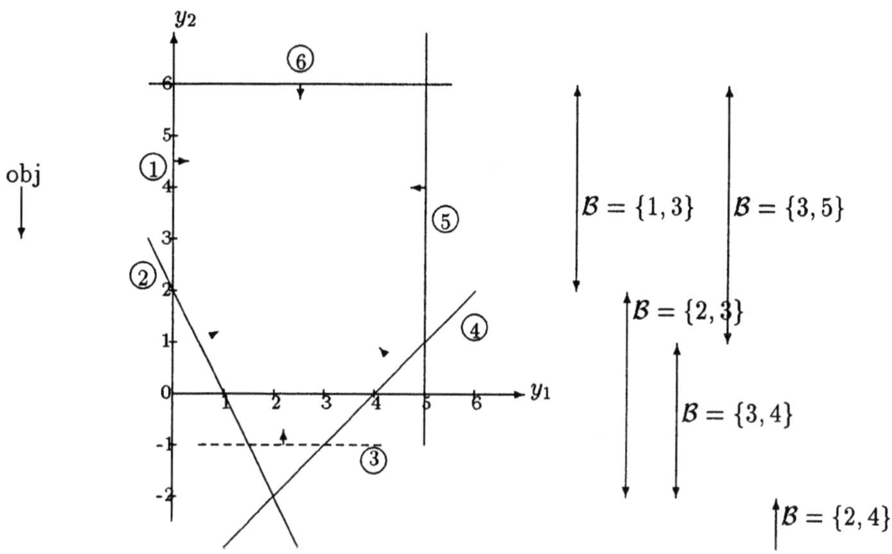

Figure 3.1 Feasible region of the dual problem in Example 3.3.4; constraint 3 is being shifted. The ranges where the different bases are optimal are denoted by the arrows.

For the bases that are not dual feasible we have the intervals

$$\mathcal{B}_3 = \{1,3\} \longrightarrow \gamma \in [-7,-3],$$
$$\mathcal{B}_4 = \{3,5\} \longrightarrow \gamma \in [-7,-2].$$

The union of the intervals of the primal optimal bases gives the complete linearity interval (Lemma 3.4.2), namely $\gamma \in [-7,1]$. Only using the optimal bases we find the subinterval $[-3,1]$. Using any of the individual bases we only find subintervals of the linearity interval. Indeed, the optimal value function is linear on $[-7,1]$, a breakpoint occurs at $\gamma = 1$, while for $\gamma < -7$ the primal problem is unbounded. Observe that the initial value $\gamma_0 = 0$ is outside the intervals implied by the (at $\gamma_0 = 0$) dual infeasible bases $\mathcal{B}_3$ and $\mathcal{B}_4$. $\diamond$

3.4.3 Example 3.2.1 (continued)

The approach outlined in this section has been applied to Example 3.2.1. Table 3.3 gives the ranges and left and right shadow costs for the objective coefficients. The ranges are obtained by considering primal optimal bases. Observe that still different linearity intervals or just a breakpoint may be obtained, since different optimal solutions may have different optimal coefficient sets.

Package	COST–ranges								
	x_{11}	x_{12}	x_{13}	x_{21}	x_{22}	x_{23}	x_{31}	x_{32}	x_{33}
CPLEX	$[1,\infty)$	$(-\infty,1]$	$[1,\infty)$	$[1,1]$	$[1,1]$	$[0,1]$	$[1,1]$	$[1,\infty)$	$[1,\infty)$
LINDO	$(-\infty,1]$	$[1,\infty)$	$[1,\infty)$	$[1,\infty)$	$[1,\infty)$	$[1,1]$	$[1,1]$	$[0,1]$	$[1,1]$
OSL	$[1,\infty)$	$(-\infty,1]$	$[1,\infty)$	$[1,1]$	$[1,1]$	$[0,1]$	$[1,1]$	$[1,\infty)$	$[1,\infty)$
PC–PROG	$[1,\infty)$	$[1,\infty)$	$[1,\infty)$	$[1,\infty)$	$[0,1]$	$[1,1]$	$[0,1]$	$[1,\infty)$	$[1,1]$
XMP	$[1,\infty)$	$[1,\infty)$	$(-\infty,1]$	$[0,1]$	$[0,1]$	$[1,\infty)$	$[1,\infty)$	$[1,\infty)$	$[1,1]$
shadow costs	2 0	2 0	2 0	3 0	3 0	3 0	3 0	3 0	3 0

	RHS–ranges					
	(3.1)	(3.2)	(3.3)	(3.4)	(3.5)	(3.6)
range	$[0,\infty)$	$[2,\infty)$	$[1,\infty)$	$[0,7]$	$[0,7]$	$[0,7]$
shadow price	0	0	0	1	1	1

Table 3.3 Ranges and prices in Example 3.2.1 using primal optimal bases.

3.5 USING OPTIMAL PARTITIONS

In Chapter 2 we showed that in each LP problem a strictly complementary solution exists (Theorem 2.3.1); such a solution uniquely determines the optimal partition of the LP problem. In this section we analyze an approach to sensitivity analysis using optimal partitions. The important result is that the linearity intervals of the optimal value function correspond to intervals where the optimal partition is constant, while in the breakpoints different partitions occur. Recalling from Section 3.3 that the optimal partition gives a complete description of the set of optimal solutions this should not be a surprise after having proved Lemmas 3.3.2 and 3.3.4.

This approach to sensitivity analysis is natural in the context of interior point methods. From Lemma 2.2.10 it follows that the limitpoint of the central path is strictly complementary, hence determines the optimal partition. Most interior point methods intrinsically follow the central path and, as shown by Güler and Ye [97], many of them actually yield a final iterate from which (at least theoretically) the optimal partition can be obtained. Mehrotra and Ye [172] propose and analyze a projection technique that yields the optimal partition in practice. Andersen and Ye [6] apply a similar technique based on [97]. In this section we show that not only we can compute linearity intervals but also the optimal partitions in the breakpoints; when computing shadow prices (costs) we automatically obtain the optimal partitions in the neighboring linearity intervals.

3.5.1 Perturbations in the right–hand side

As before we use the notation

$$b(\beta) := b + \beta\Delta b, \quad f(\beta) := z(b(\beta), c).$$

For each β we denote the corresponding optimal partition by $\pi_\beta = (B_\beta, N_\beta)$, with a strictly complementary solution $(x^{(\beta)}, y^{(\beta)}, s^{(\beta)})$.

Lemma 3.5.1 *Let the value function $f(\beta)$ be linear for $\beta \in [\beta_1, b_2]$. Then π_β is independent of β for all $\beta \in (\beta_1, \beta_2)$.*

Proof: Follows immediately from Lemma 3.3.2, observing that the optimal partition exactly identifies the optimal set. □

Let us assume that $\beta = 0$ and $\beta = 1$ are two consecutive breakpoints of the optimal value function $f(\beta)$. We show that the optimal partition in the linearity interval $0 < \beta < 1$ can be determined from the optimal partition at the breakpoint $\beta = 0$ by computing the right shadow price at $\beta = 0$. To this end we define the following primal–dual pair of LP problems[2]:

$$\begin{aligned}
&(\mathrm{P}_{\mapsto}^{\Delta b}) \quad \min_{x} \quad \left\{ c^T x \,:\, Ax = \Delta b,\ x_{N_0} \geq 0 \right\},\\
&(\mathrm{D}_{\mapsto}^{\Delta b}) \quad \max_{y,s} \quad \left\{ \Delta b^T y \,:\, A^T y + s = c,\ s_{B_0} = 0,\ s_{N_0} \geq 0 \right\}.
\end{aligned}$$

Note that in $(\mathrm{P}_{\mapsto}^{\Delta b})$ the variables x_i, $i \in B_0$, are free, hence we need to define its optimal partition $\overline{\pi} = (\overline{B}, \overline{N})$ in this case. Let $(\overline{x}, \overline{y}, \overline{s})$ be a strictly complementary solution of this pair of auxiliary problems. Since the dual variables $\overline{s}_i$

[2] The notation $\mapsto$ (and later $\leftarrow\!\!\dashv$, $\vdash\!\!\!-$ and $\to\!\!\!\!\!\rightarrow$) refers to the starting position and the direction of change. For instance, $\mapsto$ means starting in the breakpoint and increasing the parameter; $\vdash\!\!\!-$ means starting in a linearity interval and decreasing the parameter.

for $i \in B_0$ are identically zero, it is natural to let them be element of $\overline{B}$. So, we have $\overline{B} = B_0 \cup \{ i \in N_0 \ : \ \overline{s}_i = 0 \}$. We now derive the following theorem.

Theorem 3.5.2 *Let $\beta \in (0,1)$. For the primal–dual pair ($P^{\Delta b}_{\mapsto}$) and ($D^{\Delta b}_{\mapsto}$) it holds:*
(i) The optimal partition is (B_β, N_β);
(ii) $y^{(\beta)}$ is optimal in ($D^{\Delta b}_{\mapsto}$);
(iii) The optimal value $\Delta b^T y^{(\beta)}$ is the right shadow price at $\beta = 0$.

Proof: Note that (*ii*) and (*iii*) follow from Lemma 3.3.5. Let $0 < \beta < 1$ be arbitrary and consider

$$\overline{x} := \frac{x^{(\beta)} - x^{(0)}}{\beta}. \tag{3.8}$$

Since $(x^{(0)})_{N_0} = 0$ we have $\overline{x}_{N_0} \geq 0$. Obviously $A\overline{x} = \Delta b$, so $\overline{x}$ is feasible in ($P^{\Delta b}_{\mapsto}$). Observe that the dual problem ($D^{\Delta b}_{\mapsto}$) admits $(y^{(\beta)}, s^{(\beta)})$ as a feasible solution. We conclude the proof by showing that the pair $(\overline{x}, y^{(\beta)}, s^{(\beta)})$ is strictly complementary and that it determines $\pi_\beta = (B_\beta, N_\beta)$ as the optimal partition. Recall that the support of $x^{(\beta)}$ is B_β and the support of $x^{(0)}$ is B_0. So, for $i \in N_0$ we have $\overline{x}_i > 0$ if and only if $i \in N_0 \setminus N_\beta$. On the other hand, if $i \in N_0$, then we have $(s^{(\beta)})_i > 0$ if and only if $i \in N_\beta$. This proves that the given pair of solutions is strictly complementary with optimal partition $\pi_\beta = (B_\beta, N_\beta)$. The statement in (*ii*) follows immediately. Using (3.8), we obtain for $\beta \in (0,1)$

$$f(\beta) = c^T x^{(\beta)} = c^T x^{(0)} + \beta c^T \overline{x} = c^T x^{(0)} + \beta(\Delta b)^T y^{(\beta)},$$

which shows (*iii*). □

Starting from the breakpoint at $\beta = 1$ and using the optimal partition (B_1, N_1) a similar result can be obtained using the primal–dual pair of LP problems

$$(P^{\Delta b}_{\leftarrow}) \quad \min_x \quad \{ c^T x : Ax = -\Delta b, \ x_{N_1} \geq 0 \},$$
$$(D^{\Delta b}_{\leftarrow}) \quad \max_{y,s} \quad \{ -\Delta b^T y \ : \ A^T y + s = c, \ s_{B_1} = 0, \ s_{N_1} \geq 0 \}.$$

Without further proof we state the following theorem.

Theorem 3.5.3 *Let $\beta \in (0,1)$. For the primal–dual pair ($P^{\Delta b}_{\leftarrow}$) and ($D^{\Delta b}_{\leftarrow}$) it holds:*
(i) The optimal partition is (B_β, N_β);
(ii) $y^{(\beta)}$ is optimal in ($D^{\Delta b}_{\leftarrow}$);
(iii) The value $\Delta b^T y^{(\beta)}$ is the left shadow price at $\beta = 1$.

For future use we include the following result.

Lemma 3.5.4 *Let $\beta \in (0,1)$. It holds $\Delta b^T(y^{(\beta)} - y^{(0)}) > 0$ and $\Delta b^T(y^{(1)} - y^{(\beta)}) > 0$.*

Proof: Theorem 3.5.2 shows that maximizing $\Delta b^T y$ over the dual optimal face gives $y^{(\beta)}$ as an optimal solution, and $\Delta b^T y^{(\beta)}$ as the right shadow price. As a consequence of Theorem 3.5.3 minimizing $\Delta b^T y$ over the optimal face at $\beta = 0$ gives the left shadow price at $\beta = 0$; let $\overline{y}$ denote an optimal solution for this problem. Since the value function $f(\beta)$ has a breakpoint at $\beta = 0$, its left and right derivatives are different at $\beta = 0$, so we conclude $\Delta b^T \overline{y} < \Delta b^T y^{(\beta)}$. It follows that $\Delta b^T y$ is not constant on the dual optimal face. Since $y^{(0)}$ is an interior point of this face, we conclude that $\Delta b^T \overline{y} < \Delta b^T y^{(0)} < \Delta b^T y^{(\beta)}$, which implies the first result. An analogous proof using $\beta = 1$ gives the second result. □

Now we consider the case that the optimal partition associated to a linearity interval is known. We will show that the breakpoints and the corresponding optimal partitions can be found from the given partition and the perturbation vector Δb. This is done by observing that we may write the problems in Lemma 3.3.6 as LP problems.

For convenience we assume that $\beta = 0$ belongs to the linearity interval under consideration, and that the surrounding breakpoints, if they exist, occur at $\beta^- < 0$ and $\beta^+ > 0$ respectively. To determine β^- we consider the following primal–dual pair

$$(\mathrm{P}^{\Delta b}_{\leftarrow}) \quad \min_{\beta, x} \ \{ \beta \ : \ Ax - \beta\Delta b = b, \ x_{B_0} \geq 0, \ x_{N_0} = 0 \},$$

$$(\mathrm{D}^{\Delta b}_{\leftarrow}) \quad \max_{y, s} \ \{ b^T y \ : \ A^T y + s = 0, \ \Delta b^T y = -1, \ s_{B_0} \geq 0 \}.$$

Theorem 3.5.5 *For the primal–dual pair $(P^{\Delta b}_{\leftarrow})$ and $(D^{\Delta b}_{\leftarrow})$ it holds:*
(i) The optimal partition is $(B_{\beta^-}, N_{\beta^-})$;
(ii) $x^{(\beta^-)}$ is optimal in $(P^{\Delta b}_{\leftarrow})$;
(iii) The optimal value is β^-.

Proof: Items *(ii)* and *(iii)* follow from Lemma 3.3.6. The proof of *(i)* follows the same line of reasoning as the proof of Theorem 3.5.2. We construct feasible solutions for both problems and prove that they are strictly complementary with the correct partition. Since $(y^{(0)}, s^{(0)})$ is optimal in $(\mathrm{D}(b(\beta^-), c))$

(Corollary 3.3.3), we obtain the inclusion $N_0 \subseteq N_{\beta^-}$. This shows that

$$x := x^{(\beta^-)},\ \beta := \beta^-$$

is feasible for $(P^{\Delta b}_{\leftarrow})$. We will show that

$$y := \frac{y^{(\beta^-)} - y^{(0)}}{\Delta b^T (y^{(0)} - y^{(\beta^-)})} \tag{3.9}$$

is feasible for $(D^{\Delta b}_{\leftarrow})$. First we deduce from Lemma 3.5.4 that $\Delta b^T (y^{(0)} - y^{(\beta^-)})$ is positive, so y is well defined. Clearly $\Delta b^T y = -1$. Furthermore,

$$\left(\Delta b^T (y^{(0)} - y^{(\beta^-)})\right) A^T y = A^T (y^{(\beta^-)} - y^{(0)}) = s^{(0)} - s^{(\beta^-)}.$$

Since $(s^{(0)})_{B_0} = 0$ and $s^{(\beta^-)} \geq 0$, it follows that $(s^{(0)})_{B_0} - (s^{(\beta^-)})_{B_0} = -(s^{(\beta^-)})_{B_0} \leq 0$. So y is feasible for the dual problem. Since for $i \in B_0$ we have $x_i > 0$ if and only if $i \in B_{\beta^-}$, and $s_i = 0$ if and only if $i \in B_{\beta^-}$, the given pair is strictly complementary with the partition $(B_{\beta^-}, N_{\beta^-})$. This proves (i) and also (ii). To give also a proof of (iii), it follows from the linearity of the optimal value function on $[\beta^-, 0]$ that

$$b(\beta^-)^T y^{(\beta^-)} = f(\beta^-) = b^T y^{(0)} + \beta^- \Delta b^T y^{(0)},$$

or equivalently

$$b^T (y^{(\beta^-)} - y^{(0)}) = \beta^- \Delta b^T (y^{(0)} - y^{(\beta^-)}). \tag{3.10}$$

Multiplying (3.9) with b^T we obtain that the optimal value equals

$$\frac{b^T (y^{(\beta^-)} - y^{(0)})}{\Delta b^T (y^{(0)} - y^{(\beta^-)})} = \beta^-,$$

where the equality follows from (3.10). □

The breakpoint β^+ and the corresponding optimal partition are found by solving the pair of LP problems:

$$(P^{\Delta b}_{\rightarrow}) \quad \max_{\beta, x} \ \{\ \beta \ : \ Ax - \beta \Delta b = b,\ x_{B_0} \geq 0,\ x_{N_0} = 0\ \},$$

$$(D^{\Delta b}_{\rightarrow}) \quad \min_{y, s} \ \{\ -b^T y \ : \ A^T y + s = 0,\ \Delta b^T y = 1,\ s_{B_0} \geq 0\ \}.$$

Theorem 3.5.6 *For the primal–dual pair $(P^{\Delta b}_{\to|})$ and $(D^{\Delta b}_{\to|})$ it holds:*
(i) The optimal partition is $(B_{\beta+}, N_{\beta+})$;
(ii) $x^{(\beta^+)}$ is optimal in $(P^{\Delta b}_{\to|})$;
(iii) The optimal value is β^+.

To conclude this paragraph we mention that the auxiliary LP problems given here can be used to compute the optimal value function of a parametric LP problem. For instance, given some initial value β_0 with corresponding optimal partition, the part of the function to the right of β_0 is computed by alternately solving $(\mathrm{P}^{\Delta b}_{\to|})$ (for the breakpoint to the right) and $(\mathrm{D}^{\Delta b}_{\mapsto})$ (for the right shadow price in that breakpoint).

3.5.2 Perturbations in the objective

Let us now consider the effect of variations in the objective vector c on the optimal value function. By 'dualizing' the results above we obtain the appropriate results. Just as in the previous section we show that the 'surrounding' partitions of a given partition can be found by solving appropriate LP problems, which are formulated in terms of the given partition and the perturbation Δc. The proofs are based on the same idea as for their dual counterparts: one checks that natural candidate solutions for both problems are feasible indeed, and then shows that these solutions are strictly complementary with the correct partition. Therefore, we state these results without proofs. The discussion is facilitated using

$$c(\gamma) := c + \gamma\Delta c, \quad g(\gamma) := z(b, c(\gamma)),$$

where b and c are such that the pair (b, c) is feasible. For each γ we denote the corresponding optimal partition by $\pi_\gamma = (B_\gamma, N_\gamma)$ and strictly complementary solutions by $(x^{(\gamma)}, y^{(\gamma)}, s^{(\gamma)})$. We start with the case that the given partition belongs to a breakpoint. Without loss of generality we assume again that $\gamma = 0$ and $\gamma = 1$ are two consecutive breakpoints of $g(\gamma)$.

Consider the following pair of LP problems.

$$\begin{array}{lll} (\mathrm{P}^{\Delta c}_{\mapsto}) & \min\limits_{x} & \left\{ \Delta c^T x \, : \, Ax = b, \; x_{B_0} \geq 0, \; x_{N_0} = 0 \right\}, \\ (\mathrm{D}^{\Delta c}_{\mapsto}) & \max\limits_{y,s} & \left\{ b^T y \, : \, A^T y + s = \Delta c, \; s_{B_0} \geq 0 \right\}. \end{array}$$

Theorem 3.5.7 *Let $\gamma \in (0, 1)$. For the primal–dual pair $(P^{\Delta c}_{\mapsto})$ and $(D^{\Delta c}_{\mapsto})$ it holds:*

(i) The optimal partition is (B_γ, N_γ)*;*
(ii) $x^{(\gamma)}$ *is optimal in* $(P^{\Delta c}_{\mapsto})$*;*
(iii) The optimal value $\Delta c^T x^{(\gamma)}$ *is the right shadow cost at* $\gamma = 0$*.*

A similar result can be obtained for the optimal partition at $\gamma = 1$. Defining the pair of LP problems

$$(\mathrm{P}^{\Delta c}_{\mapsfrom}) \quad \max_x \quad \left\{ \Delta c^T x \; : \; Ax = b, \; x_{B_1} \geq 0, \; x_{N_1} = 0 \right\},$$
$$(\mathrm{D}^{\Delta c}_{\mapsfrom}) \quad \min_{y,s} \quad \left\{ -b^T y \; : \; A^T y + s = -\Delta c, \; s_{B_1} \geq 0 \right\},$$

one has the following theorem.

Theorem 3.5.8 *Let* $\gamma \in (0,1)$*. For the primal–dual pair* $(P^{\Delta c}_{\mapsfrom})$ *and* $(D^{\Delta c}_{\mapsfrom})$ *it holds:*
(i) The optimal partition is (B_γ, N_γ)*;*
(ii) $x^{(\gamma)}$ *is optimal in* $(P^{\Delta c}_{\mapsfrom})$*;*
(iii) The optimal value $\Delta c^T x^{(\gamma)}$ *is the left shadow price at* $\gamma = 1$*.*

Using these results we derive the following corollary.

Corollary 3.5.9 *Let* $\gamma \in (0,1)$*. It holds* $\Delta c^T (x^{(\gamma)} - x^{(0)}) < 0$ *and* $\Delta c^T (x^{(1)} - x^{(\gamma)}) < 0$*.*

The next results concern the determination of the size of the linearity interval and the optimal partition in the breakpoints, given that the optimal partition associated to the linearity interval is known. Assume that $\gamma = 0$ belongs to the linearity interval under consideration, and that the surrounding breakpoints, if they exist, occur at $\gamma^- < 0$ and $\gamma^+ > 0$ respectively. We consider the following pair of problems.

$$(\mathrm{P}^{\Delta c}_{\leftarrow}) \quad \max_x \quad \left\{ -c^T x \; : \; Ax = 0, \; \Delta c^T x = 1, \; x_{N_0} \geq 0 \right\},$$
$$(\mathrm{D}^{\Delta c}_{\leftarrow}) \quad \min_{\gamma,y,s} \quad \left\{ \gamma \; : \; A^T y + s - \gamma \Delta c = c, \; s_{B_0} = 0, \; s_{N_0} \geq 0 \right\}.$$

Theorem 3.5.10 *For the primal–dual pair* $(P^{\Delta c}_{\leftarrow})$ *and* $(D^{\Delta c}_{\leftarrow})$ *it holds:*
(i) The optimal partition is $(B_{\gamma^-}, N_{\gamma^-})$*;*
(ii) $y^{(\gamma^-)}$ *is optimal in* $(D^{\Delta c}_{\leftarrow})$*;*
(iii) The optimal value is γ^-*.*

Similarly, the breakpoint γ^+ is obtained from the pair of LP problems:

$$(\mathrm{P}^{\Delta c}_{\rightarrow}) \quad \min_{x} \quad \{ c^T x \ : \ Ax = 0, \ \Delta c^T x = -1, \ x_{N_0} \geq 0 \},$$

$$(\mathrm{D}^{\Delta c}_{\rightarrow}) \quad \max_{\gamma, y, s} \quad \{ \gamma \ : \ A^T y + s - \gamma \Delta c = c, \ s_{B_0} = 0, \ s_{N_0} \geq 0 \}.$$

Theorem 3.5.11 *For the primal-dual pair* $(P^{\Delta c}_{\rightarrow})$ *and* $(D^{\Delta c}_{\rightarrow})$ *it holds:*
(i) The optimal partition is $(B_{\gamma^+}, N_{\gamma^+})$*;*
(ii) $y^{(\gamma^+)}$ *is optimal in* $(D^{\Delta b}_{\rightarrow})$*;*
(iii) The optimal value is γ^+*.*

3.5.3 Example 3.2.1 (continued)

Let us show the results obtained with the outlined approach for Example 3.2.1. An interior point solution is given in Table 3.4. This solution is strictly complementary: B corresponds to the variables x_{ij}, $i, j = 1, \ldots, 3$ and s_i, $i = 1, \ldots, 3$, while the variables d_j, $j = 1, \ldots, 3$ are in N. Note that the number of variables with a positive value is larger than for the basic solutions. In certain applications this is unpreferable, while in others (e.g., finding critical paths in project planning) this is attractive [91]. The coefficients in the objective function and the right-hand side vector are varied one at a time, so $\Delta b = e^{(i)}$ and $\Delta c = e^{(j)}$, for all i, j. Notice that all the cost-coefficients are at a breakpoint. The intervals for the right-hand side coefficients are the same as in Table 3.3.

Method	Optimal primal solution								
	x_{11}	x_{12}	x_{13}	x_{21}	x_{22}	x_{23}	x_{31}	x_{32}	x_{33}
IPM	0.5	0.5	0.5	1.4	1.4	1.4	1.1	1.1	1.1
	COST-ranges								
range	[1,1]	[1,1]	[1,1]	[1,1]	[1,1]	[1,1]	[1,1]	[1,1]	[1,1]
shadow costs	2 0	2 0	2 0	3 0	3 0	3 0	3 0	3 0	3 0

Method	Optimal dual solution					
	(3.1)	(3.2)	(3.3)	(3.4)	(3.5)	(3.6)
IPM	0	0	0	1	1	1
	RHS-ranges					
range	[0,∞)	[2,∞)	[1,∞)	[0,7]	[0,7]	[0,7]

Table 3.4 Results from interior approach in Example 3.2.1.

3.5.4 The 100% rule

An interesting and useful, but not widely known, extension to the standard sensitivity analysis using one parameter is the *100% rule*. It was introduced by Bradley et al. [29] for multi–parameter variations in the context of optimal bases. For each of the parameters a separate sensitivity analysis is performed, keeping the other parameters constant. This gives an interval for each of the parameters. The rule says that for combined variations in all parameters the given basis remains optimal if the sum of the percentages of the variation w.r.t. the maximal variation is not larger than 100%. Stated otherwise, the optimal basis remains optimal in the 'diamond' determined by the end–points of the separate intervals. Note that the rule only gives a sufficient condition, while the actual region where the basis remains optimal might be larger. The following example clarifies the rule.

Example 3.5.12 Consider again Example 3.2.1. Let the right–hand sides of (3.1) and (3.2) be varied. We consider the solution computed by CPLEX. The dark area in Figure 3.2(Left) is given by the 100%–rule; however, also in the light–shaded area the basis from CPLEX is optimal. ◇

We show that the 100% rule can be extended to regions where the optimal partition remains constant. This is the contents of the next lemma, which considers the case of varying right–hand side coefficients.

Lemma 3.5.13 *Let the problems (P) and (D) have optimal partition* $\pi = (B, N)$ *with strictly complementary solution* $(\overline{x}, \overline{y}, \overline{s})$. *Let* $I \subseteq \{1, \ldots, n\}$. *Define*

$$b^{(i)}(\beta_i) = b + \beta_i e^{(i)}.$$

Let for each $i \in I$, $\beta_i^+ \neq 0$ *be such that the problems* $(P(b^{(i)}(\beta_i^+), c))$ *and* $(D(b^{(i)}(\beta_i^+), c))$ *have the same optimal partition* π. *Let* Δb *satisfy*

$$\Delta b_i = 0 \;\; i \notin I, \quad \frac{\Delta b_i}{\beta_i^+} \geq 0 \;\; i \in I, \quad \sum_{i \in I} \frac{\Delta b_i}{\beta_i^+} < 1.$$

Then π *is also the optimal partition of the pair*

$$(P_{\Delta b}) \quad \min_x \quad \{\, c^T x \; : \; Ax = b + \Delta b, \; x \geq 0 \,\}$$

$$(D_{\Delta b}) \quad \max_{y,s} \quad \{\, (b + \Delta b)^T y \; : \; A^T y + s = c, \; s \geq 0 \,\}.$$

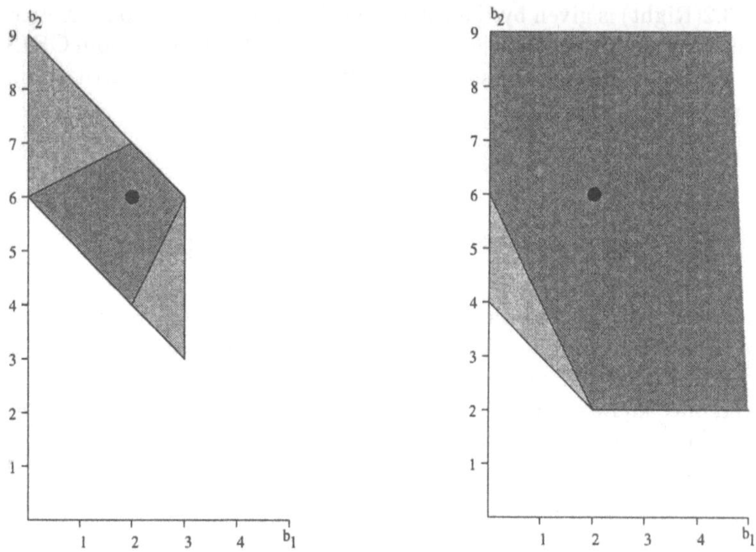

Figure 3.2 (Left) Regions for optimal basis in Example 3.5.12; the current value is $(2,6)$. (Right) Regions for optimal partition in Example 3.5.12.

Proof: Define, for $i \in I$, $f_i := \Delta b_i/\beta_i^+$ and $f := \sum_{i \in I} f_i$. Observe that $f_i \geq 0$ and $f < 1$ by assumption. Let $(x^{(i)}, y^{(i)}, s^{(i)})$ be strictly complementary solutions for $(\mathrm{P}(b^{(i)}(\beta_i^+), c))$ and $(\mathrm{D}(b^{(i)}(\beta_i^+), c))$. We show that

$$x := (1-f)\overline{x} + \sum_{i \in I} f_i x^{(i)}, \quad y := \overline{y}, \quad s := \overline{s},$$

is a strictly complementary solution of $(\mathrm{P}_{\Delta b})$ and $(\mathrm{D}_{\Delta b})$. Obviously, (y, s) is feasible in the dual problem. Also,

$$Ax = (1-f)A\overline{x} + \sum_{i \in I} f_i A x^{(i)} = (1-f)b + \sum_{i \in I} f_i (b + \beta_i^+ e^{(i)}) = b + \Delta b.$$

From $f_i \geq 0$ and $f < 1$ it follows that $x_B > 0$ and $x_N = 0$, which completes the proof. □

Example 3.5.14 Consider again Example 3.2.1. As in Example 3.5.12 the right–hand sides of constraints (3.1) and (3.2) are varied. The dark area in

Figure 3.2(Right) is given by the 100%–rule for optimal partitions. Notice, that this area is much larger than the one given by the optimal basis from CPLEX in Figure 3.2(Left). However, also in the light–shaded area this partition remains optimal. ◇

3.6 USING OPTIMAL VALUES

In Section 3.3 we showed that correct shadow prices and linearity intervals can be obtained by solving appropriate LP problems over the optimal face of the original primal or dual problem, that is, using the set of optimal solutions. However, once knowing the optimal value z^* of the LP problem, we can trivially describe the optimal faces as follows:

$$\{ \, x \; : \; Ax = b, \; x \geq 0, \; c^T x = z^* \, \},$$
$$\{ \, (y, s) \; : \; A^T y + s = c, \; s \geq 0, \; b^T y = z^* \, \}.$$

We may use this description in the results of Section 3.3. For instance, linearity intervals of $f(\beta)$ are computed by (cf. Lemma 3.3.6)

$$\begin{aligned} \beta_1 &= \min_{\beta, x} \left\{ \, \beta \; : \; Ax - \beta \Delta b = b, \; x \geq 0, \; c^T x = (b + \beta \Delta b)^T y^* \, \right\}, \\ \beta_2 &= \max_{\beta, x} \left\{ \, \beta \; : \; Ax - \beta \Delta b = b, \; x \geq 0, \; c^T x = (b + \beta \Delta b)^T y^* \, \right\}. \end{aligned}$$

where $y^* \in \mathcal{D}^*$. Similarly, left and right shadow prices are found by

$$\begin{aligned} f'_-(\beta) &= \min_{y, s} \left\{ \, \Delta b^T y \; : \; A^T y + s = c, \; s \geq 0, \; (b + \beta \Delta b)^T y = b^T y^* \, \right\}, \\ f'_+(\beta) &= \max_{y, s} \left\{ \, \Delta b^T y \; : \; A^T y + s = c, \; s \geq 0, \; (b + \beta \Delta b)^T y = b^T y^* \, \right\}. \end{aligned}$$

An advantage of the approach is that we do not need to know the optimal partition, just the optimal value. In the literature few explicit references to this idea can be found, e.g., Akgül [4], De Jong [122], Gondzio and Terlaky [83] and Mehrotra and Monteiro [171]. Similar ideas appear in Magnanti and Wong [160], who use a subproblem defined on the optimal set to compute certain cuts in Benders decomposition [22] and in Terlaky [228], who considers marginal values in ℓ_p–programming.

3.7 COMPUTATIONAL RESULTS

In De Jong [122] an extensive computational study is made to compare the three approaches to sensitivity analysis described in Sections 3.4, 3.5 and 3.6, using the NETLIB set of LP problems [69]. The approaches in the last two sections were programmed within OSL, while CPLEX was used for the standard simplex approach. In this section we report on the results obtained for an LP model which is a small representative of the type of models used in daily decision making by a major oil company. The model concerns an oil refinery and contains production, transportation and product exchange during three periods. The model has 2110 variables and 1101 constraints. Although the number of LP subproblems to be solved is quite large, *all* the sensitivity information for this model was obtained within 15 minutes on an HP9000–720 workstation.

We compare the sensitivity information from CPLEX with the results obtained from the interior approach. First consider shadow costs. Recall from Section 3.4 that different situations may occur: either the number given by CPLEX will be the shadow cost, a left or a right shadow cost (a one–sided shadow cost) or it has no meaning as a shadow cost. In the latter case, the value is in between the left and the right shadow price. In the refinery model, we found that for 1% of the coefficients the number returned by CPLEX was just a one–sided shadow cost while for 0.7% it was no shadow cost at all. Moreover, for 36.4% of the coefficients the range CPLEX reported was not equal to the complete linearity interval. For 2.9% of the coefficients a breakpoint was given. To see whether the differences in the ranges are significant, we split them up into six categories, depending on the ratio between the length of the linearity interval and the length of the range computed by CPLEX. A histogram of this is given to the left in Figure 3.3. Here, the category *breakp* means that the coefficient is in a breakpoint while CPLEX reports a range. If the linearity interval has infinite length then we use the category *inf*. A similar comparison was done for the shadow costs. We compared the CPLEX outcome with both the left and right shadow cost and categorized the largest difference (that is, either the difference between the CPLEX outcome and the left shadow cost or between CPLEX and the right shadow cost); if the current value of the objective coefficient is not a breakpoint, then the CPLEX number will be the unique shadow cost (see the right part of Figure 3.3; the category *inf* means that the LP problem becomes unbounded as soon as the objective coefficient is either increased or decreased).

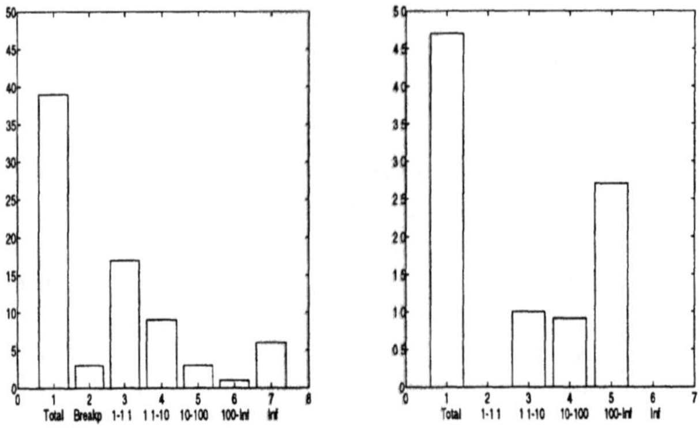

Figure 3.3 Differences in objective coefficient ranges and shadow costs (%).

For the right–hand side elements of the refinery model it appeared that CPLEX reported a one–sided shadow price for 25.4% of the coefficients and no shadow price for 2.8%. For 45.5% of the coefficients the interval returned was not equal to a complete linearity interval. Again, we made histograms to split the differences with respect to their magnitude, see Figure 3.4.

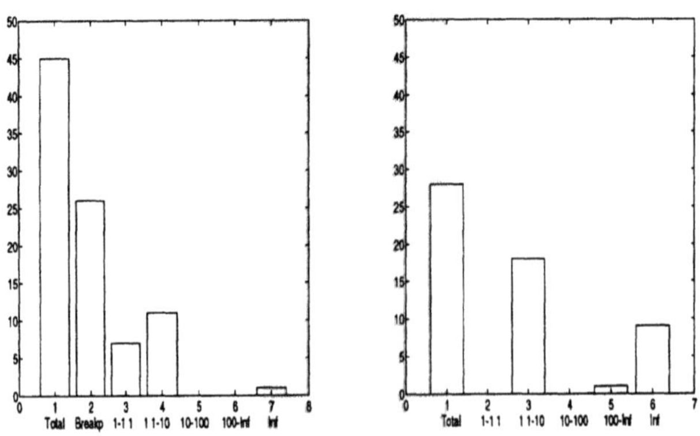

Figure 3.4 Differences in right–hand side ranges and shadow prices (%).

We conclude that the standard approach to sensitivity analysis gives different results than the one based on optimal sets, and that a commercial LP code often only gives partial sensitivity information. In De Jong [122] and Jansen et al. [110] a more detailed analysis of the differences is found in comparison with the physical meaning of the various variables and constraints.

3.8 CONCLUDING REMARKS

For long it was generally believed that sensitivity analysis based on solutions obtained with interior point methods was impossible. In this chapter we gave a unifying approach to sensitivity analysis using optimal sets instead of just one optimal solution. Within the approach we showed how to incorporate sensitivity analysis based on the simplex method (using primal/dual optimal bases), based on interior point methods (using the optimal partition), and based on the optimal value. So, sensitivity analysis using interior point solutions is possible, and sometimes even better or desirable, since it is an approach using optimal sets instead of one optimal solution. The applicability of approaches to sensitivity analysis using optimal sets has been shown with a practical LP model; we refer to Greenberg [91] for other applications.

Although discussed in many textbooks on LP the issue of sensitivity analysis deserves more attention than generally given. Specifically, problems with degeneracy should not be downplayed, instead the intimate relation between rates and ranges should be stressed. The best way to do this is via the optimal value function. We believe that practitioners of LP should be aware of the apparent difficulties with the standard approach. Furthermore, packages for LP should be able to compute (parts of) optimal value functions if desired by the user, as can be done in, e.g., AIMMS [23] and LINDO [213]. It's our experience that such functions often provide insight into the relative importance of certain changes in the problem data.

4

SENSITIVITY ANALYSIS IN QUADRATIC PROGRAMMING

4.1 INTRODUCTION

In this chapter the topic of sensitivity analysis in convex quadratic programming (CQP) is investigated, as an extension of the approach in the previous section. As will become clear, the quadratic case is more involved, but opens a wider area of application. We consider the primal CQP problem

$$(\text{QP}) \quad \min_{x} \left\{ c^T x + \frac{1}{2} x^T Q x \ : \ Ax - b, \ x \geq 0 \right\},$$

where $c, x \in \mathbb{R}^n$, $b \in \mathbb{R}^m$, A an $m \times n$ matrix with full row rank and Q a symmetric positive semidefinite $n \times n$ matrix. The Wolfe–dual of (QP) is given by

$$(\text{QD}) \quad \max_{u,y,s} \left\{ b^T y - \frac{1}{2} u^T Q u \ : \ A^T y + s - Qu = c, \ s \geq 0 \right\}.$$

Any optimal solution $(x, (u, y, s))$ of (QP) and (QD) satisfies $x^T s = 0$, i.e., complementary slackness. Furthermore, there exist optimal solutions for which $x = u$; we will only be interested in those and henceforth denote an optimal solution just by (x, y, s). While in linear programming (LP) the set of optimal solutions can be characterized with the optimal partition and strictly complementary solutions (see Chapter 3), in CQP the set of optimal solutions is characterized by *maximal complementary solutions* and the corresponding *tripartition* (Güler and Ye [97]). We define

$$\begin{aligned} B &= \{ i \ : \ x_i > 0 \text{ for an optimal solution } x \text{ of (QP) } \}, \\ N &= \{ i \ : \ s_i > 0 \text{ for an optimal solution } (x, y, s) \text{ of (QD) } \}, \\ T &= \{1, \ldots, n\} \setminus (B \cup N). \end{aligned}$$

This tripartition of the index set is denoted by $\pi := (B, N, T)$. A maximal complementary solution (x, y, s) is a solution for which

$$x_i > 0 \iff i \in B, \quad s_i > 0 \iff i \in N;$$

such a solution was shown to exist by Güler and Ye [97]. Moreover, they show that interior point methods generate such a solution (in the limit). Recall from Section 2.1 that in LP the set T is empty, which leads to the existence of a strictly complementary solution. Presence of the set T is an unpleasant feature in algorithms to solve CQP problems since it tends to slow down convergence of (interior point) algorithms (Monteiro and Tsuchiya [184]). For investigating sensitivity analysis in nonlinear programming (NLP) the existence of a strictly complementary solution is often assumed to introduce some kind of regularity, see e.g., Fiacco and McCormick [53], Fiacco [52], Harker and Pang [98]. As argued by Jittorntrum [121] in certain cases this is too restrictive.

In this section we investigate parametric versions of (QP) and (QD) as an extension of the results for LP in Section 3.5. Markowitz [162, 163] was one of the first to consider this problem, being particularly interested in applications to mean–variance analysis of investment portfolios. Other early references include Houthakker [107] and Wolfe [244] who use a parametric algorithm to solve (QP) itself (for a survey see Boot [25]). Eaves [47] also considers the parametric CQP problem. In the cited papers an extension of the simplex method and the concept of basis is used; often some type of nondegeneracy is assumed. Recent papers on mean–variance analysis by Kriens and Van Lieshout [145] and Vörös [241] show similar results but concentrate on differentiability of the optimal value function. In this chapter we do not make the assumption that the set T is empty, that (QP) is nondegenerate nor that Q is positive definite. Similarly to the analysis in Section 3.5 we give a characterization of the optimal value function using the tripartition and maximal complementary solutions. This is more general than results in Boot [25] and Bank et al. [15], who do not identify T explicitly, but merely consider N and T together. We propose an algorithm to compute the optimal value function and discuss application to calculating efficient frontiers for mean–variance models.

4.2 THE PARAMETRIC PROBLEM

We consider the parametric problem

$$(\mathrm{QP}_\lambda) \quad \min_x \left\{ (c + \lambda \Delta c)^T x + \frac{1}{2} x^T Q x \; : \; Ax = b, \; x \geq 0 \right\},$$

where $\Delta c \in \mathbb{R}^n$ is a given vector of variation. For $\lambda \in \mathbb{R}$ we denote the tripartition by $\pi_\lambda = (B_\lambda, N_\lambda, T_\lambda)$ and a maximal complementary solution by $(x^{(\lambda)}, y^{(\lambda)}, s^{(\lambda)})$. The dual is

$$(\mathrm{QD}_\lambda) \quad \max_{x,y,s} \left\{ b^T y - \frac{1}{2} x^T Q x \ : \ A^T y + s - Qx = c + \lambda \Delta c, \ s \geq 0 \right\}.$$

The optimal value of (QP_λ) is denoted by $\phi(\lambda)$, with the convention that

$$\begin{array}{ll} \phi(\lambda) = \infty & \text{if } (\mathrm{QP}_\lambda) \text{ is infeasible,} \\ \phi(\lambda) = -\infty & \text{if } (\mathrm{QP}_\lambda) \text{ is feasible and unbounded,} \end{array}$$

so $\phi : \mathbb{R} \to \mathbb{R} \cup \{-\infty, \infty\}$. Since the feasible set of (QP_λ) does not depend on λ, (QP_λ) will be either feasible or infeasible for all λ. We denote by Λ the set of values λ for which (QP_λ) has a bounded optimal value. For obvious reason we assume that Λ is nonempty. The first lemma gives some basic facts on $\phi(\lambda)$.

Lemma 4.2.1 *(i) The optimal value function $\phi(\lambda)$ is concave on Λ;*
(ii) Λ is a (possibly unbounded) closed interval;
(iii) If $\Delta c^T x \geq 0$ for all x satisfying $Ax = b, x \geq 0$, then $\phi(\lambda)$ is monotonically increasing on Λ.

Proof: (i) Let $\lambda_1, \lambda_2 \in \Lambda$ and $\alpha \in (0,1)$ be given and define $\lambda_\alpha := \alpha\lambda_1 + (1-\alpha)\lambda_2$. Then we have

$$\begin{aligned} \phi(\lambda_\alpha) &= \alpha \left((c + \lambda_1 \Delta c)^T x^{(\lambda_\alpha)} + \frac{1}{2} (x^{(\lambda_\alpha)})^T Q x^{(\lambda_\alpha)} \right) + \\ &\quad (1-\alpha) \left((c + \lambda_2 \Delta c)^T x^{(\lambda_\alpha)} + \frac{1}{2} (x^{(\lambda_\alpha)})^T Q x^{(\lambda_\alpha)} \right) \\ &\geq \alpha \phi(\lambda_1) + (1-\alpha)\phi(\lambda_2), \end{aligned}$$

where the inequality holds since the feasible set of (QP_λ) is independent of λ.
(ii) Follows from (i).
(iii) For $\lambda_1 > \lambda_2$ it holds

$$\begin{aligned} \phi(\lambda_1) - \phi(\lambda_2) &= (c + \lambda_1 \Delta c)^T x^{(\lambda_1)} + \frac{1}{2} (x^{(\lambda_1)})^T Q x^{(\lambda_1)} \\ &\quad -(c + \lambda_2 \Delta c)^T x^{(\lambda_2)} - \frac{1}{2} (x^{(\lambda_2)})^T Q x^{(\lambda_2)} \\ &\geq (c + \lambda_1 \Delta c)^T x^{(\lambda_1)} + \frac{1}{2} (x^{(\lambda_1)})^T Q x^{(\lambda_1)} \\ &\quad -(c + \lambda_2 \Delta c)^T x^{(\lambda_1)} - \frac{1}{2} (x^{(\lambda_1)})^T Q x^{(\lambda_1)} \\ &= (\lambda_1 - \lambda_2) \Delta c^T x^{(\lambda_1)} \geq 0, \end{aligned}$$

which completes the proof. □

The next lemma shows that the tripartition is constant on certain intervals.

Lemma 4.2.2 *Let λ_0 and λ_1 be such that $\pi_{\lambda_0} = \pi_{\lambda_1} = \pi$. Then $\pi_\lambda = \pi$ for all $\lambda \in [\lambda_0, \lambda_1]$.*

Proof: Without loss of generality we assume that $\lambda_0 = 0$ and $\lambda_1 = 1$. Let $(x^{(0)}, y^{(0)}, s^{(0)})$ and $(x^{(1)}, y^{(1)}, s^{(1)})$ be maximal complementary solutions for the respective problems. We define for $\lambda \in (0,1)$

$$\begin{gathered} x(\lambda) := (1-\lambda)x^{(0)} + \lambda x^{(1)}, \\ y(\lambda) := (1-\lambda)y^{(0)} + \lambda y^{(1)}, \quad s(\lambda) := (1-\lambda)s^{(0)} + \lambda s^{(1)}. \end{gathered} \tag{4.1}$$

It is easy to see that $Ax(\lambda) = b$ and $x(\lambda) \geq 0$. Also

$$A^T y(\lambda) + s(\lambda) - Qx(\lambda) = (1-\lambda)c + \lambda(c + \Delta c) = c + \lambda \Delta c.$$

So $(x(\lambda), y(\lambda), s(\lambda))$ is feasible for (QP_λ) and (QD_λ). Since $\pi_0 = \pi_1$, we have $x(\lambda)^T s(\lambda) = 0$, hence the proposed solution is optimal for (QP_λ) and (QD_λ). Using the support of $x(\lambda)$ and $s(\lambda)$ this implies

$$B \subseteq B_\lambda, \quad N \subseteq N_\lambda \text{ and } T \supseteq T_\lambda. \tag{4.2}$$

We will show that equality holds. Assuming to the contrary that $T \supset T_\lambda$, there exists a maximal complementary solution $(x^{(\lambda)}, y^{(\lambda)}, s^{(\lambda)})$ of (QP_λ) and (QD_λ) such that

$$(x^{(\lambda)})_i + (s^{(\lambda)})_i > 0 \text{ for some } i \in T, \ i \notin T_\lambda. \tag{4.3}$$

Let us now define for $\epsilon > 0$

$$\begin{gathered} \overline{x}(\epsilon) := x^{(1)} + \epsilon(x^{(1)} - x^{(0)}), \\ \overline{y}(\epsilon) := y^{(1)} + \epsilon(y^{(1)} - y^{(0)}), \quad \overline{s}(\epsilon) := s^{(1)} + \epsilon(s^{(1)} - s^{(0)}). \end{gathered}$$

For some $\overline{\epsilon} > 0$ small enough it holds

$$\overline{x}(\overline{\epsilon})_B > 0, \ \overline{x}(\overline{\epsilon})_{N \cup T} = 0, \quad \overline{s}(\overline{\epsilon})_N > 0, \ \overline{s}(\overline{\epsilon})_{B \cup T} = 0, \tag{4.4}$$

from which it follows that the proposed solutions are optimal for $(QP_{1+\overline{\epsilon}})$ and $(QD_{1+\overline{\epsilon}})$. Finally, we define

$$\begin{gathered} \tilde{x}^{(1)} := \tfrac{\overline{\epsilon}}{1-\lambda+\overline{\epsilon}} x^{(\lambda)} + \tfrac{1-\lambda}{1-\lambda+\overline{\epsilon}} \overline{x}(\overline{\epsilon}), \\ \tilde{y}^{(1)} := \tfrac{\overline{\epsilon}}{1-\lambda+\overline{\epsilon}} y^{(\lambda)} + \tfrac{1-\lambda}{1-\lambda+\overline{\epsilon}} \overline{y}(\overline{\epsilon}), \quad \tilde{s}^{(1)} := \tfrac{\overline{\epsilon}}{1-\lambda+\overline{\epsilon}} s^{(\lambda)} + \tfrac{1-\lambda}{1-\lambda+\overline{\epsilon}} \overline{s}(\overline{\epsilon}), \end{gathered}$$

which are feasible in (QP_1) and (QD_1). Also,

$$(\tilde{x}^{(1)})^T\tilde{s}^{(1)} = \frac{\bar{\epsilon}}{1-\lambda+\bar{\epsilon}}\frac{1-\lambda}{1-\lambda+\bar{\epsilon}}\left((x^{(\lambda)})^T\bar{s}(\bar{\epsilon}) + \bar{x}(\bar{\epsilon})^T s^{(\lambda)}\right).$$

Since (4.2) and (4.4) imply

$$(x^{(\lambda)})^T\bar{s}(\bar{\epsilon}) = (x^{(\lambda)})_N^T\bar{s}(\bar{\epsilon})_N = 0$$

and

$$(s^{(\lambda)})^T\bar{x}(\bar{\epsilon}) = (s^{(\lambda)})_B^T\bar{x}(\bar{\epsilon})_B = 0,$$

$(\tilde{x}^{(1)}, \tilde{y}^{(1)}, \tilde{s}^{(1)})$ is optimal for (QP_1) and (QD_1). However, if (4.3) would hold, we would have a solution of (QP_1) and (QD_1) with either $(\tilde{x}^{(1)})_i > 0$ or $(\tilde{s}^{(1)})_i > 0$ for $i \in T$, contradicting the definition of (B, N, T). Thus we conclude $T_\lambda = T$. Using (4.2) the lemma follows. □

The lemma implies the following corollary on the behavior of the optimal value function on intervals.

Corollary 4.2.3 *If* $\pi_{\lambda_0} = \pi_{\lambda_1} = \pi$ *then*
(i) On $[\lambda_0, \lambda_1]$ $\phi(\lambda)$ *is the quadratic function given by*

$$\begin{aligned}\phi(\lambda) &= \phi(0) + \frac{\lambda-\lambda_0}{\lambda_1-\lambda_0}\left(b^T(y^{(\lambda_1)} - y^{(\lambda_0)}) + (c+\lambda_0\Delta c)^T(x^{(\lambda_1)} - x^{(\lambda_0)})\right) + \\ &\quad \frac{1}{2}\left(\frac{\lambda-\lambda_0}{\lambda_1-\lambda_0}\right)^2 \Delta c^T(x^{(\lambda_1)} - x^{(\lambda_0)});\end{aligned}$$

(ii) $\phi(\lambda)$ *is linear on* $[\lambda_0, \lambda_1]$ *if and only if* $Qx^{(\lambda_0)} = Qx^{(\lambda_1)}$*; in this case*

$$\phi(\lambda) = \phi(0) + \frac{\lambda-\lambda_0}{\lambda_1-\lambda_0} b^T(y^{(\lambda_1)} - y^{(\lambda_0)}).$$

Proof: (i) Assuming $\lambda_0 = 0$ and $\lambda_1 = 1$ it follows from the proof of Lemma 4.2.2 that $x(\lambda)$ defined in (4.1) is optimal in (QP_λ) for $\lambda \in (0, 1)$. Hence

$$\begin{aligned}\phi(\lambda) &= (c+\lambda\Delta c)^T x(\lambda) + \frac{1}{2}x(\lambda)^T Q x(\lambda) \\ &= \phi(0) + \lambda\Delta c^T x^{(0)} + \lambda c^T(x^{(1)} - x^{(0)}) + \lambda^2\Delta c^T(x^{(1)} - x^{(0)}) + \\ &\quad \lambda(x^{(1)} - x^{(0)})^T Q x^{(0)} + \frac{1}{2}\lambda^2(x^{(1)} - x^{(0)})^T Q(x^{(1)} - x^{(0)}) \\ &= \phi(0) + \lambda(b^T(y^{(1)} - y^{(0)}) + c^T(x^{(1)} - x^{(0)})) + \frac{1}{2}\lambda^2\Delta c^T(x^{(1)} - x^{(0)}).\end{aligned}$$

where the last equality follows from twice using

$$A^T(y^{(1)} - y^{(0)}) + s^{(1)} - s^{(0)} = \Delta c + Q(x^{(1)} - x^{(0)}). \tag{4.5}$$

The required equality now easily follows.
(*ii*) From (*i*) we have $\phi(\lambda)$ linear on $[0, 1]$ if and only if $\Delta c^T(x^{(1)} - x^{(0)}) = 0$. Using (4.5) this is equivalent to

$$(x^{(1)} - x^{(0)})^T Q(x^{(1)} - x^{(0)}) = 0$$

which holds if and only if $Q(x^{(1)} - x^{(0)})$. Observing that

$$\begin{aligned} c^T(x^{(1)} - x^{(0)}) &= (A^T y^{(0)} + s^{(0)} - Qx^{(0)})^T(x^{(1)} - x^{(0)}) \\ &= -(Qx^{(0)})^T(x^{(1)} - x^{(0)}) = 0, \end{aligned}$$

we obtain $\phi(\lambda) = \phi(0) + \lambda\, b^T(y^{(1)} - y^{(0)})$. □

With the results obtained so far we can prove the next theorem, giving a characterization of the optimal value function in terms of tripartitions.

Theorem 4.2.4 *(i) The interval* Λ *can be partitioned in a finite set of subintervals such that the tripartition is constant on a subinterval.*
(ii) The optimal value function $\phi(\lambda)$ *is continuous, concave and piecewise quadratic on* Λ.

Proof: (*i*) Since the number of possible tripartitions is finite and the number of elements of Λ is infinite it follows from Lemma 4.2.2 that Λ can be partitioned into (open) subintervals on which the tripartition is constant, while it is different in the singletons in between the subintervals.
(*ii*) Corollary 4.2.3 implies that on each subinterval defined by a tripartition the function $\phi(\lambda)$ is quadratic. Since $\phi(\lambda)$ is concave (Lemma 4.2.1) it is continuous. □

4.3 COMPUTING THE OPTIMAL VALUE FUNCTION

Recall from Section 3.5 that in the parametric LP problem the optimal partition changes exactly where the (piecewise linear) optimal value function is not differentiable, i.e., in the breakpoints of the function. The next example shows that this is not the case in CQP.

Example 4.3.1 Let the CQP be defined by

$$Q = \begin{pmatrix} 2 & 0 \\ 0 & 1 \end{pmatrix}, \quad c = \begin{pmatrix} 0 \\ 1 \end{pmatrix}, \quad A = (2 \ \ 1), \ \ b = 2.$$

The vector of variation is taken to be $\Delta c^T = (1 \ \ 0)$. The next table gives maximal complementary solutions and the corresponding tripartitions as a function of λ.

	x_1	x_2	s_1	s_2	B	N	T
$\lambda < 0$	1	0	0	$-\lambda/2$	$\{1\}$	$\{2\}$	ϕ
$\lambda = 0$	1	0	0	0	$\{1\}$	ϕ	$\{2\}$
$0 < \lambda < 6$	$1 - \lambda/6$	$\lambda/3$	0	0	$\{1,2\}$	ϕ	ϕ
$\lambda = 6$	0	2	0	0	$\{2\}$	ϕ	$\{1\}$
$\lambda > 6$	0	2	$\lambda - 6$	0	$\{2\}$	$\{1\}$	ϕ.

Hence, $\phi(\lambda)$ is given by

$$\begin{aligned} \lambda \le 0 \qquad & \phi(\lambda) = 1 + \lambda, \\ 0 \le \lambda \le 6 \qquad & \phi(\lambda) = 1 + \lambda - \frac{1}{12}\lambda^2, \\ \lambda \ge 6 \qquad & \phi(\lambda) = 4. \end{aligned}$$

It is easy to check that $\phi(\lambda)$ is differentiable in $\lambda = 0$ and $\lambda = 6$, although the tripartition changes in these points. ◇

Let us call the *breakpoints* of $\phi(\lambda)$ those points λ where the tripartition changes; in between the breakpoints are the *curvilinearity intervals*[1]. Below we show that the breakpoints occur exactly where the *second derivative* of the optimal value function changes or does not exist. Moreover, we prove that when taking maximal complementary solutions for two parameter values in a curvilinearity interval, a convex combination of these solutions is a maximal complementary solution for the corresponding parameter value (cf. Lemma 4.2.2). We first deal with the differentiability of $\phi(\lambda)$.

Lemma 4.3.2 *If $\phi(\cdot)$ is differentiable in λ then $\Delta c^T x$ is equal for all optimal solutions x of (QP_λ). Otherwise, the left and right derivatives $\phi'_-(\lambda)$ and $\phi'_+(\lambda)$*

[1] In LP the term linearity interval is used since on these intervals the optimal value function is linear; this is not the case in CQP.

are given by

$$\phi'_-(\lambda) = \max_{x,y,s} \{ \Delta c^T x \;:\; Ax = b,\ x_{B_\lambda} \geq 0,\ x_{N_\lambda \cup T_\lambda} = 0,$$
$$A^T y + s - Qx = c + \lambda \Delta c,\ s_{N_\lambda} \geq 0,\ s_{B_\lambda \cup T_\lambda} = 0 \}, \quad (4.6)$$
$$\phi'_+(\lambda) = \min_{x,y,s} \{ \Delta c^T x \;:\; Ax = b,\ x_{B_\lambda} \geq 0,\ x_{N_\lambda \cup T_\lambda} = 0,$$
$$A^T y + s - Qx = c + \lambda \Delta c,\ s_{N_\lambda} \geq 0,\ s_{B_\lambda \cup T_\lambda} = 0 \}. \quad (4.7)$$

Proof: For arbitrary $\lambda \in \mathrm{int}(\Lambda)$ and $\epsilon \in \mathbb{R}$ sufficiently small it holds for *any* optimal solution $x(\lambda)$ of (QP$_\lambda$)

$$\begin{aligned} \phi(\lambda+\epsilon) &= (c + (\lambda+\epsilon)\Delta c)^T x^{(\lambda+\epsilon)} + \frac{1}{2}(x^{(\lambda+\epsilon)})^T Q x^{(\lambda+\epsilon)} \\ &\leq (c + \lambda\Delta c)^T x(\lambda) + \frac{1}{2} x(\lambda)^T Q x(\lambda) + \epsilon \Delta c^T x(\lambda) \\ &= \phi(\lambda) + \epsilon \Delta c^T x(\lambda), \end{aligned}$$

so the right and left derivatives at λ satisfy

$$\begin{aligned} \phi'_+(\lambda) &= \lim_{\epsilon \downarrow 0} \frac{\phi(\lambda+\epsilon) - \phi(\lambda)}{\epsilon} \leq \Delta c^T x(\lambda), \\ \phi'_-(\lambda) &= \lim_{\epsilon \uparrow 0} \frac{\phi(\lambda+\epsilon) - \phi(\lambda)}{\epsilon} \geq \Delta c^T x(\lambda). \end{aligned}$$

If $\phi(\lambda)$ is differentiable at λ then we necessarily have $\phi'(\lambda) = \Delta c^T x(\lambda)$ for any optimal solution $x(\lambda)$ of (QP$_\lambda$). Otherwise, note that the objective value of each x feasible in (QP) as a function of λ is linear with slope $\Delta c^T x$. Since $\phi(\cdot)$ is assumed to be not differentiable in λ there must be different optimal solutions with different slope; using the concavity of $\phi(\cdot)$ (Lemma 4.2.1) it is obvious that the left and right derivative of $\phi(\lambda)$ are obtained for those solutions optimal at λ having the largest respectively smallest value of $\Delta c^T x$. This implies the definitions (4.6) and (4.7). □

Before proceeding, we quote the following well–known result.

Lemma 4.3.3 *Let (x^*, y^*, s^*) and $(\overline{x}^*, \overline{y}^*, \overline{s}^*)$ both be optimal solutions of (QP) and (QD). Then $Qx^* = Q\overline{x}^*$ and $c^T x^* = c^T \overline{x}^*$.*

Proof: Follows easily using the convexity of the objective function of (QP) in x. □

The support of a vector u is defined as $\sigma(u) := \{ i \,:\, u_i > 0 \}$.

Lemma 4.3.4 *Assume that $\lambda = 0$ is a breakpoint of the optimal value function with tripartition $\pi_0 = (B_0, N_0, T_0)$; further assume that the curvilinearity interval to the right of zero contains $\lambda = 1$ with tripartition $\pi_1 = (B_1, N_1, T_1)$. Let (x^*, s^*) belong to a strictly complementary solution of (4.7) with $\lambda = 0$. Then $\sigma(x^*) \subseteq B_1$ and $\sigma(s^*) \subseteq N_1$.*

Proof: On $(0, 1]$ the tripartition is constant, hence linear combinations of maximal complementary solutions for two values in this interval are optimal in between these two values. Taking limit to zero implies the existence of $\underline{x}$, optimal in (QP_0) with $\sigma(\underline{x}) \subseteq B_1$ and $\Delta c^T \underline{x} = \phi'_+(0) = \Delta c^T x^*$. Since x^* and $\underline{x}$ are both optimal in (QP_0), Lemma 4.3.3 implies

$$\begin{aligned}(x^*)^T s^{(1)} &= (x^*)^T(c + \Delta c + Qx^{(1)} - A^T y^{(1)}) \\ &= c^T \underline{x} + \Delta c^T \underline{x} + \underline{x}^T Q x^{(1)} - \underline{x}^T A^T y^{(1)} = \underline{x}^T s^{(1)} = 0.\end{aligned}$$

Analogously, one shows that $(s^*)^T x^{(1)} = 0$. Consider now for $\epsilon \in (0, 1)$

$$x(\epsilon) = (1-\epsilon)x^* + \epsilon\, x^{(1)}, \quad y(\epsilon) = (1-\epsilon)y^* + \epsilon\, y^{(1)}, \quad s(\epsilon) = (1-\epsilon)s^* + \epsilon\, s^{(1)},$$

then $x(\epsilon)$ and $s(\epsilon)$ are feasible and complementary in (QP_ϵ) and (QD_ϵ), hence optimal. So it holds $\sigma(x^*) \subseteq B_1$ and $\sigma(s^*) \subseteq N_1$. □

Obviously, a similar result can be obtained for the interval to the left of a breakpoint, using the solution of (4.6).

Corollary 4.3.5 *The breakpoints of the optimal value function occur exactly where its second derivative changes or does not exist.*

Proof: Suppose to the contrary that the optimal value function is quadratic on $[-1, 1]$, while the tripartition is π_0 for $\lambda = 0$, π_{-1} on $[-1, 0)$ and π_1 on $(0, 1]$. Since the optimal value function is differentiable in $\lambda = 0$, it follows from the proof of Lemma 4.3.4 that $x(\epsilon) := (1-\epsilon)x^{(-1)} + \epsilon x^{(0)}$ is optimal in $(\mathrm{QP}_{\epsilon-1})$ for $\epsilon \in (0, 1)$, while $x(\epsilon) := (1-\epsilon)x^{(0)} + \epsilon x^{(1)}$ is optimal in (QP_ϵ) for $\epsilon \in (0, 1)$. Since the derivative of the optimal value function is linear on $[-1, 1]$ it follows that $(x^{(-1)} + x^{(1)})/2$ is optimal in (QP_0). This implies $B_1 \cup B_{-1} \subseteq B_0$. Similarly one shows $N_1 \cup N_{-1} \subseteq N_0$. Combining this result with Lemma 4.3.4 it follows $B_1 = B_{-1} = B_0$ and $N_1 = N_{-1} = N_0$ contradicting the assumption. □

In Section 3.5 we showed that in LP solving the analogon of (4.7) in a breakpoint rendered the optimal partition in the curvilinearity interval to the right of it for free. This is not the case in CQP; instead, we still have to solve another CQP problem. Observe from Lemma 4.3.3 that the solution of (4.7) is directionally differentiable. This motivates the following lemma.

Lemma 4.3.6 *Consider the situation as in Lemma 4.3.4. Let* (x^*, s^*) *belong to a strictly complementary solution of (4.7) for* $\lambda = 0$*, and define* $B := \sigma(x^*)$*,* $N := \sigma(s^*)$*,* $T := \{1, \ldots, n\} \setminus (B \cup N)$*. Consider*

$$\min_{\xi,\rho,\eta} \quad \{ \, \Delta c^T \xi + \xi^T Q \xi \; : \; A\xi = 0, \; A^T \eta + \rho - Q\xi = \Delta c \, , \quad \xi_N = 0, \; \rho_B = 0, \xi_T \geq 0, \rho_T \geq 0 \, \}, \tag{4.8}$$

and let (ξ^*, ρ^*, η^*) *belong to a maximal complementary solution of it. Then* $B_1 = B \cup \{i : \xi_i^* > 0\}$*,* $N_1 = N \cup \{i : \rho_i^* > 0\}$*, and* $T_1 = \{1, \ldots, n\} \setminus (B_1 \cup N_1)$*.*

Proof: It is easy to check that for a feasible solution of (4.8)

$$\Delta c^T \xi + \xi^T Q \xi = \xi^T \rho = \xi_T^T \rho_T \geq 0.$$

The dual of (4.8) is given by

$$\max_{\delta,\xi,\gamma,\zeta} \{ \, -\Delta c^T \delta - \xi^T Q \xi \; : \; A\delta = 0, \; A^T \zeta + \gamma + Q\delta - 2Q\xi = \Delta c, \; \gamma_B = 0, \; \delta_N = 0, \; \gamma_T \geq 0, \; \delta_T \geq 0 \, \},$$

and for a feasible solution it holds

$$-\Delta c^T \delta - \xi^T Q \xi = -\delta^T \gamma - (\delta - \xi)^T Q (\delta - \xi) \leq 0.$$

Consequently, the optimal value of (4.8), if it exists, is zero. Consider the assignment

$$\xi = \delta = x^{(1)} - x^*, \; \rho = \gamma = s^{(1)} - s^*, \; \eta = \zeta = y^{(1)} - y^*,$$

which satisfies the first two linear equalities in (4.8). Using the fact that $B \subseteq B_1$ and $N \subseteq N_1$ (Lemma 4.3.4) it follows

$$\xi_N = x_N^{(1)} - x_N^* = 0, \quad \xi_T = x_T^{(1)} - x_T^* = x_T^{(1)} \geq 0,$$

and

$$\rho_B = s_B^{(1)} - s_B^* = 0, \quad \rho_T = s_T^{(1)} - s_T^* = s_T^{(1)} \geq 0,$$

and so $\xi_T^T \rho_T = 0$, implying that the assignment is an optimal solution. The fact that $(x^{(1)}, s^{(1)})$ is maximal complementary implies that the assignment must be maximal complementary in (4.8) as well, implying the result. □

In case a parameter value is not a breakpoint of the optimal value function the curvilinearity interval must be computed. This can be done by solving appropriate LP problems.

Lemma 4.3.7 *Let $(\underline{\lambda}, \overline{\lambda})$ be an interval on which the tripartition is constant and given by $\pi = (B, N, T)$, and such that $\pi_{\underline{\lambda}} \neq \pi$ and $\pi_{\overline{\lambda}} \neq \pi$. Then*

$$\begin{aligned}
\underline{\lambda} = \min_{\lambda,x,y,s} \{ \lambda \ : \ & Ax = b,\ x_B \geq 0,\ x_{N \cup T} = 0, \\
& A^T y + s - Qx - \lambda \Delta c = c,\ s_N \geq 0,\ s_{B \cup T} = 0 \}, \quad (4.9) \\
\overline{\lambda} = \max_{\lambda,x,y,s} \{ \lambda \ : \ & Ax = b,\ x_B \geq 0,\ x_{N \cup T} = 0, \\
& A^T y + s - Qx - \lambda \Delta c = c,\ s_N \geq 0,\ s_{B \cup T} = 0 \} \quad (4.10)
\end{aligned}$$

Proof: We prove (4.9). For any $\lambda \in (\underline{\lambda}, \overline{\lambda})$ the pair $(x^{(\lambda)}, y^{(\lambda)}, s^{(\lambda)})$ is feasible in (4.9). If the minimum value of the LP problem in (4.9) would be smaller than $\underline{\lambda}$ than a linear combination of the corresponding solution and a solution at $\lambda \in (\underline{\lambda}, \overline{\lambda})$ would give a contradiction with the definition of (B, N, T). The proof of (4.10) is similar. □

It need not be the case that a strictly complementary solution of (4.9) or (4.10) gives the tripartition in the corresponding breakpoint; only *some* of the primal and dual variables can be shown to be either in its B or N part. Instead, the CQP problem itself needs to be solved here, while using the information on some of the variables.

All the ingredients for the computation of the optimal value function have been obtained at this stage. The algorithm is outlined as follows (note: by 'solve' we mean 'compute a maximal complementary solution of')

1. Solve (QP$_\lambda$) for some initial λ_0;
2. Solve (4.10) to obtain $\overline{\lambda}$, a breakpoint;
3. Solve (QP$_{\overline{\lambda}}$) to compute the tripartition in $\overline{\lambda}$;
4. Solve (4.7) to obtain the slope and a support;
5. Solve (4.8) to obtain the tripartition of the next interval and proceed with Step 2.

Evidently, if the problem in Step 2 or Step 4 is unbounded the algorithm stops. The part of the optimal value function to the left of λ_0 can be computed analogously.

4.4 APPLICATION TO MEAN–VARIANCE MODELS

An important application of the parametric CQP problem is the computation of an efficient frontier for mean–variance models, introduced by Markowitz [163] for portfolio analysis. Given assets with expected return r_i and covariances v_{ij}, the problem is to find portfolios of the assets that have minimal variance given a level of total return, and maximal return given a level of total variance. Mathematically, let x_i be the proportion invested in asset i and consider the basic mean–variance problem

$$\min_x \left\{ \frac{1}{2}x^T V x \ : \ e^T x = 1, \ r^T x = \lambda, \ x \geq 0 \right\},$$

which can be viewed as a (right–hand side) parametric CQP problem. The efficient frontier is the function relating the (square root of the) objective function (variance or standard deviation of the portfolio) to λ (expected return of the portfolio).

The theoretical results in the previous section can in an obvious way be adjusted to this situation. This has some useful practical consequences. First, the efficient frontier is piecewise quadratic; the quadratic parts correspond to intervals where the set of assets that can occur in *some* efficient portfolio is the same, while there exist linearly varying 'maximal complementary portfolios' over such an interval. Maybe even more important, the tripartition shows which of the assets will never appear in an efficient portfolio given a specified return.

4.5 CONCLUDING REMARKS

In this chapter we extended the approach to sensitivity analysis using a description of the optimal set (as described in Chapter 3 for LP) to CQP. Instead of strictly complementary solutions we need to use maximal complementary solutions to describe the optimal value function, and to be able to compute it efficiently. As in the linear case this approach gives a better hindsight on the optimal value function and its derivatives. A further computational investigation will have to show how computationally demanding the proposed algorithm is. To do this, currently very efficient (interior point) codes for CQP are available: for instance, LOQO [237] and the 1996 release of CPLEX [35].

5

PRIMAL–DUAL AFFINE SCALING METHODS FOR LINEAR PROBLEMS

5.1 INTRODUCTION

The introduction of Karmarkar's polynomial time algorithm for linear programming (LP) [126] in 1984 is responsible for an enormous production of research papers on LP in the last decade. Two of the early papers were Barnes [16] and Vanderbei et al. [238], who independently proposed the *affine scaling algorithm* as a simplification of Karmarkar's algorithm. While Karmarkar performed a projective scaling in each iteration, this algorithm used a simpler affine scaling. Convergence of the iterates to optimality was shown (under nondegeneracy assumptions), but no complexity proof was given. Implementations soon showed that the affine scaling algorithm was very efficient, see Adler et al. [1]. After the publication of Vanderbei's paper, the editor of Mathematical Programming received a letter [44] from the Russian researcher Dikin, revealing that the affine scaling algorithm was already proposed by him in 1967 [42], see also [43]. Around the same time a paper by Gill et al. [71] showed that Karmarkar's algorithm has close connections with the logarithmic barrier method, introduced by Frisch [62] in 1955 and extensively investigated by Fiacco and McCormick [53] among others.

All the methods mentioned above are either primal or dual algorithms, i.e., methods generating either primal or dual feasible solutions in *each* iteration of the method. In 1987, Monteiro and Adler [180] and Kojima et al. [139] independently proposed a *primal–dual logarithmic barrier algorithm*, that works with the primal and dual of the LP problem simultaneously. Monteiro et al. [182] then gave a simplified version of the algorithm in [180] which they considered to be a primal–dual version of Dikin's affine scaling algorithm. This algorithm is normally referred to as the *primal–dual affine scaling* algorithm.

In this chapter we introduce and analyze a different generalization of Dikin's algorithm to the primal–dual setting. The generalization is more natural and leads to a better complexity bound than the one in [182]. The algorithm is named the *primal–dual Dikin–affine scaling* algorithm. In Jansen et al. [113] the polynomial complexity of this algorithm was first shown for LP. In the analysis a homogeneous potential function was used, viz. $f : \mathbb{R}^n_{++} \to \mathbb{R}$ defined by

$$f(v) := n \ln \|v\|^2 - \sum_{i=1}^{n} \ln v_i^2 - n \ln n,$$

where $v := \sqrt{xs}$ represents a point in the v–space (Section 2.1). This function was introduced by Tanabe [226] and used by Todd and Ye [231] and Kojima et al. [136], among others. Later Ling [150] simplified and improved the analysis in [113] using a different way to measure proximity to the central path. Surprisingly, he showed that the complexity of the algorithm remains unchanged when working with large neighborhoods of the central path.

After a short introduction in primal (or dual) and primal–dual affine scaling and logarithmic barrier methods in Section 5.2 we will derive the new algorithm in an LP context in Section 5.3. We prove the polynomial complexity of the algorithm for the class of *linear complementarity problems* (LCPs) with P_* (or sufficient) matrices, for which interior point methods are as difficult to analyze as for LP (cf. Kojima et al. [136]). The class of LCPs contains LP and convex quadratic programming (CQP) as special cases. We further extend the algorithm and its applicability in several ways. In Section 5.4 we analyze the use of corrector steps, which aim at diminishing the second order effect when using Newton's method. We show that the complexity bound can asymptotically be improved to the best one currently known for LCPs.

5.2 LOGARITHMIC BARRIER AND AFFINE SCALING

We give a short survey of logarithmic and affine scaling methods to later place the primal–dual Dikin–affine scaling method in a proper context. More elaborate surveys can be found in Gonzaga [87] and Den Hertog [100]. Consider the primal and dual LP problem in standard form:

$$\text{(P)} \qquad \min_x \left\{ c^T x \; : \; Ax = b, \; x \geq 0 \right\},$$

$$\text{(D)} \qquad \max_{y,s} \left\{ b^T y \; : \; A^T y + s = c, \; s \geq 0 \right\},$$

where $c, x \in \mathbb{R}^n$, $b \in \mathbb{R}^m$ and A is an $m \times n$ matrix. We make the standard assumptions that A has full rank and that there exists a positive primal-dual pair (Definition 2.1.4). The primal and dual feasible sets are denoted by $\mathcal{P}$ and $\mathcal{D}$. The logarithmic barrier algorithm for (P) was introduced in 1955 by Frisch [62] and extensively investigated in the 1960s and 1970s (Fiacco and McCormick [53]). The primal logarithmic barrier method uses the logarithmic barrier function $f_P : \mathbb{R}^n_{++} \times \mathbb{R}_{++} \to \mathbb{R}$:

$$f_P(x; \mu) = \frac{c^T x}{\mu} - \sum_{i=1}^{n} \ln(x_i).$$

The barrier parameter μ is used to balance the original objective with the distance to the inequality constraints $x \geq 0$. The optimality conditions for minimizing $f_P(x;\mu)$ over $\mathcal{P}$, for fixed μ, are given by the system

$$\begin{aligned} Ax &= b, \quad x > 0, \\ A^T y + s &= c, \quad s > 0, \\ xs &= \mu e. \end{aligned} \tag{5.1}$$

The existence of a solution to (5.1) is guaranteed from Theorem 2.1.5, and is denoted by $(x(\mu), y(\mu), s(\mu))$. The sets

$$\{\, x(\mu) \;:\; \mu > 0 \,\} \quad \text{and} \quad \{\, (y(\mu), s(\mu)) \;:\; \mu > 0 \,\}$$

are the central paths of (P) and (D), alternatively defined in primal-dual form as the set

$$\mathcal{C} = \left\{ (x, s) \;:\; x \in \mathcal{P},\; s \in \mathcal{D},\; xs = \frac{x^T s}{n} e \right\}.$$

In the latter form, the central path was introduced and investigated by Bayer and Lagarias [17], Megiddo [168] and Sonnevend [218]. As $x^T s \to 0$, this path leads to a strictly complementary solution of (P) and (D) (cf. Lemma 2.2.10), hence it can be used as a guideline to optimality. The generic *path-following* method works as follows. Given a value μ compute an approximate minimizer of $f_P(x;\mu)$; update μ and proceed. Different methods are obtained by varying the updating scheme for μ, the method for minimizing the barrier function and the criterion that judges approximation to the exact minimizer. The latter aspect is related to the use of neighborhoods of the central path. These neighborhoods include

$$\begin{aligned} \mathcal{N}_2(\beta) &= \left\{ (x,s) \;:\; x \in \mathcal{P},\; s \in \mathcal{D},\; \|xs - \mu e\| \leq \beta\mu, \;\text{ where }\; \mu = \frac{x^T s}{n} \right\} \\ \mathcal{N}_\infty(\beta) &= \left\{ (x,s) \;:\; x \in \mathcal{P},\; s \in \mathcal{D},\; \|xs - \mu e\|_\infty \leq \beta\mu \;\text{ where }\; \mu = \frac{x^T s}{n} \right\}. \end{aligned}$$

for some $\beta \in (0,1)$. Algorithms based on the larger $\mathcal{N}_\infty$ neighborhood are called *long-step algorithms*, and those based on the smaller $\mathcal{N}_2$ are called *short-step algorithms*. Among all existing path-following (infeasible or feasible) algorithms for LP, the theoretical iteration complexity to obtain an ϵ-approximate solution with a short-step algorithm is $\mathcal{O}(\sqrt{n}\ln 1/\epsilon)$ (e.g., Renegar [206], Gonzaga [84], Roos and Vial [210], Kojima et al. [139] and Monteiro and Adler [180]). The complexity of the long-step algorithms is at least $\mathcal{O}(nL)$ (e.g., Gonzaga [85], Mizuno et al. [179], Den Hertog [100], Anstreicher and Bosch [9] and Jansen et al. [116]). In contrast, long-step algorithms outperform short-step ones by a big margin in practice (e.g., Lustig et al. [156, 157] and Mehrotra [170]). Nesterov [191] discusses several long-step strategies in a general (nonlinear) setting.

In the primal logarithmic barrier (path-following) method the search-direction is obtained by minimizing the second order Taylor approximation of $f_P(x;\mu)$ over $\mathcal{P}$ at a given iterate $\overline{x} > 0$, i.e., by Newton's method. The search-direction obtained in this way is given by

$$\Delta x = \overline{X}(I - \overline{X}A^T(A\overline{X}^2A^T)^{-1}A\overline{X})(\mu e - \overline{X}c). \tag{5.2}$$

Dikin's primal affine scaling method does not use the central path as a guideline. Instead, the search-direction is obtained by solving a subproblem in which the nonnegativity constraints $x \geq 0$ are replaced by an ellipsoidal constraint:

$$\min_x \{ \, c^T x \; : \; Ax = b, \; \|\overline{X}^{-1}(\overline{x} - x)\| \leq 1 \, \}.$$

Taking its solution x^+ as the new iterate, the *affine scaling direction* is

$$\Delta x = x^+ - \overline{x} = -\overline{X}(I - \overline{X}A^T(A\overline{X}^2A^T)^{-1}A\overline{X})\overline{X}c. \tag{5.3}$$

Note that the search-direction in (5.3) can be obtained from (5.2) by putting $\mu = 0$, i.e., by not using the *centering direction*

$$\Delta x = \overline{X}(I - \overline{X}A^T(A\overline{X}^2A^T)^{-1}A\overline{X})\mu e. \tag{5.4}$$

Hence, the main difference between affine scaling and path-following algorithms is that in the former the direction only depends on the current iterate, while in the latter target-points (on the central path) are used. Den Hertog and Roos [103], Yamashita [248] and Gonzaga [86], among others, have shown that in many interior point methods the search-direction used is a linear combination of the affine scaling direction and the centering direction.

We noted before that the minimizer of the logarithmic barrier function $f_P(x;\mu)$ over $\mathcal{P}$ is characterized by the system (5.1). This system is also the KKT–system for minimizing the primal–dual logarithmic barrier function

$$f_{PD}(x, s; \mu) = \frac{x^T s}{\mu} - \sum_{i=1}^{n} \ln(x_i s_i).$$

over both $\mathcal{P}$ and $\mathcal{D}$. Using (5.1) Monteiro and Adler [180] and Kojima et al. [139] proposed a primal–dual algorithm in which the search–directions $\Delta x, \Delta y, \Delta s$ are determined from

$$\begin{aligned} A\Delta x &= 0, \\ A^T \Delta y + \Delta s &= 0, \\ \overline{x}\Delta s + \overline{s}\Delta x &= \mu e - \overline{x}\,\overline{s}, \end{aligned} \tag{5.5}$$

where $(\overline{x}, \overline{y}, \overline{s})$ is the current iterate. The last equation is motivated from the desire to compute Δx and Δs such that

$$(\overline{x} + \Delta x)(\overline{s} + \Delta s) = \mu e.$$

Note that in the last equation of (5.5) the term $\Delta x \Delta s$ is not included (the effect of this linearization will be studied in detail in Chapter 8). A short–step version of the method was shown to be polynomial in [139, 180]. In Jansen et al. [116] the algorithm was interpreted and analyzed completely using the primal–dual logarithmic barrier function; moreover, long–step algorithms of this type were investigated. By analogy with the primal affine scaling direction Monteiro et al. [182] proposed to simplify the direction of the primal–dual algorithm by setting $\mu = 0$ in (5.5), and to call the resulting direction the *primal–dual affine scaling* direction. This terminology has become standard since 1987. The method was shown to have a polynomial complexity bound of $\mathcal{O}(n(\ln(1/\epsilon))^2)$ iterations to obtain an ϵ–accurate solution.

In the next section we show that in the primal–dual case Dikin's affine scaling idea of replacing the entangling inequality constraints by an ellipsoid and optimizing the objective (the duality gap) can also be done in the primal–dual setting. As we will see, this gives rise to a new affine scaling algorithm, named hereafter *primal–dual Dikin–affine scaling* algorithm. The search direction used is different from the one in Monteiro et al. [182]. We will show that the algorithm has a complexity bound of $\mathcal{O}(n \ln 1/\epsilon)$ iterations, which is better than that for the primal–dual affine scaling algorithm. This improvement is due to the fact that the new direction contains a centering effect as opposed to the primal–dual affine scaling direction. Moreover, we will show that it can be argued that the primal–dual affine scaling direction in [182] is obtained without scaling at all.

5.3 THE PRIMAL–DUAL DIKIN–AFFINE SCALING ALGORITHM

In this section we introduce and motivate the primal–dual Dikin–affine scaling method for LP as derived and analyzed in Jansen et al. [113]. Then we analyze the complexity of the algorithm for the more general class of LCPs with *matrices in* P_*.

5.3.1 Deriving the search–direction

Let (x, s) be a positive primal–dual pair, that is

$$\begin{aligned} Ax &= b, \quad x > 0, \\ A^T y + s &= c, \quad s > 0. \end{aligned}$$

Let $\Delta x, \Delta y, \Delta s$ denote search–directions in the respective spaces. For the equality constraints, feasibility is maintained if

$$A\Delta x = 0, \quad A^T \Delta y + \Delta s = 0.$$

Observe that Δx and Δs are orthogonal. Following Dikin's idea [42], we replace the nonnegativity conditions by requiring the next iterates $(x + \Delta x, y + \Delta y, s + \Delta s)$ to belong to a suitable ellipsoid. In the primal–dual setting we define it as follows

$$\|x^{-1}\Delta x + s^{-1}\Delta s\| \leq 1,$$

and call it the (primal–dual) *Dikin–ellipsoid*. The duality gap (complementarity) after a step is given by

$$(x + \Delta x)^T (s + \Delta s) = x^T s + x^T \Delta s + s^T \Delta x.$$

Minimizing the new duality gap over the Dikin–ellipsoid amounts to solving the following optimization problem:

$$\min_{\Delta x, \Delta y, \Delta s} \quad \left\{ s^T \Delta x + x^T \Delta s \ : \ A\Delta x = 0, \ A^T \Delta y + \Delta s = 0 \ , \right.$$
$$\left. \|x^{-1}\Delta x + s^{-1}\Delta s\| \leq 1 \right\}. \tag{5.6}$$

We proceed by showing that this problem uniquely determines the search–direction. We apply a common rescaling of the variables (see e.g., Gonzaga [87]). Let the vectors d and v be given by

$$d := \sqrt{xs^{-1}}, \quad v := \sqrt{xs}. \tag{5.7}$$

Using d we rescale both x and s as follows: $d^{-1}x = ds = v$; likewise we define

$$p_x := d^{-1}\Delta x, \quad p_s := d\Delta s. \tag{5.8}$$

Note that the orthogonality of Δx and Δs implies that p_x and p_s are orthogonal as well. The important property of this scaling is that it maps both x and s to the same vector v. As we will see this implies that the scaled search–directions can be expressed as orthogonal complements of a vector (this property has been generalized to a nonlinear setting by Nesterov and Todd [194]). Now we may write

$$\begin{aligned} x\Delta s + s\Delta x &= xd^{-1}d\Delta s + sdd^{-1}\Delta x = v(p_x + p_s), \\ x^{-1}\Delta x + s^{-1}\Delta s &= x^{-1}dd^{-1}\Delta x + s^{-1}d^{-1}d\Delta s = v^{-1}(p_x + p_s). \end{aligned}$$

The affine constraints can be reformulated as

$$ADp_x = 0, \quad DA^T\Delta y + p_s = 0,$$

where $D = \text{Diag}\,(d)$. Clearly p_x and p_s are uniquely characterized as the components of the vector

$$p_v := p_x + p_s,$$

in the null space and the row space of the matrix AD, respectively. Therefore, we may reformulate (5.6) in terms of the vector p_v:

$$\min_{p_v} \left\{ v^T p_v \ : \ \|v^{-1}p_v\| \le 1 \right\}. \tag{5.9}$$

The solution of this problem is given by

$$p_v = -\frac{v^3}{\|v^2\|},$$

with objective value $v^T p_v = -\|v^2\|$. The solution of the original problem is then:

$$\Delta x = DP_{AD}p_v, \quad \Delta s = D^{-1}Q_{AD}p_v,$$

where P_{AD} and Q_{AD} denote the orthogonal projections onto the null space and the row space of AD, respectively.

Observe that these directions are uniquely determined by the system

$$\begin{aligned} A\Delta x &= 0, \\ A^T\Delta y + \Delta s &= 0, \\ x\Delta s + s\Delta x &= -\frac{x^2s^2}{\|xs\|}. \end{aligned} \tag{5.10}$$

Note that the directions differ from the primal–dual affine scaling directions: recall from (5.5) with $\mu = 0$ that the latter have $-xs$ as the right–hand side of the third equation. The relation between the two directions becomes more apparent by noting that, up to some scalar factor, the primal–dual affine scaling directions are given by $DP_{AD}q_v$ for the x–space and $D^{-1}Q_{AD}q_v$ for the s–space, where

$$q_v := -\frac{v}{\|v\|}.$$

Note that q_v is the solution of the problem

$$\min_{q_v} \left\{ v^T q_v \ : \ \|q_v\| \leq 1 \right\}. \tag{5.11}$$

Recall that Theorem 2.1.5 implies that we may identify any positive primal–dual pair with a positive vector v, where $v = \sqrt{xs}$; this mapping is one–to–one. So both types of affine scaling directions are actually obtained in the v–space (see Section 2.1). Comparing problems (5.9) and (5.11) one sees that the objectives are the same (minimize the complementarity). The Dikin–affine scaling algorithm is obtained using the maximal ellipsoid contained in the v–space; observe that this is exactly the way in which Dikin defined the affine scaling direction for the primal problem. The primal–dual affine scaling direction is obtained using a sphere in the v–space, which corresponds to a steepest descent step[1]. The analogy just described justifies the name *affine scaling direction* for the new direction more than for the old direction. In fact, we may say that the primal–dual affine scaling directions are obtained without scaling at all. In Figure 5.1 we show the difference between the two directions in a two–dimensional v–space. In the v–space we measure distance to the central path by the ratio

$$\omega(v) = \frac{\min(v)}{\max(v)}.$$

This measure was proposed by Ling [150]. Using this measure we define the following large neighborhood:

$$\mathcal{N}_\omega(\rho) := \left\{ (x,s) \ : \ x \in \mathcal{P}, \ s \in \mathcal{D}, \ v = \sqrt{xs}, \ \omega(v) \geq \rho \right\}. \tag{5.12}$$

Note that $0 \leq \omega(v) \leq 1$. On the central path it holds $\omega(v) = 1$. Observe, that the Dikin–affine scaling direction has a centering effect in the v–space, while the primal–dual affine scaling direction keeps $\omega(v)$ unchanged.

[1] See Gonzaga [87] for an extensive discussion on the difference between steepest descent (Cauchy) and scaled steepest descent (Dikin).

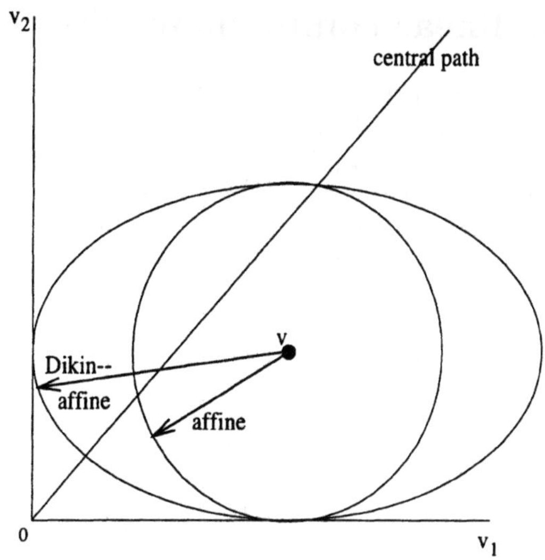

Figure 5.1 Primal–dual affine and Dikin–affine scaling directions in the v–space.

In [190] Nazareth gives a different perspective on the Dikin–affine scaling direction. Let us consider the centering conditions

$$\begin{aligned} Ax &= b, \quad x > 0, \\ A^T y + s &= c, \quad s > 0, \\ \tfrac{1}{x_i s_i} &= \tfrac{1}{\mu}, \quad i = 1, \ldots, n. \end{aligned} \tag{5.13}$$

Obviously, the system is equivalent to (5.1). Nazareth showed that the Newton direction for (5.13) is given by

$$\begin{aligned} A\Delta x &= 0, \\ A^T \Delta y + \Delta s &= 0, \\ x\Delta s + s\Delta x &= \frac{xs}{\mu}(\mu e - xs). \end{aligned}$$

The resulting direction is a linear combination of the primal–dual affine scaling direction and the primal–dual Dikin–affine scaling direction. As μ converges to zero it approaches the primal–dual Dikin–affine scaling direction.

5.3.2 The linear complementarity problem

We perform the analysis of the primal–dual Dikin–affine scaling algorithm introduced above for the class of LCPs, which contains LP as a special case. As proposed by Ling [150] we use $\omega(v)$ as a measure of proximity to the central path, hence we analyze a large neighborhood algorithm. We extend the analysis in Jansen et al. [112], allowing *nonmonotone* linear mappings in the definition of the LCP; the class of LCPs considered is that of P_* or sufficient matrices, defined below (cf. [136]).

Let M be a given $n \times n$ matrix and $q \in \mathbb{R}^n$. The LCP is defined as follows.

(LCP) Find $(x,s) \in \mathbb{R}^{2n}$ such that $s = Mx + q,\ (x,s) \geq 0$ and $x^T s = 0$.

We denote the sets of feasible and interior–feasible points of (LCP) by:

$$\begin{aligned} \mathcal{F} &= \{ (x,s) \in \mathbb{R}^{2n} : s = Mx + q,\ (x,s) \geq 0 \}, \\ \mathcal{F}^0 &= \{ (x,s) \in \mathbb{R}^{2n} : s = Mx + q,\ (x,s) > 0 \}. \end{aligned}$$

We assume that $\mathcal{F}^0$ is nonempty, or stated otherwise, that an interior point exists. For an introduction in complementarity problems and traditional solution methods we refer to the book of Cottle et al. [34]. The monograph by Kojima et al. [136] analyzes certain interior point methods for complementarity problems. Special classes for matrices M have been distinguished to guarantee the existence of a solution. These include PSD (M positive semidefinite), P (M has positive principal minors), P_* (see below) and CS and RS (column–sufficient resp. row–sufficient). Some known implications are

$$PSD \subset P_* \subset CS, \quad P \subset P_*, \quad P_* = CS \cap RS,$$

see e.g., Cottle [34], Den Hertog et al. [106], Väliaho [235] and Kojima et al. [136]. In this section we will be interested in the class P_* (i.e., the class of sufficient matrices).

Definition 5.3.1 (P_*–matrices) *Let $\kappa \geq 0$. The matrix $M \in \mathbb{R}^{n\times n}$ is in $P_*(\kappa)$ if*

$$(1+4\kappa) \sum_{i \in I_+(\xi)} \xi_i (M\xi)_i + \sum_{i \in I_-(\xi)} \xi_i (M\xi)_i \geq 0, \quad \forall\, \xi \in \mathbb{R}^n,$$

where

$$I_+(\xi) = \{ i : \xi_i (M\xi)_i > 0 \}, \qquad I_-(\xi) = \{ i : \xi_i (M\xi)_i < 0 \}.$$

The matrix M is in P_* if it is in $P_*(\kappa)$ for some κ.

Throughout this section, we impose the following condition on the matrix M.

Condition 5.3.2 *There exists a constant $\kappa \geq 0$ such that M is in $P_*(\kappa)$.*

If $\kappa = 0$ then the complementarity problem is called monotone, since for any $(x, s) \in \mathcal{F}$ and $(\overline{x}, \overline{s}) \in \mathcal{F}$ it holds

$$(x - \overline{x})^T(s - \overline{s}) = (x - \overline{x})^T M(x - \overline{x}) \geq 0.$$

In practice, it might be hard to compute the actual value of κ; fortunately, an implementation of the algorithm does not need its knowledge, although the theoretical complexity depends on κ.

Observe that the skew–symmetric self–dual reformulation of the LP problem we developed in Chapter 2 is in fact an LCP. Letting $(\overline{\text{SP}})$ be as in Section 2.3 we may define

$$M := \begin{pmatrix} 0 & A & \overline{b} & -b \\ -A^T & 0 & -\overline{c} & c \\ -\overline{b}^T & \overline{c}^T & 0 & -\alpha \\ b^T & -c^T & \alpha & 0 \end{pmatrix}, \quad q := \begin{pmatrix} 0 \\ 0 \\ \beta \\ 0 \end{pmatrix}.$$

In this case M is skew–symmetric which implies that $\xi^T M\xi = 0$ for all $\xi \in \mathbb{R}^{n+m+2}$, so M is in $P_*(0)$. Also, the CQP problem

$$\min_x \left\{ c^T x + \frac{1}{2} x^T Q x \ : \ Ax \geq b, \ x \geq 0 \right\},$$

where Q is a symmetric positive semidefinite matrix, can be written as an LCP. The dual problem is given by

$$\max_{x,y,s} \left\{ b^T y - \frac{1}{2} x^T Q x \ : \ A^T y - Qx \leq c, \ y \geq 0 \right\}.$$

The corresponding LCP is defined by

$$M := \begin{pmatrix} Q & -A^T \\ A & 0 \end{pmatrix}, \quad q := \begin{pmatrix} c \\ -b \end{pmatrix}.$$

In this case we have $\xi^T M\xi = \xi^T_{\{1,\ldots,n\}} Q \xi_{\{1,\ldots,n\}} \geq 0$, so again M is in $P_*(0)$.

Observe from these two examples that complementarity problems offer a natural framework for primal–dual algorithms.

5.3.3 The algorithm and its convergence analysis

The primal–dual Dikin–affine scaling method is easily extended to the LCP. Let a strictly feasible pair $(x, s) \in \mathcal{F}^0$ be given. We determine the search–direction $(\Delta x, \Delta s)$ from which the next iterate follows by

$$x(\theta) = x + \theta\Delta x, \quad s(\theta) = s + \theta\Delta s,$$

for some step size $\theta \in (0, 1)$. The componentwise product $x(\theta)s(\theta)$ can be expressed as

$$x(\theta)s(\theta) = (x + \theta\Delta x)(s + \theta\Delta s) = xs + \theta\,(s\Delta x + x\Delta s) + \theta^2\Delta x\Delta s.$$

For the search–direction we consider (cf. (5.10))

$$\begin{array}{lll} \min\limits_{\Delta x, \Delta s} & (xs)^T(x^{-1}\Delta x + s^{-1}\Delta s) & \\ \text{s.t.} & -M\Delta x + \Delta s & = 0, \\ & \|x^{-1}\Delta x + s^{-1}\Delta s\| & \leq 1. \end{array} \tag{5.14}$$

Using the same construction as in Section 5.3.1 the search–direction is obtained from the system of equations

$$\begin{array}{rcl} -M\Delta x + \Delta s & = & 0, \\ x\Delta s + s\Delta x & = & -\frac{(xs)^2}{\|xs\|}. \end{array} \tag{5.15}$$

Since M is in $P_*(\kappa)$ this system has a unique solution (cf. Kojima et al. [136, Lemma 4.1]). The algorithm is formally described in Figure 5.2.

As in Section 5.3.1 we may reformulate the system (5.15) in terms of the v–space using the usual scaling (see (5.7)-(5.8)), which gives

$$\begin{array}{rcl} -DMDp_x + p_s & = & 0, \\ p_x + p_s & = & p_v, \end{array} \tag{5.16}$$

where $p_v = -v^3/\|v^2\|$. We now prove the following lemma. Similar lemmas can be found in, e.g., Mizuno et al. [179], Kojima et al. [136] and Jansen et al. [112]. The lemma (and its variants) is important to prove the polynomiality of many primal–dual algorithms ([133, 138, 139, 135, 136, 175, 180], etc.).

Lemma 5.3.3 *Let p_x, p_s, and p_v be as defined above. Then, it holds*
(i) $\|p_v\| \leq \|v\|_\infty \leq \|v\|$;
(ii) $-\kappa\|p_v\|^2 \leq \Delta x^T\Delta s = p_x^T p_s \leq \|p_v\|^2/4$;
(iii) $\|\Delta x\Delta s\|_\infty = \|p_x p_s\|_\infty \leq (1 + 4\kappa)\|p_v\|^2/4$.

Input
  $(x^{(0)}, s^{(0)})$: the initial pair of interior–feasible solutions;
Parameters
  ϵ is the accuracy parameter;
  θ is the step size;
begin
  $x := x^{(0)}$; $s := s^{(0)}$;
  while $x^T s > \epsilon$ **do**
    calculate Δx and Δs from (5.15);
    $x := x + \theta \Delta x$;
    $s := s + \theta \Delta s$;
  end
end.

Figure 5.2 Primal–dual Dikin–affine scaling algorithm.

Proof: (*i*) Obvious from the definition $p_v = -v^3/||v^2||$.
(*ii*) The vectors p_x and p_s satisfy system (5.16). Applying Lemma 3.4 in Kojima et al. [136] gives

$$p_x^T p_s \geq -\kappa ||p_v||^2. \tag{5.17}$$

Note that the cited lemma applies since the $P_*(\kappa)$ property is preserved by pre– and post–multiplication with a positive definite diagonal matrix (cf. [136, Theorem 3.5]). Defining $q_v = p_x - p_s$, it holds

$$p_x^T p_s = \frac{1}{4}(||p_v||^2 - ||q_v||^2) \leq \frac{1}{4}||p_v||^2. \tag{5.18}$$

(*iii*) Using (5.17) and (5.18) we obtain

$$||q_v||^2 \leq (1 + 4\kappa)||p_v||^2.$$

Since $p_x p_s = (p_v^2 - q_v^2)/4$ it holds

$$||p_x p_s||_\infty \leq \frac{1}{4}\max(||p_v||_\infty^2, ||q_v||_\infty^2) \leq \frac{1}{4}\max(||p_v||^2, ||q_v||^2) \leq \frac{1}{4}(1 + 4\kappa)||p_v||^2.$$

This completes the proof. □

Defining $v(\theta)^2 = x(\theta)s(\theta)$ we derive

$$v(\theta)^2 = (x + \theta\Delta x)(s + \theta\Delta s) = v^2 + \theta v p_v + \theta^2 p_x p_s. \tag{5.19}$$

We first give an estimate for the new complementarity, then show that $v(\theta)$ can be kept sufficiently close to the central path.

Lemma 5.3.4 *If $\theta \leq 2/\sqrt{n}$ then the new complementarity satisfies*

$$x(\theta)^T s(\theta) \leq \left(1 - \frac{\theta}{2\sqrt{n}}\right) x^T s.$$

Proof: With (5.19) and the definition of p_v we have

$$x(\theta)^T s(\theta) = e^T v^2 - \theta \frac{e^T v^4}{\|v^2\|} + \theta^2 p_x^T p_s \leq e^T v^2 - \theta \frac{\|v\|^2}{\sqrt{n}} + \theta^2 \frac{1}{4}\|v\|^2,$$

where the inequality follows from the Cauchy–Schwarz inequality and Lemma 5.3.3. Using the bound on θ the lemma is proved. □

We proceed with a condition on the step size that guarantees feasibility of the new iterates.

Lemma 5.3.5 *If $\overline{\theta}$ is such that $v(\theta)^2 > 0$ for all $0 \leq \theta \leq \overline{\theta}$ then $x(\overline{\theta}) > 0$ and $s(\overline{\theta}) > 0$.*

Proof: If $\overline{\theta}$ satisfies the hypothesis of the lemma then $x(\theta)$ and $s(\theta)$ cannot vanish for any $\theta \in [0, \overline{\theta}]$. Hence, by continuity, $x(\theta)$ and $s(\theta)$ must be positive for any such θ. □

In the analysis we use the large neighborhood $\mathcal{N}_\omega(\rho)$ of the central path defined by the ratio $\omega(v) = \min(v)/\max(v)$ in the v–space. Specifically, we require for all iterates that $\omega(v) \geq \rho$ for some $\rho \in (0,1)$. This neighborhood is essentially the same as the large neighborhood $\mathcal{N}_\infty(\beta)$ introduced in Section 5.2, which can be seen by rewriting its condition in terms of the v–space as

$$(1-\beta)\frac{e^T v^2}{n} \leq v_i^2 \leq (1+\beta)\frac{e^T v^2}{n}, \quad i = 1, \ldots, n.$$

So, if v is in $\mathcal{N}_\infty(\beta)$ it satisfies $\omega(v) \geq \sqrt{(1-\beta)/(1+\beta)}$. Naturally, the complexity of the algorithm will depend on the value of ρ used. In fact, if $(x^{(0)}, s^{(0)})$

is an arbitrary interior–feasible starting point and $v^{(0)} := \sqrt{x^{(0)}s^{(0)}}$, we may take $\rho := \omega(v^{(0)})$. The next theorem makes clear that, with a suitable step size, the new iterates not only stay feasible, but also that the ratio $\omega(v)$ remains bounded by ρ.

Theorem 5.3.6 *If $(x, s) \in \mathcal{F}^0$, $0 < \rho < 1$, $\omega(v) \geq \rho$, $\theta \geq 0$ and*

$$\theta \leq \min\left(\frac{2\omega(v)}{\sqrt{1+4\kappa}}\left(\sqrt{1+\frac{\omega(v)^2}{n(1+4\kappa)}} - \frac{\omega(v)}{\sqrt{n(1+4\kappa)}}\right), \frac{\omega(v)^2\sqrt{n}}{2}, \frac{4\rho^2(1-\rho^2)}{(1+4\kappa)(1+\rho^2)\sqrt{n}}\right) \quad (5.20)$$

then $(x(\theta), s(\theta)) \in \mathcal{F}^0$ and $\omega(v(\theta)) \geq \rho$.

Proof: The hypothesis of the theorem provides three upper bounds for the step size θ. As we will see below, the first upper bound guarantees feasibility of the new iterates, the last guarantees that $\omega(v(\theta)) \geq \rho$, both under the premise that the second bound holds.

Recall from (5.19) and the definition of p_v that

$$v(\theta)^2 = v^2 - \theta\frac{v^4}{\|v^2\|} + \theta^2 p_x p_s.$$

Consider the function

$$\phi(t) = t - \theta\frac{t^2}{\|v^2\|}. \quad (5.21)$$

One easily verifies that ϕ is monotonically increasing on the interval $[0, \max(v)^2]$ if $\theta \leq \|v^2\|/(2\max(v)^2)$. Note that

$$\frac{\|v^2\|}{2\max(v)^2} \geq \frac{\sqrt{n}\min(v)^2}{2\max(v)^2} = \frac{\sqrt{n}\,\omega(v)^2}{2}; \quad (5.22)$$

hence if we enforce the second upper bound in (5.20) the largest and smallest coordinates of $v(\theta)$ can be estimated, upon also using Lemma 5.3.3, as follows:

$$\begin{aligned}\max(v(\theta))^2 &\leq \max(v)^2 - \theta\frac{\max(v)^4}{\|v^2\|} + \theta^2\frac{(1+4\kappa)\max(v)^2}{4},\\ \min(v(\theta))^2 &\geq \min(v)^2 - \theta\frac{\min(v)^4}{\|v^2\|} - \theta^2\frac{(1+4\kappa)\max(v)^2}{4}.\end{aligned} \quad (5.23)$$

Lemma 5.3.5 implies that the new iterates will be feasible if $\min(v(\theta)^2) > 0$. After dividing (5.23) by $\min(v)^2$ this is certainly true if θ satisfies

$$1 - \frac{\theta \min(v)^2}{\|v^2\|} - \frac{\theta^2(1+4\kappa)}{4\omega(v)^2} \geq 0.$$

Using the inequality in (5.22) this certainly holds if

$$1 - \frac{\theta}{\sqrt{n}} - \frac{\theta^2(1+4\kappa)}{4\omega(v)^2} \geq 0.$$

Elementary calculations make clear that this condition is satisfied due to the first upper bound on θ in (5.20). So the new iterates are feasible. Using again the monotonicity of (5.21) and the fact that

$$\min(v)^2 = \omega(v)^2 \max(v)^2 \geq \rho^2 \max(v)^2,$$

we derive

$$\min(v(\theta))^2 \geq \rho^2 \max(v)^2 - \theta \frac{\rho^4 \max(v)^4}{\|v^2\|} - \theta^2 \frac{(1+4\kappa)\max(v)^2}{4}.$$

Now $\omega(v(\theta)) \geq \rho$ will certainly hold if

$$\max(v)^2 - \theta \frac{\max(v)^4}{\|v^2\|} + \theta^2 \frac{(1+4\kappa)\max(v)^2}{4} \leq \max(v)^2 - \theta \frac{\rho^2 \max(v)^4}{\|v^2\|} - \theta^2 \frac{(1+4\kappa)\max(v)^2}{4\rho^2}.$$

By rearranging the inequality we see that it is equivalent to

$$\theta \frac{1+4\kappa}{4} \left(1 + \frac{1}{\rho^2}\right) \leq \frac{\max(v)^2}{\|v^2\|}(1-\rho^2).$$

Finally, using

$$\frac{\max(v)^2}{\|v^2\|} \geq \frac{1}{\sqrt{n}},$$

the third bound on θ in (5.20) is obtained. This completes the proof. □

Now we are ready to derive the complexity of the algorithm.

Theorem 5.3.7 *Let $\epsilon > 0$, $(x^{(0)}, s^{(0)}) \in \mathcal{F}^0$ be given, such that $\omega(v^{(0)}) \geq \rho$ for some $0 < \rho < 1$, and let θ satisfy the conditions of Lemma 5.3.4 and Theorem 5.3.6. The primal–dual Dikin–affine scaling algorithm stops with a feasible solution (x^*, s^*) for which $(x^*)^T s^* \leq \epsilon$ and $\omega(v^*) \geq \rho$ holds, after at most*

$$\frac{2\sqrt{n}}{\theta} \ln \frac{(x^{(0)})^T s^{(0)}}{\epsilon}$$

iterations.

Proof: By Lemma 5.3.4 in each iteration the duality gap reduces by at least the factor

So, after k steps the (error in) complementarity will be less than ϵ if

$$\left(1 - \frac{\theta}{2\sqrt{n}}\right)^k (x^{(0)})^T s^{(0)} \leq \epsilon.$$

Taking logarithms gives

$$k \ln \left(1 - \frac{\theta}{2\sqrt{n}}\right) \leq \ln \frac{\epsilon}{(x^{(0)})^T s^{(0)}},$$

which is certainly true if

$$-k \frac{\theta}{2\sqrt{n}} \leq \ln \frac{\epsilon}{(x^{(0)})^T s^{(0)}}.$$

From this the bound follows. The statement $\omega(v^*) \geq \rho$ is contained in Theorem 5.3.6. □

To derive the actual complexity it should be checked which of the conditions on θ in Lemma 5.3.4 and Theorem 5.3.6 is strongest. From the bounds it can be seen that for n sufficiently large the third bound in Theorem 5.3.6 will dominate the step size computation. For instance, we may derive the following corollary.

Corollary 5.3.8 *If $(x^{(0)}, s^{(0)}) \in \mathcal{F}^0$ is such that $\omega(v^{(0)}) \geq \rho$ and n is sufficiently large then the primal–dual Dikin–affine scaling algorithm requires at most*

$$\mathcal{O}\left(\frac{n(1+4\kappa)}{\rho^2(1-\rho^2)} \ln \frac{(x^{(0)})^T s^{(0)}}{\epsilon}\right)$$

iterations.

Specifically, it follows that for 'standard' large neighborhood algorithms in which $\rho = \Omega(1)$ the algorithm stops after at most $\mathcal{O}(n(1+4\kappa)\ln(x^{(0)})^T s^{(0)}/\epsilon)$ iterations, which is the same as for other algorithms as in Mizuno et al. [179] and Kojima et al. [133]. Observe, that the theoretical step size depends on κ, which might be hard to compute. However, in practice we may compute the step size with a line search by requiring $\omega(v) = \rho$ for all iterates, which means that the actual step size is at least as large as the theoretical one.

5.4 USING CORRECTOR STEPS

5.4.1 Motivation

To compute a search–direction interior point methods for LCPs typically use (nonlinear) systems of the form

$$\begin{aligned} Mx - s &= -q, \quad x > 0,\ s > 0, \\ xs &= \overline{v}^2, \end{aligned} \tag{5.24}$$

for some $\overline{v} \in \mathbb{R}^n_{++}$ (cf. Chapter 8). As in the previous section we assume that M is in $P_*(\kappa)$ for some $\kappa \geq 0$ and that $\mathcal{F}^0$ is nonempty. Let $(\overline{x}, \overline{s}) \in \mathcal{F}^0$ be the current iterate. Similar to the derivation of (5.5) a direction can be computed from the (linear) system

$$\begin{aligned} M\Delta x - \Delta s &= 0, \\ \overline{x}\Delta s + \overline{s}\Delta x &= \overline{v}^2 - \overline{x}\,\overline{s}. \end{aligned}$$

Obviously, the new iterate obtained in this way will not satisfy equation (5.24) due to the linearization. Moreover, the major computational effort per iteration is to compute a factorization of a $n \times n$ matrix, which makes it important to take the utmost advantage of it. This led Monteiro et al. [182] and Mehrotra [170] to the idea of using *corrector steps* to diminish the error that is made by the linearization. The use of one such corrector step appeared to be very useful in practice and is related to the predictor–corrector algorithm (see e.g., Mehrotra [170] and Lustig et al. [156, 157]). The use of several correctors per iteration has been investigated computationally by Mehrotra [170], Carpenter et al. [32] and Gondzio [82] among others. Recently, Hung and Ye [109] developed an r–order predictor–corrector primal–dual algorithm, similar to the ones mentioned, and analyzed the theoretical behavior of the algorithm, hereto inspired by Zhang and Zhang [257]. Their algorithm uses a large neighborhood, namely

$$\mathcal{N}_\infty^-(\beta) = \left\{(x,s) \in \mathcal{F}^0 : xs \geq (1-\beta)\mu e \quad \text{where} \quad \mu = \frac{x^T s}{n}\right\},$$

where β is a fixed constant in $(0,1)$. Consequently, the algorithm is a large neighborhood long–step algorithm. They showed their algorithm to need at most $\mathcal{O}(n^{(r+1)/(2r)} \ln 1/\epsilon)$ iterations where $r \in [1,n]$ equals one plus the number of corrections per iteration. Each iteration requires $\mathcal{O}(n^3 + rn^2)$ arithmetic operations. Note that if $r = n$ then the iteration bound is asymptotically $\mathcal{O}(\sqrt{n} \ln 1/\epsilon)$, so the complexity approaches the bound for short–step algorithms.

In this section, we prove that an r–order version of the primal–dual Dikin–affine scaling algorithm also possesses iteration complexity $\mathcal{O}(n^{(r+1)/(2r)} \ln 1/\epsilon)$, where $r \in [1,n]$. This implies that the affine scaling algorithm is not only polynomial but can also asymptotically achieve the best complexity bound as r and n increase. In the analysis we use the neighborhood $\mathcal{N}_\infty(\beta)$, defined in Section 5.2. Recall from the previous section that this neighborhood is essentially equivalent to the one using $\omega(v)$. The results in this section were derived in Jansen et al. [117] for LP and monotone LCPs and is extended here to LCPs with a P_*–matrix.

5.4.2 A generic r–order algorithm

We describe a generic high–order version of primal–dual interior point algorithms, where (for the moment) the search–direction is determined by a vector $h^{(1)}$. We use the neighborhood $\mathcal{N}_\infty(\beta)$ and denote the order of the algorithm by the positive integer r. We use the notation

$$v^2 = xs, \quad \mu = \frac{x^T s}{n} = \frac{e^T v^2}{n}, \quad \text{and} \quad w = \frac{v^2}{\mu}, \tag{5.25}$$

and also

$$v(\theta)^2 = x(\theta)s(\theta), \quad \mu(\theta) = \frac{x(\theta)^T s(\theta)}{n},$$

to refer to the complementarity after a step of size θ. Algorithms fitting in the general scheme are specified by the choice of $h^{(1)}$, for instance

$$\begin{aligned} h^{(1)} = \gamma\mu e - v^2 \quad \text{for some} \quad 0 < \gamma \leq 1 &\quad : \quad \text{primal–dual path–following} \\ h^{(1)} = -v^2 &\quad : \quad \text{primal–dual affine scaling} \\ h^{(1)} = -\frac{v^4}{\|v^2\|} &\quad : \quad \text{primal–dual Dikin–affine scaling.} \end{aligned}$$

The second choice was studied by Monteiro et al. [182] and the first choice was analyzed in Hung and Ye [109] for LP. Later in this section we consider the third choice for $h^{(1)}$. The generic algorithm is described in Figure 5.3.

Input
$(x^{(0)}, s^{(0)}) \in \mathcal{N}_\infty(\beta)$: the initial pair of interior–feasible solutions;
Parameters
$\epsilon > 0$ is the accuracy parameter;
$r \in [1, n]$ is the order;
$\beta \in (0, 1)$ is the accuracy parameter;
$h^{(1)}$ specifies the algorithm;
begin
$x := x^{(0)}$; $s := s^{(0)}$;
while $x^T s > \epsilon$ **do**
solve the first order direction $(\Delta x^{(1)}, \Delta s^{(1)})$ from

$$\begin{aligned} M\Delta x^{(1)} - \Delta s^{(1)} &= 0, \\ x\Delta s^{(1)} + s\Delta x^{(1)} &= h^{(1)}; \end{aligned} \tag{5.26}$$

for $j = 2, 3, \ldots, r$, solve the jth order direction $(\Delta x^{(j)}, \Delta s^{(j)})$ from

$$\begin{aligned} M\Delta x^{(j)} - \Delta s^{(j)} &= 0, \\ x\Delta s^{(j)} + s\Delta x^{(j)} &= -\sum_{t=1}^{j-1} \Delta x^{(t)} \Delta s^{(j-t)}; \end{aligned} \tag{5.27}$$

choose a step size $\theta > 0$ such that

$$x(\theta) = x + \sum_{j=1}^{r} \theta^j \Delta x^{(j)}, \quad s(\theta) = s + \sum_{j=1}^{r} \theta^j \Delta s^{(j)},$$

is in $\mathcal{N}_\infty(\beta)$ and $\mu(\theta)$ as small as possible;
$x := x(\theta)$, $s := s(\theta)$;
end
end.

Figure 5.3 Generic r-order algorithm.

5.4.3 Complexity analysis

To stay in the neighborhood defined by β, we choose θ as large as possible but still satisfying the inequalities

$$(1-\beta)\mu(\zeta)e \le v(\zeta)^2 \le (1+\beta)\mu(\zeta)e, \quad \forall\, 0 \le \zeta \le \theta. \tag{5.28}$$

Although in practice one will choose a different value for θ in each iteration (namely, by checking for condition (5.28)), the analysis guarantees the existence of a fixed value for θ that gives the desired complexity bound. Without loss of generality we assume that $\mu = 1$ (otherwise we perform a scaling to accomplish this), hence (5.25) implies

$$w = \frac{v^2}{\mu} = v^2, \tag{5.29}$$

and we have the bound

$$(1-\beta)e \le w \le (1+\beta)e. \tag{5.30}$$

We first derive some general lemmas, without making an explicit choice for $h^{(1)}$.

Lemma 5.4.1 *Using the directions defined by (5.26) and (5.27) in Figure 5.3 it holds*

$$v(\theta)^2 = x(\theta)s(\theta) = v^2 + \theta h^{(1)} + \sum_{j=r+1}^{2r} \left(\theta^j \sum_{t=j-r}^{r} \Delta x^{(t)} \Delta s^{(j-t)} \right). \tag{5.31}$$

Proof: We prove the lemma by induction on r. For $r = 1$ it holds

$$\begin{aligned} x(\theta)s(\theta) &= v^2 + \theta(s\Delta x^{(1)} + x\Delta s^{(1)}) + \theta^2 \Delta x^{(1)} \Delta s^{(1)} \\ &= v^2 + \theta h^{(1)} + \theta^2 \Delta x^{(1)} \Delta s^{(1)}, \end{aligned}$$

which is the desired equality. Assume the result is true for $r-1$. Denote

$$\Delta^{(r-1)}x := \sum_{j=1}^{r-1} \theta^j \Delta x^{(j)}, \qquad \Delta^{(r-1)}s := \sum_{j=1}^{r-1} \theta^j \Delta s^{(j)}.$$

Then for r we have

$$x(\theta)s(\theta) = (x + \Delta^{(r-1)}x + \theta^r \Delta x^{(r)})(s + \Delta^{(r-1)}s + \theta^r \Delta s^{(r)})$$

$$\begin{aligned}
&= (x+\Delta^{(r-1)}x)(s+\Delta^{(r-1)}s)+\theta^r(x\Delta s^{(r)}+s\Delta x^{(r)})+\\
&\quad 2\theta^r\left(\Delta x^{(r)}\Delta^{(r-1)}s+\Delta s^{(r)}\Delta^{(r-1)}x\right)+\theta^{2r}\Delta x^{(r)}\Delta s^{(r)}\\
&= v^2+\theta h^{(1)}+\sum_{j=r}^{2(r-1)}\theta^j\sum_{t=j-(r-1)}^{r-1}\Delta x^{(t)}\Delta s^{(j-t)}-\theta^r\sum_{t=1}^{r-1}\Delta x^{(t)}\Delta s^{(r-t)}+\\
&\quad \theta^r\left(\Delta x^{(r)}\sum_{j=1}^{r-1}\theta^j\Delta s^{(j)}+\Delta s^{(r)}\sum_{j=1}^{r-1}\theta^j\Delta x^{(j)}\right)+\theta^{2r}\Delta x^{(r)}\Delta s^{(r)}.
\end{aligned}$$

The right–hand side is a polynomial in θ. It is easy to see that the coefficient of θ^j is zero for $2\le j<r$. For $j=r$, the coefficient is also zero due to cancellation of terms. The coefficient for θ^j, $r+1\le j\le 2(r-1)$, is

$$\sum_{t=j-r+1}^{r-1}\left(\Delta x^{(t)}\Delta s^{(j-t)}\right)+\Delta x^{(r)}\Delta s^{(j-r)}+\Delta s^{(r)}\Delta x^{(j-r)}=\sum_{t=j-r}^{r}\Delta x^{(t)}\Delta s^{(j-t)}.$$

For θ^{2r-1} we find

$$\Delta x^{(r)}\Delta s^{(r-1)}+\Delta s^{(r)}\Delta x^{(r-1)}=\sum_{t=2r-1-r}^{r}\Delta x^{(t)}\Delta s^{(2r-1-t)},$$

and for θ^{2r} we have the coefficient

$$\Delta x^{(r)}\Delta s^{(r)}=\sum_{t=2r-r}^{r}\Delta x^{(t)}\Delta s^{(2r-t)}.$$

Combining gives the required equality for r and completes the proof. □

Using (5.29) and (5.31) we observe that condition (5.28) is certainly satisfied if

$$\left\|\sum_{j=r+1}^{2r}\left(\theta^j\sum_{t=j-r}^{r}\Delta x^{(t)}\Delta s^{(j-t)}\right)\right\|_\infty\le(1+\beta)\mu(\theta)-(w_i+\theta h_i^{(1)}),\quad i=1,\ldots,n, \tag{5.32}$$

and also

$$\left\|\sum_{j=r+1}^{2r}\left(\theta^j\sum_{t=j-r}^{r}\Delta x^{(t)}\Delta s^{(j-t)}\right)\right\|_\infty\le w_i+\theta h_i^{(1)}-(1-\beta)\mu(\theta),\quad i=1,\ldots,n. \tag{5.33}$$

For ease of notation we introduce the vectors

$$\xi^{(j)} := \sum_{t=j-r}^{r} \Delta x^{(t)} \Delta s^{(j-t)}$$

for $j = r+1, \ldots, 2r$. Using the Cauchy–Schwarz inequality we may bound $\mu(\theta)$ from below by

$$\mu(\theta) = 1 + \theta \frac{e^T h^{(1)}}{n} + \frac{1}{n} \sum_{j=r+1}^{2r} \theta^j e^T \xi^{(j)} \geq 1 + \theta \frac{e^T h^{(1)}}{n} - \frac{1}{\sqrt{n}} \sum_{j=r+1}^{2r} \theta^j \|\xi^{(j)}\|,$$

and from above by

$$\mu(\theta) \leq 1 + \theta \frac{e^T h^{(1)}}{n} + \frac{1}{\sqrt{n}} \sum_{j=r+1}^{2r} \theta^j \|\xi^{(j)}\|. \tag{5.34}$$

The main task in the complexity analysis is to analyze the effect of the higher–order corrections. For this purpose, it appears that we can use very weak bounds. Observe that

$$\left\| \sum_{j=r+1}^{2r} \theta^j \xi^{(j)} \right\|_\infty \leq \sum_{j=r+1}^{2r} \theta^j \|\xi^{(j)}\|. \tag{5.35}$$

Combining (5.35) with the bounds on $\mu(\theta)$ it follows that (5.32) and (5.33) are certainly satisfied if θ satisfies

$$\sum_{j=r+1}^{2r} \theta^j \|\xi^{(j)}\| \leq \frac{\sqrt{n}}{1+\beta+\sqrt{n}} \left((1+\beta)\left(1 + \theta \frac{e^T h^{(1)}}{n}\right) - (w_i + \theta h_i^{(1)}) \right), \quad i = 1, \ldots, n,$$

and also

$$\sum_{j=r+1}^{2r} \theta^j \|\xi^{(j)}\| \leq \frac{\sqrt{n}}{1-\beta+\sqrt{n}} \cdot \left(w_i + \theta h_i^{(1)} - (1-\beta)\left(1 + \theta \frac{e^T h^{(1)}}{n}\right) \right), \quad i = 1, \ldots, n.$$

We will proceed as follows. First, we show that the terms $\|\xi^{(j)}\|$ can be bounded in terms of $\|h^{(1)}\|$. Then we consider bounds for the right–hand side quantities in equations (5.35) and (5.35). Let $d = (xs^{-1})^{1/2}$, so the usual primal–dual scaling vector. Define $p^{(j)} := d\Delta s^{(j)} + d^{-1}\Delta x^{(j)}$, $1 \le j \le r$. Observe that

$$p^{(1)} = w^{-1/2}h^{(1)}. \tag{5.36}$$

Now for $j \ge 1$, we have

$$\begin{aligned} \|d\Delta s^{(j)}\|^2 + \|d^{-1}\Delta x^{(j)}\|^2 &= \|d\Delta s^{(j)} + d^{-1}\Delta x^{(j)}\|^2 \\ &\quad -2(\Delta s^{(j)})^T\Delta x^{(j)} \le (1+2\kappa)\|p^{(j)}\|^2, \end{aligned}$$

where the inequality follows from Lemma 5.3.3. So

$$\max(\|d^{-1}\Delta x^{(j)}\|, \|d\Delta s^{(j)}\|) \le \sqrt{1+2\kappa}\|p^{(j)}\|. \tag{5.37}$$

In the following lemma a bound on the norm of $p^{(j)}$ is derived.

Lemma 5.4.2 *For $j = 1, 2, ..., r$, it holds*

$$\|p^{(j)}\| \le \frac{(1+2\kappa)^{j-1}\phi(j)}{(1-\beta)^{(j-1)/2}}\|p^{(1)}\|^j, \tag{5.38}$$

where the integer sequence $\phi(j)$ is defined recursively as follows:

$$\phi(1) = 1, \quad \phi(j) = \sum_{t=1}^{j-1}\phi(t)\phi(j-t), \quad j = 2, 3, ...$$

Proof: The proof is by induction on j. Note that (5.38) trivially holds for $j = 1$. Assume (5.38) holds for all ℓ with $1 \le \ell < j$. We will show it to hold for j, as follows.

$$\begin{aligned} \|p^{(j)}\| &= \left\| w^{-1/2}\left(\sum_{t=1}^{j-1}\Delta x^{(t)}\Delta s^{(j-t)}\right)\right\| \\ &\le \left\|w^{-1/2}\right\|_\infty \left\|\sum_{t=1}^{j-1}\Delta x^{(t)}\Delta s^{(j-t)}\right\| \\ &\le \frac{1}{(1-\beta)^{1/2}}\sum_{t=1}^{j-1}\|d^{-1}\Delta x^{(t)}\|\,\|d\Delta s^{(j-t)}\| \\ &\le \frac{1+2\kappa}{(1-\beta)^{1/2}}\sum_{t=1}^{j-1}\|p^{(t)}\|\,\|p^{(j-t)}\|. \end{aligned}$$

In the second inequality we use (5.30); in the third we invoke (5.37). Using the induction hypothesis we obtain

$$\begin{aligned}
\|p^{(j)}\| &\le \frac{1+2\kappa}{(1-\beta)^{1/2}} \sum_{t=1}^{j-1} \left(\frac{(1+2\kappa)^{t-1}\phi(t)}{(1-\beta)^{(t-1)/2}} \|p^{(1)}\|^t \cdot \right. \\
&\quad \left. \frac{(1+2\kappa)^{j-t-1}\phi(j-t)}{(1-\beta)^{(j-t-1)/2}} \|p^{(1)}\|^{j-t} \right) \\
&= \frac{(1+2\kappa)^{j-1}}{(1-\beta)^{(j-1)/2}} \left(\sum_{t=1}^{j-1} \phi(t)\phi(j-t) \right) \|p^{(1)}\|^j \\
&= \frac{(1+2\kappa)^{j-1}\phi(j)}{(1-\beta)^{(j-1)/2}} \|p^{(1)}\|^j .
\end{aligned}$$

□

Next, we establish an upper bound for $\|\xi^{(j)}\|$ in terms of $p^{(1)}$.

Lemma 5.4.3 *For $j = r+1, ..., 2r$, it holds*

$$\|\xi^{(j)}\| = \left\| \sum_{t=j-r}^{r} \Delta x^{(t)} \Delta s^{(j-t)} \right\| \le \frac{(1+2\kappa)^{j-1}\|p^{(1)}\|^j}{(1-\beta)^{j/2-1}} \frac{16^r}{8r}.$$

Proof: Using Lemma 5.4.2 and (5.37) we have for $j = r+1, \ldots, 2r$

$$\begin{aligned}
\left\| \sum_{t=j-r}^{r} \Delta x^{(t)} \Delta s^{(j-t)} \right\| &\le \sum_{t=j-r}^{r} \|d^{-1}\Delta x^{(t)}\| \, \|d\Delta s^{(j-t)}\| \\
\le (1+2\kappa) \sum_{t=j-r}^{r} &\left(\frac{(1+2\kappa)^{t-1}\phi(t)}{(1-\beta)^{(t-1)/2}} \|p^{(1)}\|^t \frac{(1+2\kappa)^{j-t-1}\phi(j-t)}{(1-\beta)^{(j-t-1)/2}} \|p^{(1)}\|^{j-t} \right) \\
= \frac{(1+2\kappa)^{j-1}\|p^{(1)}\|^j}{(1-\beta)^{j/2-1}} &\sum_{t=j-r}^{r} \phi(t)\phi(j-t) \\
\le \frac{(1+2\kappa)^{j-1}\|p^{(1)}\|^j}{(1-\beta)^{j/2-1}} &\phi(2r).
\end{aligned}$$

The inequality in the lemma follows from the solution of the recurrence relation for $\phi(j)$ in Lemma 5.4.2, which is given by

$$\phi(j) = \frac{1}{j} \binom{2(j-1)}{j-1} \le \frac{1}{j} 2^{2j-2}.$$

□

In the primal–dual Dikin–affine scaling algorithm the vector $h^{(1)}$ is given by (recall $v^2 = w$ by assuming $\mu = 1$)

$$h^{(1)} = -\frac{w^2 e}{||w||}.$$

Since $(1-\beta)e \le w \le (1+\beta)e$, we have (recall (5.36))

$$||p^{(1)}||^2 = \frac{||w^{3/2}e||^2}{||w||^2} \le ||w^{1/2}||_\infty^2 \le 1+\beta.$$

This estimation enables us to make the bound in Lemma 5.4.3 independent of j, since the functions

$$\left(\frac{1+\beta}{1-\beta}\right)^{j/2} \quad \text{and} \quad (1+2\kappa)^{j-1}$$

are increasing in j. Hence it holds for $j = r+1, \ldots, 2r$

$$\left\| \sum_{t=j-r}^{r} \Delta x^{(t)} \Delta s^{(j-t)} \right\| \le (1+2\kappa)^{2r-1} \left(\frac{1+\beta}{1-\beta}\right)^r \frac{(1-\beta)16^r}{8r}. \tag{5.39}$$

We introduce

$$\overline{\mu}(\theta) := 1 + \theta\frac{e^T h^{(1)}}{n} = 1 - \theta\frac{e^T w^2}{n||w||} = 1 - \theta\frac{||w||}{n}.$$

In the next lemma we obtain bounds for the right–hand sides of conditions (5.35) and (5.35) for our choice of $h^{(1)}$.

Lemma 5.4.4 *Let $\overline{\mu}(\theta) = 1 - \theta||w||/n$, $\theta \le \sqrt{n}/2$ and $1-\beta \le w_i \le 1+\beta$, $\forall i$. Then the following bounds hold:*

$$\min_i \left[\left(w_i - \theta\frac{w_i^2}{||w||}\right) - (1-\beta)\overline{\mu}(\theta)\right] \ge \frac{\theta\beta(1-\beta)}{\sqrt{n}},$$

and

$$\min_i \left[(1+\beta)\overline{\mu}(\theta) - \left(w_i - \theta\frac{w_i^2}{||w||}\right)\right] \ge \frac{\theta\beta(1-\beta)}{\sqrt{n}}.$$

Proof: To prove the first inequality we use the fact that $w_i - \theta w_i^2/\|w\|$ as a function of w_i attains its minimal value at one of the bounds $1-\beta$ or $1+\beta$. In the first case we have

$$
\begin{aligned}
1-\beta &- \frac{\theta(1-\beta)^2}{\|w\|} - (1-\beta)\overline{\mu}(\theta) \\
&= (1-\beta)\left(1 - \frac{\theta(1-\beta)}{\|w\|} - \left(1 - \frac{\theta\|w\|}{n}\right)\right) \\
&= \theta(1-\beta)\left(\frac{\|w\|}{n} - \frac{1-\beta}{\|w\|}\right) \\
&\geq \theta(1-\beta)\left(\frac{1}{\sqrt{n}} - \frac{1-\beta}{\sqrt{n}}\right) \\
&= \frac{\theta\beta(1-\beta)}{\sqrt{n}},
\end{aligned}
$$

where the inequality follows from the fact that $\|w\| \geq \sqrt{n}$. In the second case we have

$$
\begin{aligned}
1+\beta &- \frac{\theta(1+\beta)^2}{\|w\|} - (1-\beta)\overline{\mu}(\theta) \\
&= 1+\beta - \frac{\theta(1+\beta)^2}{\|w\|} - (1-\beta)\left(1 - \frac{\theta\|w\|}{n}\right) \\
&\geq 2\beta - \frac{\theta(1+\beta)^2}{\sqrt{n}} + \frac{(1-\beta)\theta}{\sqrt{n}} \\
&= 2\beta - \frac{\theta\beta}{\sqrt{n}}(3+\beta) \\
&= \frac{\theta\beta}{\sqrt{n}}\left(\frac{2\sqrt{n}}{\theta} - (3+\beta)\right) \\
&\geq \frac{\theta\beta(1-\beta)}{\sqrt{n}},
\end{aligned}
$$

where the last inequality follows from $\theta \leq \sqrt{n}/2$. This completes the proof of the first inequality in the lemma. The second inequality can be proved in a similar way, where instead of using $\|w\| \geq \sqrt{n}$, the upper bound on $\|w\|$ from Lemma A.2 should be used. This is left to the reader. □

We combine the bounds obtained so far to give a (stronger) condition on θ such that it certainly satisfies (5.35) and (5.35). Notice that for $n \geq 2$ it holds

$$\frac{\sqrt{n}}{1+\beta+\sqrt{n}} > \frac{1}{4}, \quad \frac{\sqrt{n}}{1-\beta+\sqrt{n}} > \frac{1}{4}.$$

Using (5.39) and the bounds in Lemma 5.4.4 we now require θ to satisfy

$$r\,\theta^{r+1}(1+2\kappa)^{2r-1}\left(\frac{1+\beta}{1-\beta}\right)^r \frac{(1-\beta)16^r}{8r} \leq \frac{\theta\beta(1-\beta)}{4\sqrt{n}}, \tag{5.40}$$

to replace conditions (5.35) and (5.35). Hence we obtain the (sufficient) condition that θ satisfies

$$\theta^r \leq \frac{2}{16^r}\frac{1}{(1+2\kappa)^{2r-1}}\left(\frac{1-\beta}{1+\beta}\right)^r \frac{\beta}{\sqrt{n}}. \tag{5.41}$$

We can now derive the complexity for the primal–dual Dikin–affine scaling algorithm with corrector steps.

Theorem 5.4.5 *Let be given an initial point $(x^{(0)}, s^{(0)}) \in \mathcal{F}^0$ and in $\mathcal{N}_\infty(\beta)$ for some $\beta \in (0,1)$ and let $r \in [1,n]$ be an integer. Then after at most*

$$\mathcal{O}\left(n^{(r+1)/(2r)}\frac{(1+2\kappa)^{2-1/r}}{(1-\beta)\sqrt[r]{\beta}}\ln\frac{(x^{(0)})^T s^{(0)}}{\epsilon}\right)$$

iterations the r-order primal-dual Dikin-affine scaling algorithm has generated a feasible pair (x^*, s^*) such that $(x^*)^T s^* \leq \epsilon$ and $(x^*, s^*) \in \mathcal{N}_\infty(\beta)$; each iteration of the algorithm uses $\mathcal{O}(n^3 + rn^2)$ arithmetic operations.

Proof: From (5.28) and (5.41), it immediately follows that in any iteration we can choose the step size

$$\theta = n^{-1/(2r)}\frac{1}{(1+2\kappa)^{2-1/r}}\frac{1-\beta}{16(1+\beta)}\sqrt[r]{2\beta}.$$

Using (5.34), the definition of $h^{(1)}$ and Cauchy–Schwarz we have (recall $\mu = 1$, so $e^T w = n$),

$$\mu(\theta) \leq 1 - \frac{\theta}{\sqrt{n}} + \frac{1}{\sqrt{n}}\sum_{j=r+1}^{2r}\theta^j\,\|\xi^{(j)}\|.$$

The last term can be bounded with (5.39), giving

$$\mu(\theta) \leq 1 - \frac{\theta}{\sqrt{n}} + \frac{1}{\sqrt{n}} r\theta^{r+1}(1+2\kappa)^{2r-1}\left(\frac{1+\beta}{1-\beta}\right)^r \frac{(1-\beta)16^r}{8r}.$$

Using inequality (5.40) we obtain

$$\mu(\theta) \leq 1 - \frac{\theta}{\sqrt{n}} + \frac{\theta\beta(1-\beta)}{4n} \leq 1 - \frac{\theta}{2\sqrt{n}},$$

for $n \geq 2$. Combining this with the bound on θ implies that no more

$$\frac{2\sqrt{n}}{n^{-1/(2r)}}(1+2\kappa)^{2-1/r}\frac{16(1+\beta)}{1-\beta}\frac{1}{\sqrt[r]{2\beta}}\ln\frac{(x^{(0)})^T s^{(0)}}{\epsilon}.$$

iterations are needed to obtain a solution (x^*, s^*) for which $(x^*)^T s^* \leq \epsilon$. This implies the iteration bound. In each iteration, one factorization of an $n \times n$ matrix has to be performed, together with r backsubstitutions. This gives the number of arithmetic operations per iteration. □

The theorem shows that asymptotically only $\mathcal{O}(\sqrt{n}\ln(x^{(0)})^T s^{(0)}/\epsilon)$ iterations are required with $\mathcal{O}(n^3)$ arithmetic operations per iteration. In practice, however, computational experience with analogous techniques have shown that the use of one corrector is very effective, while using more than a few correctors is not worth the effort (e.g., Carpenter et al. [32]). In Chapter 7 we will see the same phenomenon.

5.5 CONCLUDING REMARKS

In this chapter we derived an interior point method, which gives a generalization of Dikin's primal affine scaling method to a primal–dual setting. The algorithm, called the primal–dual Dikin–affine scaling algorithm, was originally developed for LP but could be extended to linear complementarity problems (LCPs). The method outperforms the classical primal–dual affine scaling method by Monteiro et al. [182] from a theoretical point of view: it is more natural and has a better complexity bound. The main reason for this is its property of simultaneously driving the iterates to optimality as well as to the central path, when started at an arbitrary interior–feasible solution.

To enhance the complexity of the algorithm we studied the effect on complexity of the use of corrector steps and showed how to derive the (asymptotically) best known bound for LCPs.

6

PRIMAL–DUAL AFFINE SCALING METHODS FOR NONLINEAR PROBLEMS

6.1 INTRODUCTION

(Monotone) Nonlinear complementarity problems ((M)NCPs) form a large class of mathematical programming problems with many applications. For instance, any convex programming problem can be modeled as an MNCP. This class of problems is closely related to the class of variational inequalities, which play an important role in the study of equilibria in, e.g., economics, transportation planning and game–theory. We refer to Cottle et al. [34] and Pang [198] for surveys on complementarity problems. A survey on variational inequalities is provided by Harker and Pang [98].

The study of interior point methods for LP also led to the use of barrier methods for nonlinear convex programming problems, see Nesterov and Nemirovskii [193], Jarre [118], Den Hertog et al. [100, 101, 105], Kortanek and Zhu [143], Monteiro and Adler [181], Zhu [258], etc. An important difference between applying these algorithms to linear versus nonlinear problems is that for the latter the convergence rate has only been established for problems satisfying certain Lipschitz conditions. A general and unifying analysis was provided by Nesterov and Nemirovskii [193] for the study of central path–following methods for NLP problems satisfying the *self–concordance condition.* Jarre [118] and Den Hertog et al. [105] used the *relative Lipschitz condition*, which was later shown (see e.g. [120]) to be essentially equivalent to self–concordance. Zhu et al. [143, 225, 258] used the *scaled Lipschitz condition* to analyze path–following methods. These conditions are applicable to monotone mappings only.

Several interior point algorithms for LCPs and NCPs have been developed as well, see e.g., [94, 95, 136, 137, 138, 183, 199, 246, 252, 253] and the references therein. Only in the beginning 1990s it appeared that many of them are related to fundamental work of McLinden [166]. The global convergence of these algorithms has been shown using the existence of the central path and relying on a barrier function. For NCPs the study of the convergence rate has been restricted to mappings satisfying a smoothness condition, similarly as for convex programming. Chapter 7 in [193] is completely devoted to the study of variational inequalities, and uses a generalization of the self-concordance condition for convex programming. Some drawbacks of the analysis in [193] are that it focuses on central path-following methods, that the type of search-direction to be used is prescribed (Newton's direction w.r.t. a self-concordant function), that the algorithm is purely primal, and that small neighborhoods of the central path are used. These negative aspects have been improved upon in subsequent papers. Nesterov [191] uses various strategies with potential functions to be able to make long steps; Nesterov and Todd [194] use a symmetric primal-dual conic reformulation of nonlinear convex problems for the analysis of primal-dual potential reduction methods. Potra and Ye [202, 203] deal with a primal-dual potential reduction algorithm for solving MNCPs, and study the global and local convergence rate and the complexity of the algorithm. Their smoothness condition can be regarded as a generalization of Zhu's condition [258]. Kortanek et al. [142] employ this algorithm to solve entropy optimization problems. However, we are not aware of results concerning other search-directions, neither of large neighborhood algorithms nor an analysis for problems with nonmonotone mappings.

In this chapter a complexity analysis for a family of primal-dual affine scaling algorithms for NCPs is developed. The family is constructed by varying the amount of scaling in the primal-dual (Dikin-)affine scaling direction, and is characterized by a scaling parameter ν. Special cases include the primal-dual affine scaling direction with $\nu = 0$ and the primal-dual Dikin-affine scaling direction with $\nu = 1$. Neither the algorithm nor its analysis use a barrier function. The theoretical analysis we provide shows that within the family the Dikin-affine scaling method has the best complexity bound.

We use the wide neighborhood $\mathcal{N}_\omega(\rho)$ of the central path as introduced in (5.12) and used in Section 5.3.1. An implication is that we may not hope for a complexity bound better than $\mathcal{O}(n \ln 1/\epsilon)$, and that we need to consider separate components of the vector of complementarity products instead of its norm. The latter, and the fact that we allow for nonmonotone mappings, require us to reconsider the usability of the smoothness conditions mentioned above for this type of analysis. We introduce a new condition (Condition 6.2.2), which is trivially satisfied in the linear case. An advantage of this condition is that it does not require monotonicity of the mapping while this seems to be indispensable for the scaled Lipschitz condition and self-concordance condition, even if the mapping is linear.

6.2 PROBLEM STATEMENT AND SEARCH-MAPPINGS

Let us consider the NCP:

$$(NCP) \quad \text{Find } (x,s) \in \mathbb{R}^{2n} \text{ such that } s = f(x),\ (x,s) \geq 0 \text{ and } x^T s = 0.$$

Here f is a C^1-mapping from $\mathbb{R}^n_+$ to $\mathbb{R}^n$. We denote the sets of feasible and interior-feasible points of (NCP) by:

$$\begin{aligned} \mathcal{F} &= \{ (x,s) \in \mathbb{R}^{2n} \ : \ s = f(x),\ (x,s) \geq 0 \}, \\ \mathcal{F}^0 &= \{ (x,s) \in \mathbb{R}^{2n} \ : \ s = f(x),\ (x,s) > 0 \}. \end{aligned}$$

We assume that $\mathcal{F}^0$ is not empty, stated otherwise, that an interior point exists. Obviously, the LCP is the special case where the mapping f is given by

$$f(x) = Mx + q,$$

for $M \in \mathbb{R}^{n \times n}$ and $q \in \mathbb{R}^n$. Throughout this chapter we impose the following condition on the mapping f (recall Definition 5.3.1).

Condition 6.2.1 *There exists a constant $\kappa \geq 0$ such that the Jacobian $\nabla f(x)$ of the mapping f is in $P_*(\kappa)$ for all $x \geq 0$.*

Notice that if $\kappa = 0$ then f is a monotone mapping. In practice, it might be hard to compute the actual value of κ; fortunately, our algorithm does not need its explicit knowledge. As in Section 5.3 we use the following notation:

$$v = (xs)^{1/2}, \quad \omega(v) = \min(v)/\max(v).$$

Let us now define a search-mapping and derive some of its properties. Suppose that we have an interior-feasible point $(x,s) \in \mathcal{F}^0$. Given displacements Δx and Δs, we define

$$x(\theta) = x + \theta \Delta x, \tag{6.1}$$

$$g(\theta) = f(x + \theta\Delta x) - f(x) - \theta \nabla f(x) \Delta x, \tag{6.2}$$

$$\Delta s(\theta) = \Delta s + g(\theta)/\theta, \tag{6.3}$$

$$s(\theta) = s + \theta\Delta s + g(\theta) = s + \theta \Delta s(\theta). \tag{6.4}$$

The mapping $s(\theta)$ was introduced in [140] as a modification of the one in [181] for the convex programming problem. The mapping $g(\theta)$ contains the second order effect introduced by the displacement. For a linear mapping f the term $g(\theta)$ vanishes and we obtain

$$x(\theta) = x + \theta\Delta x, \quad s(\theta) = s + \theta\Delta s,$$

which means a usual linear search–mapping; absence of $g(\theta)$ accounts for the fact that no smoothness conditions are required in the linear case. Hence, one of the major tasks in this section will be to accommodate the analysis of Section 5.3 to the presence of $g(\theta)$.

Obviously, it holds $(x(0), s(0)) = (x, s)$. We require our search–directions to satisfy the following (linear) system of equations

$$-\nabla f(x)\Delta x + \Delta s = 0. \tag{6.5}$$

Then it follows

$$\begin{aligned} s(\theta) - f(x(\theta)) &= s + \theta\Delta s + f(x + \theta\Delta x) - f(x) - \theta\nabla f(x)\Delta x - f(x + \theta\Delta x) \\ &= s - f(x) + \theta(-\nabla f(x)\Delta x + \Delta s) \\ &= 0, \end{aligned} \tag{6.6}$$

for every $\theta \geq 0$, i.e., feasibility is preserved by construction. Consequently, if we can find θ such that $(x(\theta), s(\theta)) > 0$ then $(x(\theta), s(\theta))$ is also interior–feasible. The term $g(\theta)$ is continuous and higher–order in θ, i.e., $\lim_{\theta\to 0} \|g(\theta)\|/\theta = 0$; hence we have

$$\left.\frac{\partial s(\theta)}{\partial \theta}\right|_{\theta=0} = \Delta s.$$

The componentwise complementarity product $x(\theta)s(\theta)$ can be expressed as follows:

$$x(\theta)s(\theta) = (x+\theta\Delta x)(s+\Delta s(\theta)) = xs+\theta\,(s\Delta x + x\Delta s(\theta))+\theta^2\Delta x\Delta s(\theta). \tag{6.7}$$

Using as before the primal–dual scaling vector defined by $d = (xs^{-1})^{1/2}$ the search–directions in scaled space are

$$p_x = d^{-1}\Delta x, \quad p_s = d\Delta s = D\nabla f(x)\Delta x. \tag{6.8}$$

Let us define the vector p_v as

$$p_v = p_x + p_s. \tag{6.9}$$

For convenience in further discussions we define the mappings:

$$\begin{aligned} p_s(\theta) &= d\Delta s(\theta) = p_s + d\, g(\theta)/\theta, \\ p_v(\theta) &= p_x + p_s(\theta) = p_v + d\, g(\theta)/\theta, \\ v(\theta) &= (x(\theta)s(\theta))^{1/2}. \end{aligned} \tag{6.10}$$

Using these definitions we may write

$$\begin{aligned} s\Delta x + x\Delta s(\theta) &= v p_v(\theta), \\ \Delta x \Delta s(\theta) &= p_x p_s(\theta). \end{aligned}$$

Hence equation (6.7) can be rewritten as

$$v(\theta)^2 = x(\theta)s(\theta) = v^2 + \theta v p_v(\theta) + \theta^2 p_x p_s(\theta). \tag{6.11}$$

Our next task is to prove results analogous to Lemma 5.3.3 for the vectors Δx and $\Delta s(\theta)$. For this purpose we introduce the following smoothness condition and impose it on the mapping f.

Condition 6.2.2 *Let $x \in \mathbb{R}^n_{++}$, $s = f(x) \in \mathbb{R}^n_{++}$ be given. Let p_s and $p_s(\theta)$ be as defined above. There exist $\Theta > 0$ and $\gamma \geq 0$ such that*

$$\|p_s(\theta) - p_s\| \leq \gamma\, \theta\, \|p_s\|$$

for every $\theta \in (0, \Theta]$.

The following identity is useful in the analysis:

$$p_v(\theta) - p_v = p_s(\theta) - p_s.$$

Notice that the inequality $\|p_s(\theta) - p_s\| \leq \gamma\theta\|p_s\|$ is equivalent to

$$\left\| d\left(\frac{f(x + \theta\Delta x) - f(x)}{\theta} - \nabla f(x)\Delta x \right) \right\| \leq \gamma\theta \|d\nabla f(x)\Delta x\|, \tag{6.12}$$

or stated otherwise

$$\left\| d\frac{g(\theta)}{\theta} \right\| \leq \gamma\theta\|d\Delta s\|.$$

Note that the condition depends not only on the mapping f but also on the displacement $(\Delta x, \Delta s)$ used in an algorithm. If the mapping f is linear, however, the above condition holds with $\Theta = +\infty$ and $\gamma = 0$, independent of the search–directions. In Section 6.5 we will show how Condition 6.2.2 is related to other smoothness conditions on the mapping f. We derive the following lemma.

Lemma 6.2.3 *Let p_s, $p_s(\theta)$, p_v and $p_v(\theta)$ be as defined above and let Condition 6.2.2 hold. Then for any $\theta \in (0, \Theta]$ it holds*
(i) $\|p_v(\theta)\| \le (1+\gamma\theta\sqrt{1+2\kappa})\|p_v\|$;
(ii) $|p_v(\theta) - p_v| \le \gamma\theta\sqrt{1+2\kappa}\|p_v\|e \le \gamma\theta\sqrt{1+2\kappa}\sqrt{n}|p_v|/\omega(|p_v|)$.

Proof: Applying Lemma 5.3.3(*ii*) to relations (6.9) and (6.5) yields

$$\|p_s\|^2 = \|p_v\|^2 - 2p_x^T p_s - \|p_x\|^2 \le (1+2\kappa)\|p_v\|^2. \tag{6.13}$$

Since Condition 6.2.2 holds, we have

$$\|p_v(\theta)\| \le \|p_v\| + \|p_s(\theta) - p_s\| \le \|p_v\| + \gamma\theta\|p_s\| \le (1+\gamma\theta\sqrt{1+2\kappa})\|p_v\|$$

which is the assertion of (*i*). Similarly,

$$|p_v(\theta) - p_v| = |p_s(\theta) - p_s| \le \gamma\theta\|p_s\|e \le \gamma\theta\sqrt{1+2\kappa}\|p_v\|e.$$

Finally, we derive

$$\|p_v\|e = \|p_v\| p_v^{-1} p_v \le \|p_v\|\,|p_v^{-1}|\,|p_v| \le \frac{\sqrt{n}\max(|p_v|)}{\min(|p_v|)}|p_v| = \frac{\sqrt{n}}{\omega(|p_v|)}|p_v|.$$

This completes the proof of (*ii*). □

The following lemma serves a similar goal as Lemma 5.3.3 in the linear case.

Lemma 6.2.4 *Let p_x, p_s, $p_s(\theta)$, p_v and $p_v(\theta)$ be as defined above and let Condition 6.2.2 hold. Then, for any $\theta \in (0, \Theta]$, it holds*
(i) $-(1+\gamma\theta)(1+2\kappa)\|p_v\|^2 \le \Delta x^T \Delta s(\theta)$
$= p_x^T p_s(\theta) \le (1+\theta\gamma\sqrt{1+2\kappa})^2\|p_v\|^2/4$;
(ii) $\|\Delta x \Delta s(\theta)\|_\infty = \|p_x p_s(\theta)\|_\infty$
$\le \left((1+\theta\gamma\sqrt{1+2\kappa})^2/4 + (1+\gamma\theta)(1+2\kappa)\right)\|p_v\|^2$.

Proof: By (6.2)-(6.5) and (6.10), we have

$$p_s(\theta) = d\Delta s(\theta) = d\left(\frac{f(x+\theta\Delta x) - f(x)}{\theta}\right).$$

Using also (6.8) we obtain

$$\Delta x^T \Delta s(\theta) = p_x^T p_s(\theta) = \frac{1}{\theta} p_x^T d(f(x+\theta\Delta x) - f(x)).$$

From (6.12) we derive

$$\|d(f(x+\theta\Delta x)-f(x))\| \leq (1+\gamma\theta)\theta\|d\nabla f(x)\Delta x\| = (1+\gamma\theta)\theta\|p_s\|.$$

Just as in (6.13) it holds

$$\|p_x\|^2 = \|p_v\|^2 - 2p_x^T p_s - \|p_s\|^2 \leq (1+2\kappa)\|p_v\|^2. \tag{6.14}$$

Consequently,

$$\begin{aligned} |\Delta x^T \Delta s(\theta)| &\leq \frac{1}{\theta}\|p_x\|\,\|d(f(x+\theta\Delta x)-f(x))\| \leq \frac{1}{\theta}\|p_x\|\,(1+\gamma\theta)\theta\|p_s\| \\ &\leq (1+\gamma\theta)(1+2\kappa)\|p_v\|^2, \end{aligned} \tag{6.15}$$

where the last inequality follows from (6.13) and (6.14). This proves the left inequality in (*i*); for the right we proceed as in the proof of Lemma 5.3.3. By (6.10) we have

$$p_v(\theta) = p_x + p_s(\theta).$$

Letting $q_v(\theta) = p_x - p_s(\theta)$, we obtain the following bound:

$$p_x^T p_s(\theta) = \frac{1}{4}(\|p_v(\theta)\|^2 - \|q_v(\theta)\|^2) \leq \frac{1}{4}\|p_v(\theta)\|^2 \leq \frac{1}{4}(1+\theta\gamma\sqrt{1+2\kappa})^2\|p_v\|^2, \tag{6.16}$$

where the last inequality follows from Lemma 6.2.3. This shows the bounds in (*i*). For (*ii*) observe that combining (6.15) and (6.16) leads to the bound

$$\frac{1}{4}\|q_v(\theta)\|^2 \leq \frac{1}{4}(1+\gamma\theta\sqrt{1+2\kappa})^2\|p_v\|^2 + (1+\gamma\theta)(1+2\kappa)\|p_v\|^2.$$

Hence, using

$$p_x p_s(\theta) = \frac{1}{4}(p_v(\theta)^2 - q_v(\theta)^2)$$

we have

$$\begin{aligned} \|p_x p_s(\theta)\|_\infty &\leq \frac{1}{4}\max(\|p_v(\theta)\|_\infty^2, \|q_v(\theta)\|_\infty^2) \leq \frac{1}{4}\max(\|p_v(\theta)\|^2, \|q_v(\theta)\|^2) \\ &\leq \frac{1}{4}(1+\gamma\theta\sqrt{1+2\kappa})^2\|p_v\|^2 + (1+\gamma\theta)(1+2\kappa)\|p_v\|^2. \end{aligned}$$

This completes the proof of the lemma. □

The lemmas derived above enable the analysis of primal–dual algorithms for NCPs. Before proceeding we mention that for MNCPs the bounds in Lemma 6.2.4 can be improved using

$$\Delta x^T \Delta s(\theta) = \frac{1}{\theta^2}(\theta\Delta x)^T(f(x+\theta\Delta x) - f(x)) \geq 0,$$

which is the monotonicity property.

6.3 A FAMILY OF AFFINE SEARCH–DIRECTIONS

Up to this point the analysis was general in the sense that we did not specify our search–directions explicitly. We now derive the family of affine scaling directions introduced in Jansen et al. [112]. The directions are obtained by minimizing the complementarity (suitably scaled) over the Dikin–ellipsoid, which is the idea of Dikin's primal affine scaling algorithm [42]. Consider the problem

$$\min_{x,s} \left\{ x^T s \, : \, (x,s) \in \mathcal{F} \right\}.$$

The NCP satisfying Condition 6.2.1 is equivalent to the above problem in the sense that (x, s) is a solution of the NCP if and only if it is a minimum solution of the above problem with objective value zero. According to the search–mapping defined by (6.1) and (6.4), the complementarity after a step is given by

$$x(\theta)^T s(\theta) = x^T s + \theta \left(s^T \Delta x + x^T \Delta s(\theta) \right) + \theta^2 \Delta x^T \Delta s(\theta).$$

It follows from definition (6.2) of $g(\theta)$ that

$$\left. \frac{\partial (x(\theta)^T s(\theta))}{\partial \theta} \right|_{\theta=0} \equiv \lim_{\theta \to 0} \frac{x(\theta)^T s(\theta) - x^T s}{\theta} = s^T \Delta x + x^T \Delta s. \qquad (6.17)$$

This relation points to the following idea for the determination of the search–direction $(\Delta x, \Delta s)$. Let ν be a fixed nonnegative constant (the order of scaling in the algorithm). Taking account of the equation $-\nabla f(x) \Delta x + \Delta s = 0$, we consider the following subproblem, which for $\nu = 1$ is essentially the same as (5.14)

$$\begin{aligned} \min_{\Delta x, \Delta s} \quad & \left\{ ((xs)^\nu)^T (x^{-1} \Delta x + s^{-1} \Delta s) \, : \, -\nabla f(x) \Delta x + \Delta s = 0 \, , \right. \\ & \left. \| x^{-1} \Delta x + s^{-1} \Delta s \| \leq 1 \right\}. \end{aligned}$$

For $\nu = 1$ the solution of this subproblem minimizes the derivative (6.17). Proceeding as in Section 5.3 it follows that the solution satisfies the KKT–system:

$$\begin{aligned} -\nabla f(x) \Delta x + \Delta s &= 0, \\ s \Delta x + x \Delta s &= -\frac{v^{2\nu+2}}{\| v^{2\nu} \|}. \end{aligned}$$

The reader may observe that in case of LP or CQP for $\nu = 0$ this system exactly determines the primal–dual affine scaling direction of Monteiro et al.

[182]. From this point on p_v will have the following definition:

$$p_v := \frac{v^{2\nu+1}}{||v^{2\nu}||}. \tag{6.18}$$

Using the definitions in (6.8), the above optimality system can be rewritten as

$$-D\nabla f(x)Dp_x + p_s = 0, \tag{6.19}$$

$$p_x + p_s = p_v. \tag{6.20}$$

Since the Jacobian $\nabla f(x)$ is in $P_*(\kappa)$ the system has a unique solution for every $x \in \mathbb{R}^n$ (cf. [136, Lemma 4.1]).

6.4 CONVERGENCE ANALYSIS

6.4.1 General results

We analyze the behavior of the family of primal–dual affine scaling algorithms as follows. First, we derive general results for $\nu \geq 0$. These regard the complementarity and the feasibility of the iterates after a step. Throughout this section, we impose the following condition on the mapping f which is a special case of Condition 6.2.2.

Condition 6.4.1 *There exists a constant $\pi \in [0,1)$ and a value Θ such that for all p_s and $p_s(\theta)$ the algorithm computes, it holds*

$$||p_s(\theta) - p_s|| \leq \frac{\pi\omega(v)^{2\nu+1}}{\sqrt{n}}\theta\,||p_s||$$

for every $\theta \in (0, \Theta]$.

Obviously, we impose the condition for those p_s and $p_s(\theta)$ generated in the algorithm under consideration. Hence, in this section we will further assume that p_v is given by (6.18) for certain constant $\nu \geq 0$, and that $p_x, p_s, p_s(\theta)$ etc., are obtained from solving (6.19)–(6.20). From Lemma 6.2.3 and Lemma 6.2.4 we obtain the following result, which is a key for the behavior of $v(\theta)^2 = x(\theta)s(\theta)$.

Lemma 6.4.2 *Suppose that Condition 6.4.1 holds. Then for every* $\theta \in (0, \Theta]$ *we have*
(i) $\|p_v(\theta)\| \leq (1 + \theta\pi\omega(v)^{2\nu+1}\sqrt{1+2\kappa}/\sqrt{n})\|v\|_\infty$
$\leq (1 + \theta\pi\omega(v)^{2\nu+1}\sqrt{1+2\kappa}/\sqrt{n})\|v\|$;
(ii) $-(1 + \theta\pi\sqrt{1+2\kappa})v^{2\nu+2}/\|v^{2\nu}\| \leq vp_v(\theta)$
$\leq -(1 - \theta\pi\sqrt{1+2\kappa})v^{2\nu+2}/\|v^{2\nu}\|$;
(iii) $p_x^T p_s(\theta) \leq \left(1 + \theta\pi\omega(v)^{2\nu+1}\sqrt{1+2\kappa}/\sqrt{n}\right)\|v\|^2/4$;
(iv) $\|p_x p_s(\theta)\|_\infty \leq \big((1 + \theta\pi\omega(v)^{2\nu+1}\sqrt{1+2\kappa}/\sqrt{n})^2/4 +$
$+(1 + \theta\pi\omega(v)^{2\nu+1}/\sqrt{n})(1+2\kappa)\big)\|v\|_\infty^2$.

Proof: The vector p_v is given by $p_v = -v^{2\nu+1}/\|v^{2\nu}\|$. Hence we have

$$\begin{aligned}
\|p_v\| &= \frac{\|v^{2\nu+1}\|}{\|v^{2\nu}\|} \leq \|v\|_\infty \leq \|v\|,\\
\omega(|p_v|) &= \omega(v)^{2\nu+1},\\
|p_v| &= -p_v.
\end{aligned}$$

Substituting these in Lemmas 6.2.3 and 6.2.4 we complete the proof. □

Let us introduce some notation:

$$\begin{aligned}
\overline{\pi} &:= \pi\sqrt{1+2\kappa},\\
\overline{\vartheta} &:= \omega(v)^{2\nu+1}/(2\sqrt{n}),\\
\overline{\gamma} &:= \pi\omega(v)^{2\nu+1}\sqrt{1+2\kappa}/\sqrt{n} = 2\overline{\pi}\overline{\vartheta}.
\end{aligned} \tag{6.21}$$

Notice that $\overline{\vartheta}$ and $\overline{\gamma}$ depend on v; however, in this paragraph we are concerned with the behavior in one iteration so v can be considered to be fixed. Later we will derive uniform bounds for these quantities. We are now ready to show that the complementarity can be reduced with a suitable step size θ.

Lemma 6.4.3 *Let* $\overline{\pi}, \overline{\vartheta}, \overline{\gamma}$ *be as in (6.21). After a step it holds*
(i) If $0 \leq \nu \leq 1$ *and* $\theta \leq \min(\Theta, 1/(2\overline{\pi}), 1/(\sqrt{n}(1+\overline{\vartheta})^2))$ *then*

$$x(\theta)^T s(\theta) = \|v(\theta)\|^2 \leq \left(1 - \frac{\theta}{4\sqrt{n}}\right)\|v\|^2;$$

(ii) If $\nu \geq 1$ *and* $\theta \leq \min(\Theta, 1/(2\overline{\pi}), \omega(v)^{2\nu-2}/(\sqrt{n}(1+\overline{\vartheta})^2))$ *then*

$$x(\theta)^T s(\theta) = \|v(\theta)\|^2 \leq \left(1 - \frac{\omega(v)^{2\nu-2}}{4\sqrt{n}}\theta\right)\|v\|^2.$$

Proof: Since $\theta \leq 1/(2\overline{\pi})$ we have $1+\theta\overline{\gamma} \leq 1+\overline{\vartheta}$. Using (6.11) and Lemma 6.4.2 the new complementarity is bounded by

$$e^T v(\theta)^2 \leq \|v\|^2 - \theta(1-\theta\overline{\pi})\frac{e^T v^{2\nu+2}}{\|v^{2\nu}\|} + \theta^2\frac{1}{4}(1+\theta\overline{\gamma})^2\|v\|^2.$$

Applying Lemma A.3 and the third bound assumed on θ it holds

$$e^T v(\theta)^2 \quad \leq \quad \|v\|^2 - \theta\frac{\|v\|^2}{2\sqrt{n}} + \theta\frac{1}{\sqrt{n}(1+\overline{\vartheta})^2}\frac{1}{4}(1+\overline{\vartheta})^2\|v\|^2 = \left(1-\frac{\theta}{4\sqrt{n}}\right)\|v\|^2.$$

The proof of (*ii*) is similar. □

In the following lemma we derive a bound $\overline{\theta}$ such that $(x(\theta), s(\theta)) \in \mathcal{F}^0$ for every $\theta \leq \overline{\theta}$, i.e., the new iterate is interior–feasible.

Lemma 6.4.4 *Let $\nu \geq 0$ be a given constant and let $\overline{\pi}, \overline{\vartheta}, \overline{\gamma}$ be as in (6.21); define also*

$$\overline{\eta}^2 := \frac{1}{4}(1+\overline{\vartheta})^2 + \left(1+2\frac{\overline{\vartheta}}{\sqrt{n}(1+\overline{\vartheta})^2}\right)(1+2\kappa). \tag{6.22}$$

Suppose that Condition 6.4.1 holds. If $(x,s) \in \mathcal{F}^0$ and

$$\begin{aligned} 0 \quad \leq \quad & \theta < \min\left(\Theta, \frac{1}{2\overline{\pi}}, \frac{1}{\sqrt{n}(1+\overline{\vartheta})^2}, \frac{2\sqrt{n}\,\omega(v)^{2\nu}}{3(1+\nu)}\right., \\ & \left.\frac{\omega(v)}{\overline{\eta}}\left(\sqrt{1+\frac{9\omega(v)^2}{16n\overline{\eta}^2}} - \frac{3\omega(v)}{4\sqrt{n}\overline{\eta}}\right)\right) \end{aligned} \tag{6.23}$$

then $(x(\theta), s(\theta)) \in \mathcal{F}^0$.

Proof: From the fact that the search–direction satisfies (6.5) we have feasibility from (6.6). We still need to show that $(x(\theta), s(\theta))$ is interior–feasible. The first upper bound $\theta \leq \Theta$ follows from Condition 6.4.1. The second bound on θ implies $1+\theta\overline{\gamma} \leq 1+\overline{\vartheta}$. Using also the third bound on θ and Lemma 6.4.2(*iv*) we get

$$\|p_x p_s(\theta)\|_\infty \leq \overline{\eta}^2\|v\|_\infty^2.$$

Relation (6.11) and Lemma 6.4.2 imply then for every $\theta \in (0, \Theta]$

$$\begin{aligned} v(\theta)^2 \quad &\leq \quad v^2 - \theta(1-\theta\overline{\pi})\frac{v^{2\nu+2}}{\|v^{2\nu}\|} + \theta^2\overline{\eta}^2\|v\|_\infty^2 e, \\ v(\theta)^2 \quad &\geq \quad v^2 - \theta(1+\theta\overline{\pi})\frac{v^{2\nu+2}}{\|v^{2\nu}\|} - \theta^2\overline{\eta}^2\|v\|_\infty^2 e. \end{aligned}$$

Let $\alpha > 0$ and consider the function

$$\phi(t) = t - \theta\alpha \frac{t^{\nu+1}}{\|v^{2\nu}\|}.$$

For every α, $\phi(\cdot)$ is monotonically increasing on the interval $[0, \max(v)^2]$ if $\theta \leq \|v^{2\nu}\|/((1+\nu)\alpha\max(v)^{2\nu})$. Note that

$$\begin{aligned}
\frac{\|v^{2\nu}\|}{(1+\nu)(1-\theta\overline{\pi})\max(v)^{2\nu}} &\geq \frac{\|v^{2\nu}\|}{(1+\nu)(1+\theta\overline{\pi})\max(v)^{2\nu}} \\
&\geq \frac{2\sqrt{n}\min(v)^{2\nu}}{3(1+\nu)\max(v)^{2\nu}} \\
&= \frac{2\sqrt{n}\,\omega(v)^{2\nu}}{3(1+\nu)},
\end{aligned}$$

hence if we enforce the fourth upper bound in (6.23) the largest coordinate $\max(v(\theta))$ of $v(\theta)$ and the smallest coordinate $\min(v(\theta))$ can be estimated as follows:

$$\max(v(\theta))^2 \leq \max(v)^2 - \theta(1-\theta\overline{\pi})\frac{\max(v)^{2\nu+2}}{\|v^{2\nu}\|} + \theta^2\overline{\eta}^2\max(v)^2, \quad (6.24)$$

$$\min(v(\theta))^2 \geq \min(v)^2 - \theta(1+\theta\overline{\pi})\frac{\min(v)^{2\nu+2}}{\|v^{2\nu}\|} - \theta^2\overline{\eta}^2\max(v)^2. \quad (6.25)$$

By the continuity of the mapping v, dividing relation (6.25) by $\min(v)^2$ gives the following condition to ensure $\min(v(\theta)) > 0$:

$$1 - \theta(1+\theta\overline{\pi})\frac{\min(v)^{2\nu}}{\|v^{2\nu}\|} - \theta^2\frac{\overline{\eta}^2}{\omega(v)^2} > 0.$$

Since

$$\frac{\min(v)^{2\nu}}{\|v^{2\nu}\|} \leq \frac{\min(v)^{2\nu}}{\sqrt{n}\min(v)^{2\nu}} = \frac{1}{\sqrt{n}},$$

and $1 + \theta\overline{\pi} \leq 3/2$ the condition certainly holds if

$$1 - \theta\frac{3}{2\sqrt{n}} - \theta^2\frac{\overline{\eta}^2}{\omega(v)^2} \geq 0.$$

The last upper bound in the lemma ensures the inequality above. This completes the proof of the lemma. □

6.4.2 Polynomial convergence for $\nu > 0$

We now prove the polynomiality of the class of primal–dual affine scaling algorithms for $\nu > 0$. As in Section 5.3 each algorithm in this class generates a sequence of iterates $\{(x^{(k)}, s^{(k)}) : k = 1, 2, \ldots\}$ satisfying $\omega(v^{(k)}) \geq \rho$ for some constant $\rho \in (0, 1)$. Suppose that the current point $(x, s) \in \mathcal{F}^0$ satisfies $\omega(v) \geq \rho$. Our algorithm determines the next point along the curve $(x(\theta), s(\theta))$ given by (6.1) and (6.4) by choosing a step size θ. The following theorem ensures the existence of $\overline{\theta} > 0$ for which $(x(\theta), s(\theta)) \in \mathcal{F}^0$ and $\omega(v(\theta)) \geq \rho$ for every $\theta \in (0, \overline{\theta}]$.

Theorem 6.4.5 *Let $\nu > 0$ be a given constant, let $\overline{\pi}, \overline{\vartheta}, \overline{\gamma}$ be as in (6.21) and $\overline{\eta}^2$ as in (6.22). Suppose that Condition 6.4.1 holds. If $(x, s) \in \mathcal{F}^0$, $\omega(v) \geq \rho$, and θ satisfies (6.23) and*

$$0 \leq \theta \leq \min\left(\frac{1-\rho^{2\nu}}{2\overline{\pi}(1+\rho^{2\nu})}, \frac{\rho^2(1-\rho^{2\nu})}{2\overline{\eta}^2(1+\rho^2)\sqrt{n}}\right), \tag{6.26}$$

then $(x(\theta), s(\theta)) \in \mathcal{F}^0$ and $\omega(v(\theta)) \geq \rho$.

Proof: The part $(x(\theta), s(\theta)) \in \mathcal{F}^0$ is obvious from Lemma 6.4.4. Hence we only need to show that $\omega(v(\theta)) \geq \rho$, i.e., $\rho^2 \max(v(\theta))^2 \leq \min(v(\theta))^2$. Using the relation $\min(v) = \omega(v)\max(v) \geq \rho\max(v)$, the same discussion for finding the bounds (6.24) and (6.25) leads to the following relation:

$$\min(v(\theta))^2 \geq \rho^2\left(\max(v)^2 - \theta(1+\theta\overline{\pi})\frac{\rho^{2\nu}\max(v)^{2\nu+2}}{\|v^{2\nu}\|} - \theta^2\frac{\overline{\eta}^2}{\rho^2}\max(v)^2\right). \tag{6.27}$$

Hence, from (6.24) and (6.27) we derive a sufficient condition for θ as follows:

$$\begin{aligned}
&\rho^2\left(\max(v)^2 - \theta(1-\theta\overline{\pi})\frac{\max(v)^{2\nu+2}}{\|v^{2\nu}\|} + \theta^2\overline{\eta}^2\max(v)^2\right) \\
&\quad\leq \rho^2\left(\max(v)^2 - \theta(1+\theta\overline{\pi})\frac{\rho^{2\nu}\max(v)^{2\nu+2}}{\|v^{2\nu}\|} - \theta^2\frac{\overline{\eta}^2}{\rho^2}\max(v)^2\right).
\end{aligned}$$

Rearranging this inequality gives

$$\theta\frac{\overline{\eta}^2(1+\rho^2)}{\rho^2} \leq \frac{\max(v)^{2\nu}}{\|v^{2\nu}\|}\left((1-\theta\overline{\pi}) - (1+\theta\overline{\pi})\rho^{2\nu}\right).$$

Since $\|v^{2\nu}\| \leq \sqrt{n}\max(v)^{2\nu}$, we obtain the bound

$$\theta \leq \frac{\rho^2\left((1-\theta\overline{\pi}) - (1+\theta\overline{\pi})\rho^{2\nu}\right)}{\overline{\eta}^2(1+\rho^2)\sqrt{n}}.$$

Using the first bound in (6.26) we find that θ will certainly satisfy this inequality if

$$\theta \leq \frac{\rho^2(1-\rho^{2\nu})}{2\overline{\eta}^2(1+\rho^2)\sqrt{n}}.$$

Thus we obtain the theorem. □

We are now in the position to derive the complexity of our algorithms.

Theorem 6.4.6 *Let $\nu > 0$ be a given constant and let $0 < \rho < 1$ be given. Suppose that Condition 6.4.1 is satisfied. Let $\epsilon > 0$, $(x^{(0)}, s^{(0)}) \in \mathcal{F}^0$, such that $\omega(v^{(0)}) \geq \rho$, be given and let θ satisfy the conditions in Lemma 6.4.3, (6.23) and (6.26). Then the primal-dual affine scaling algorithm with order of scaling ν stops with a solution (x^*, s^*) for which $(x^*)^T s^* \leq \epsilon$ and $\omega(v^*) \geq \rho$ holds, after at most*

$$\mathcal{O}\left(\frac{\sqrt{n}}{\theta}\ln\frac{(x^{(0)})^T s^{(0)}}{\epsilon}\right)$$

iterations if $0 < \nu \leq 1$, and after at most

$$\mathcal{O}\left(\frac{\sqrt{n}}{\rho^{2\nu-2}\theta}\ln\frac{(x^{(0)})^T s^{(0)}}{\epsilon}\right)$$

iterations if $\nu \geq 1$.

Proof: Follows from Lemma 6.4.3 and Theorem 6.4.5. □

To be more specific about the complexity we have to check which of the various conditions on the step size θ is strongest and how θ depends on the input parameters. It is easy to verify the following bounds for the quantities in (6.21)

$$\begin{array}{rcccl} 0 & \leq & \overline{\pi} & \leq & \sqrt{1+2\kappa} < 1+\kappa, \\ \rho^{2\nu+1}/(2\sqrt{n}) & \leq & \overline{\vartheta} & \leq & 1/(2\sqrt{n}), \\ 0 & \leq & \overline{\gamma} & \leq & \sqrt{1+2\kappa}/\sqrt{n}, \end{array}$$

where we use $\omega(v) \geq \rho$. Using also $n \geq 2$ we get from (6.22)

$$\frac{5}{4} \leq \overline{\eta}^2 \leq \frac{1}{4}\left(1+\frac{1}{2\sqrt{2}}\right)^2 + \left(1+2\frac{1/(2\sqrt{2})}{\sqrt{2}}\right)(1+2\kappa) < 3(1+\kappa). \quad (6.28)$$

We analyze the bounds in (6.23) consecutively. Notice,

$$\frac{1}{2\overline{\pi}} \geq \frac{1}{2\sqrt{1+2\kappa}},$$

$$\frac{1}{\sqrt{n}(1+\overline{\vartheta})^2} \geq \frac{1}{\sqrt{n}(1+1/(2\sqrt{2}))^2} > \frac{1}{2\sqrt{n}},$$

$$\frac{2\sqrt{n}\omega(v)^{2\nu}}{3(1+\nu)} \geq \frac{2\sqrt{n}\rho^{2\nu}}{3(1+\nu)}.$$

We have

$$\frac{3\omega(v)}{4\sqrt{n}\overline{\eta}} \leq \frac{3}{4\sqrt{2}\sqrt{5/4}} < 1;$$

since the function $\phi(t) = \sqrt{1+t^2} - t$ is monotonically decreasing we obtain

$$\frac{\omega(v)}{\overline{\eta}}\left(\sqrt{1+\frac{9\omega(v)^2}{16n\overline{\eta}^2}} - \frac{3\omega(v)}{4\sqrt{n}\overline{\eta}}\right) \geq \frac{\omega(v)}{\overline{\eta}}(\sqrt{2}-1) \geq \frac{\rho}{5\sqrt{1+\kappa}}.$$

For the bounds in (6.26) it holds

$$\frac{1-\rho^{2\nu}}{2\overline{\pi}(1+\rho^{2\nu})} \geq \frac{1-\rho^{2\nu}}{2(1+\kappa)(1+\rho^{2\nu})},$$

$$\frac{\rho^2(1-\rho^{2\nu})}{2\overline{\eta}^2(1+\rho^2)\sqrt{n}} \geq \frac{\rho^2(1-\rho^{2\nu})}{6(1+\kappa)(1+\rho^2)\sqrt{n}}.$$

Finally, for the bound in Lemma 6.4.3(*ii*) we have

$$\frac{\omega(v)^{2\nu-2}}{\sqrt{n}(1+\overline{\vartheta})^2} > \frac{\rho^{2\nu-2}}{2\sqrt{n}}.$$

Thus we obtain the following result as a corollary of Theorem 6.4.6.

Corollary 6.4.7 *Let us take the situation as in Theorem 6.4.6 and specify* $\rho = 1/\sqrt{2}$. *Then*
(i) If $0 < \nu \leq 1$ *and* $n \geq 2$ *we may choose*

$$\theta = \min\left(\Theta, \frac{1-2^{-\nu}}{18\sqrt{n}(1+\kappa)}\right),$$

hence the number of iterations required by the algorithm is

$$\mathcal{O}\left(\max\left(\frac{\sqrt{n}}{\Theta}, \frac{n(1+\kappa)}{1-2^{-\nu}}\right)\ln\frac{(x^{(0)})^T s^{(0)}}{\epsilon}\right).$$

(ii) If $\nu = 1$ *and* $n \geq 2$ *then we may choose*

$$\theta = \min\left(\Theta, \frac{1}{36\sqrt{n}(1+\kappa)}\right),$$

hence the number of iterations required by the algorithm is

$$\mathcal{O}\left(\max\left(\frac{\sqrt{n}}{\Theta}, n(1+\kappa)\right)\ln\frac{(x^{(0)})^T s^{(0)}}{\epsilon}\right).$$

(iii) If $\nu > 1$ and n sufficiently large we may choose

$$\theta = \min\left(\Theta, \frac{1}{2^{\nu}\sqrt{n}(1+\kappa)}\right)$$

hence the number of iterations required by the algorithm is

$$\mathcal{O}\left(\max\left(\frac{2^{\nu}\sqrt{n}}{\Theta}, 4^{\nu}n(1+\kappa)\right)\ln\frac{(x^{(0)})^T s^{(0)}}{\epsilon}\right).$$

Observe that for $\nu \geq 1$ the complexity bound gets worse as ν increases; similarly, for $\nu < 1$ the complexity bound inclines as ν decreases to zero. From a theoretical point of view the complexity is 'optimal' if $\nu = 1$, i.e., for the primal–dual Dikin–affine scaling algorithm. This behavior is also apparent from the computational results we will present in Chapter 7.

6.4.3 Polynomial convergence for $\nu = 0$

Now we show that, with suitable step size, the primal–dual affine scaling algorithm of Monteiro et al. [182] can be applied to NCPs satisfying Condition 6.2.1, with a polynomial complexity bound. We believe that this is the first proof of polynomial convergence of this affine scaling algorithm for NCPs. Mizuno and Nagasawa [178] already showed (using a potential reduction approach) that the complexity of the algorithm is not affected using large neighborhoods of the central path.

So assume that $\nu = 0$. It is easily verified that Lemma 6.4.3 and Lemma 6.4.4 still apply. Theorem 6.4.5, however, is not valid for $\nu = 0$. In fact, taking the limit in (6.26) as ν tends to zero one obtains that the step size θ becomes zero. Hence, it cannot be guaranteed that the iterates remain in the neighborhood determined by a given $\rho \in (0,1)$. Below we show that by a step $\omega(v)$ may well decrease, but that the decrease can be bounded from below. This is the contents of the next lemma.

Lemma 6.4.8 *Let $\overline{\pi}, \overline{\vartheta}$ be as in (6.21) and $\overline{\eta}$ as in (6.22). If $(x,s) \in \mathcal{F}^0$ and*

$$0 \le \theta < \min\left(\Theta, \frac{1}{2\overline{\pi}}, \frac{1}{\sqrt{n}(1+\overline{\vartheta})^2}, \frac{2\sqrt{n}}{3}, \frac{\omega(v)}{\overline{\eta}}\left(\sqrt{1+\frac{9\omega(v)^2}{16n\overline{\eta}^2}} - \frac{3\omega(v)}{4\sqrt{n}\overline{\eta}}\right)\right) \tag{6.29}$$

then $(x(\theta), s(\theta)) \in \mathcal{F}^0$ and

$$1+\omega(v(\theta))^2 \ge \frac{1+\omega(v)^2}{1+\theta^2(\overline{\eta}^2\sqrt{n}+2\overline{\pi})/(\sqrt{n}-\theta(1+\theta\overline{\pi}))}. \tag{6.30}$$

Proof: From Lemma 6.4.4 it is clear that (6.29) guarantees feasibility of $(x(\theta), s(\theta))$. So it remains to show (6.30). Observe that (6.24) and (6.25) still hold for $\nu = 0$. Hence, using the notation $\omega^2 := \omega(v)^2 =: \alpha/\beta$ with α and β such that $\alpha e \le v^2 \le \beta e$, one has

$$\begin{aligned}
\omega(v(\theta))^2 &\ge \frac{(1-\theta(1+\theta\overline{\pi})/\sqrt{n})\alpha - \theta^2\overline{\eta}^2\beta}{(1-\theta(1-\theta\overline{\pi})/\sqrt{n})\beta + \theta^2\overline{\eta}^2\beta} \\
&= \frac{\omega^2(\sqrt{n}-\theta(1+\theta\overline{\pi})) - \theta^2\overline{\eta}^2\sqrt{n}}{(\sqrt{n}-\theta(1+\theta\overline{\pi})) + \theta^2\overline{\eta}^2\sqrt{n} + 2\theta^2\overline{\pi}} \\
&\ge \frac{\omega^2(\sqrt{n}-\theta(1+\theta\overline{\pi})) - \theta^2\overline{\eta}^2\sqrt{n} - 2\theta^2\overline{\pi}}{(\sqrt{n}-\theta(1+\theta\overline{\pi})) + \theta^2\overline{\eta}^2\sqrt{n} + 2\theta^2\overline{\pi}}.
\end{aligned}$$

Adding one to both sides and rearranging terms gives (6.30). □

Now we are ready to prove the polynomial complexity. We denote by $(x^{(k)}, s^{(k)})$ the iterate after k iterations and use $\omega_k := \omega(\sqrt{x^{(k)}s^{(k)}})$.

Theorem 6.4.9 *Let an initial interior point $(x^{(0)}, s^{(0)}) \in \mathcal{F}^0$, with $\omega_0 \le 1$ and $0 < \epsilon < (x^{(0)})^T s^{(0)}/2$ be given. We define parameters L and τ as follows:*

$$L := \ln\frac{(x^{(0)})^T s^{(0)}}{\epsilon}, \qquad \tau := \frac{80(1+\kappa)}{\omega_0^2} + \frac{2}{nL}.$$

Let t be the smallest real number in the interval $(\tau, \tau + 1/(4nL^2)]$ such that $K := 4tnL^2$ is integral. If $\theta := 1/(t\sqrt{n}L) \le \Theta$, then after $O(nL^2/(\omega_0^2))$ iterations the algorithm yields a solution (x^*, s^*) such that $(x^*)^T s^* \le \epsilon$ and $\omega(\sqrt{x^* s^*})^2 \ge \omega_0^2/2$.

Proof: For the moment we assume that in each iteration the step size $\theta = 1/(t\sqrt{n}L)$ satisfies the conditions of Lemma 6.4.8. Later we will justify this assumption. Taking logarithms in (6.30) and substituting the given value of θ we obtain

$$\begin{aligned}\ln\frac{1+\omega_0^2}{1+\omega_k^2} &\leq k\ln\left(1+\frac{\theta^2(\overline{\eta}^2\sqrt{n}+2\overline{\pi})}{\sqrt{n}-\theta(1+\theta\overline{\pi})}\right)\leq k\frac{\theta^2(\overline{\eta}^2\sqrt{n}+2\overline{\pi})}{\sqrt{n}-\theta(1+\theta\overline{\pi})}\\ &\leq k\frac{1/(t^2nL^2)(\overline{\eta}^2+2\overline{\pi})\sqrt{n}}{\sqrt{n}-2/(t\sqrt{n}L)}=k\frac{\overline{\eta}^2+2\overline{\pi}}{tL(tnL-2)},\end{aligned}$$

where we used $\overline{\pi} < \overline{\pi}\sqrt{n}$ and $\theta\overline{\pi} < 1$ in the third inequality. Hence we certainly have $\omega_k^2 \geq \omega_0^2/2$ if

$$k\frac{\overline{\eta}^2+2\overline{\pi}}{tL(tnL-2)}\leq \ln\frac{1+\omega_0^2}{1+\omega_0^2/2}. \tag{6.31}$$

Since $\phi(\sigma) := \ln((1+\sigma)/(1+\sigma/2))$ is concave and $\phi(0) = 0$, $\phi(1) \geq 1/4$, it holds

$$\ln\frac{1+\omega_0^2}{1+\omega_0^2/2}\geq\frac{\omega_0^2}{4}.$$

As a consequence, (6.31) is certainly satisfied if

$$k\leq\frac{\omega_0^2tL(tnL-2)}{4(\overline{\eta}^2+2\overline{\pi})}. \tag{6.32}$$

We conclude that if the total number of iterations satisfies (6.32) then the inequality $\omega_k^2 \geq \omega_0^2/2$ holds. Since Lemma 6.4.3 is (assumed to be) valid and we employ a fixed step parameter θ, the algorithm stops after at most k iterations, where

$$k\geq\frac{4\sqrt{n}}{\theta}\ln\frac{(x^{(0)})^Ts^{(0)}}{\epsilon}=4tnL^2,$$

after which $(x^{(k)})^Ts^{(k)} \leq \epsilon$. Note that the definition of t guarantees that $4tnL^2$ is integral. This number of iterations respects (6.32) if

$$4tnL^2\leq\omega_0^2tL\frac{tnL-2}{4(\overline{\eta}^2+2\overline{\pi})}.$$

Dividing by ω_0^2tL and rearranging terms gives the condition

$$t\geq\frac{16(\overline{\eta}^2+2\overline{\pi})}{\omega_0^2}+\frac{2}{nL}. \tag{6.33}$$

which is satisfied by the value assigned to t in the theorem, since $\overline{\eta}^2 \leq 3(1+\kappa)$ and $\overline{\pi} \leq 1+\kappa$.

It remains to show that in each iteration of the algorithm the specified step size θ satisfies condition (6.29) of Lemma 6.4.8, to justify the assumption made at the beginning of this proof. First, observe that $\theta < \Theta$ by assumption. From (6.33) we have $t \geq 32\overline{\pi}/\omega_0^2$. The condition $\theta \leq 1/(2\overline{\pi})$ is equivalent to $t \geq 2\overline{\pi}/(\sqrt{n}L)$, hence certainly satisfied if $\omega_0^2 \leq 16\sqrt{n}L$, which is guaranteed by the assumption on L. The third condition in (6.29) is satisfied if

$$\theta \leq \frac{1}{\sqrt{n}(1+1/(2\sqrt{n}))},$$

which is satisfied since $t \geq 80$ from the definition of τ. The fourth condition is trivially guaranteed, so it remains to deal with the condition that for each k

$$\theta < \frac{\omega_k}{\overline{\eta}}\left(\sqrt{1+\frac{9\omega_k^2}{16n\overline{\eta}^2}}-\frac{3\omega_k}{4\sqrt{n}\overline{\eta}}\right).$$

Using $n \geq 2$ and $\overline{\eta} \geq \sqrt{5/4}$ (see (6.28)), we have

$$\frac{3\omega_k}{4\sqrt{n}\overline{\eta}} < \frac{\omega_k}{2} \leq \frac{1}{2}.$$

Therefore, since

$$\sqrt{1+\sigma^2}-\sigma > \frac{1}{2} \quad \text{if} \quad 0 \leq \sigma < \frac{3}{4},$$

it is sufficient to show that $\theta \leq \omega_k/(2\overline{\eta})$ for each k. As we have seen before, with the given step size we have $\omega_k \geq \omega_0/(\sqrt{2})$ for each k. So is it sufficient that θ satisfies $\theta \leq \omega_0/(2\sqrt{2}\overline{\eta})$ or even $\theta \leq \omega_0/(2\sqrt{6}\sqrt{1+\kappa})$. This amounts to $\omega_0 t\sqrt{n}L \geq 2\sqrt{6}\sqrt{1+\kappa}$, which is easily shown to hold using the definition of τ and the assumption on L. Hence the proof of the theorem is complete. □

6.5 SMOOTHNESS CONDITIONS

In the literature on interior point methods for NLP problems some smoothness conditions on the functions involved have been proposed, with the purpose to bound the second order effect not taken into account by Newton's method. In this section we show how these conditions are generalized to NCPs and indicate their use for analyzing algorithms in wide neighborhoods. All the conditions given below only apply to monotone mappings, which is the major difference with Condition 6.2.2.

6.5.1 Zhu's scaled Lipschitz condition

In Zhu [258] the scaled Lipschitz condition was introduced to analyze a path-following algorithm for convex programming problems. It was used by Kortanek et al. [142] and Potra and Ye [203] for an analysis of a primal–dual potential reduction method for entropy optimization problems and by Sun et al. [225] for min–max saddle problems. Potra and Ye [202] modified the condition for use in the analysis of interior point methods for MNCPs. Let us first give the definition.

Definition 6.5.1 (Scaled Lipschitz) *Let G be a closed convex domain in $\mathbb{R}^n$, with nonempty interior $Q := \mathrm{int}(G)$. A single–valued monotone operator $f : Q \to \mathbb{R}^n$ satisfies the scaled Lipschitz condition if there is a nondecreasing function $\psi(\alpha)$ such that*

$$\|X(f(x+h) - f(x) - \nabla f(x)h)\| \leq \psi(\alpha) h^T \nabla f(x) h$$

for all $x > 0$ and h satisfying $\|x^{-1}h\| \leq \alpha$.

We show that any mapping satisfying the scaled Lipschitz condition also satisfies Condition 6.2.2 with certain values for γ and Θ.

Theorem 6.5.2 *Let the mapping f satisfy the scaled Lipschitz condition. Then there exist values for γ and Θ such that f satisfies Condition 6.2.2.*

Proof: For θ and Δx satisfying

$$\|\theta x^{-1} \Delta x\| \leq \alpha, \tag{6.34}$$

we have (recall (6.12))

$$\begin{aligned}
\|p_s(\theta) - p_s\| &= \left\| d \left(\frac{f(x+\theta\Delta x) - f(x)}{\theta} - \nabla f(x) \Delta x \right) \right\| \\
&\leq \frac{1}{\theta} \|v^{-1}\|_\infty \|x(f(x+\theta\Delta x) - f(x) - \theta \nabla f(x) \Delta x)\| \\
&\leq \frac{1}{\theta} \|v^{-1}\|_\infty \psi(\alpha) \theta^2 \Delta x^T \nabla f(x) \Delta x \\
&\leq \theta \|v^{-1}\|_\infty \psi(\alpha) \|d^{-1} \Delta x\| \, \|d \nabla f(x) \Delta x\| \\
&\leq \theta \|v^{-1}\|_\infty \psi(\alpha) \|v\|_\infty \|x^{-1} \Delta x\| \, \|d \nabla f(x) \Delta x\| \\
&\leq \theta \frac{1}{\omega(v)} \psi(\alpha) \frac{\alpha}{\Theta} \|p_s\|.
\end{aligned}$$

The last inequality follows from the fact that (6.34) should specifically be satisfied for Θ. We compute Θ and γ from

$$\frac{1}{\omega(v)}\psi(\alpha)\frac{\alpha}{\Theta} = \gamma,$$

which yields $\Theta = \alpha\psi(\alpha)/(\omega(v)\gamma)$. This completes the proof. □

Observe that Theorem 6.5.2 leaves some freedom in the choice of γ and Θ; however, Θ and γ influence the complexity at the same rate. Definition 6.5.1 of the scaled Lipschitz condition implies that $h^T\nabla f(x)h \geq 0$ for every $x > 0$ and h with $\|x^{-1}h\| \leq \alpha$. Hence, using this condition seems not to be possible for nonmonotone mappings. In fact, even if f is linear, i.e. given by $f(x) = Mx+q$, the condition does not necessarily hold for the matrices M in P_*. On the other hand, Condition 6.2.2 does not need the monotonicity and holds for any linear mapping with a matrix in P_*, which is an advantage of the condition.

6.5.2 Self–concordance and relative Lipschitz condition

The most important (and also most general) smoothness condition is self–concordance, introduced by Nesterov and Nemirovskii [193], later used by Jarre [119], Den Hertog [100] and Den Hertog et al. [101], and Nesterov and Todd [194], among others. The crux of the condition is that it bounds the first and the third order derivative of a convex function in its second order derivative.

Definition 6.5.3 (Self–concordance) *Let G be a closed convex domain in a finite-dimensional real vector space E and let $a, \vartheta \geq 0$. A function F : $\operatorname{int}(G) \rightarrow \mathbb{R}$ is called a-self-concordant if $F \in C^3$ is a convex function on $\operatorname{int}(G)$ that for all $y \in \operatorname{int}(G)$ and $h \in E$ satisfies the condition*

$$\left|\nabla^3 F(y)[h,h,h]\right| \leq 2\,a^{-1/2}\left(\nabla^2 F(y)[h,h]\right)^{3/2};$$

F is called strongly a–self–concordant if it is a–self–concordant and is unbounded above for a sequence of points converging to the boundary of G; F is called an (a,ϑ)–self–concordant barrier for G if F is strongly a–self–concordant and moreover for all $y \in \operatorname{int}(G)$ and $h \in E$

$$|\nabla F(y)[h]| \leq \sqrt{\vartheta}\left(\nabla^2 F(y)[h,h]\right)^{1/2}.$$

A main difference with the scaled Lipschitz condition is that in the analysis of interior point methods self-concordance does not apply to the mapping itself; instead a self-concordant barrier for the domain is needed. In our case the domain is $\mathbb{R}^n_+$; it is easy to verify that the function $-\sum_{i=1}^n \ln x_i$ is a self-concordant barrier for this domain. We also need the following definition, relating the mapping defining the complementarity problem to a self-concordant barrier for the domain.

Definition 6.5.4 (β–compatibility) *A C^2-smooth monotone operator $f : \mathbb{R}^n_+ \to \mathbb{R}^n$ is called β-compatible with $F(x) = -\sum_{i=1}^n \ln x_i$ if for all $x > 0$ and $h^{(i)} \in \mathbb{R}^n$, $i = 1, 2, 3$, the following inequality holds:*

$$|\nabla^2 f(x)[h^{(1)}, h^{(2)}, h^{(3)}]| \leq 3^{3/2}\beta \prod_{i=1}^{3} \{\nabla f(x)[h^{(i)}, h^{(i)}]^{1/3} \|x^{-1}h^{(i)}\|^{1/3}\}.$$

If the mapping f is β–compatible with the function $-\sum_{i=1}^n \ln x_i$ then the barrier function $f_t(x) := (1+\beta)^2\{tf(x) + x^{-1}\}$ is strongly self–concordant for all $t > 0$ [193, Prop. 7.3.2]. A similar property can be obtained if the mapping f is strongly self-concordant itself. For all $\alpha > 0$ and $\beta > 0$, if the mappings ϕ and ψ are self-concordant then the mapping $\alpha\phi + \beta\psi$ is also self-concordant with some parameter. Self–concordant mappings satisfy the following condition, which is called the relative Lipschitz condition, introduced by Jarre [118], see also [193, Prop. 7.2.1] and [120, Sec. 2.1.4]).

Definition 6.5.5 (Relative Lipschitz) *Let G be a closed convex domain in $\mathbb{R}^n$, with nonempty interior $Q := \text{int}(G)$. A single-valued monotone operator $f : Q \to \mathbb{R}^n$ satisfies the relative Lipschitz condition if for all $x, y \in Q$ for which $\tau := \sqrt{(y-x)^T \nabla f(x)(y-x)} \leq 1$ the inequality*

$$|h^T(\nabla f(y) - \nabla f(x))h| \leq \left(\frac{1}{(1-\tau)^2} - 1\right) h^T \nabla f(x) h.$$

holds for all $h \in \mathbb{R}^n$.

In the following lemma, we will show that β–compatibility and the relative Lipschitz condition can be used to bound the inner product $p_x^T p_s(\theta)$, which plays an important role in the complexity analysis of the primal–dual algorithms for NCPs, see Lemma 6.2.4.

Lemma 6.5.6 *Let x and $s = f(x)$ be in $\mathcal{F}^0$ and $\Delta x, \Delta s$ displacements satisfying (6.5). Let p_x and $p_s(\theta)$ be as in (6.10) and let θ satisfy*

$$\theta\|x^{-1}\Delta x\| \le \frac{1}{2} \quad \text{and} \quad \theta\sqrt{\frac{\Delta x^T \Delta s}{\min(v)^2}} \le \frac{1}{2}. \tag{6.35}$$

If the mapping f is β-compatible with $F(x) = -\sum_{i=1}^n \ln x_i$, then

$$p_x^T p_s(\theta) \le \Delta x^T \Delta s + 63\beta \left(\frac{\Delta x^T \Delta s}{\min(v)^2} + \|x^{-1}\Delta x\|^2\right) \min(v)^2.$$

Proof: Using Taylor's expansion, we have

$$g(\theta) = f(x+\theta\Delta x) - f(x) - \theta\nabla f(x)\Delta x = \frac{1}{2}\theta^2\nabla^2 f(y)[\Delta x, \Delta x, \cdot],$$

where $y = x + \theta\lambda\Delta x$ for some $\lambda \in \mathbb{R}^n_+$ with $\lambda_i \le 1$. Then it holds

$$\begin{aligned} p_x^T p_s(\theta) &= \Delta x^T\left(\Delta s + \frac{g(\theta)}{\theta}\right) = \Delta x^T \Delta s + \frac{1}{2}\theta\nabla^2 f(y)[\Delta x, \Delta x, \Delta x] \\ &\le \Delta x^T \Delta s + \frac{3^{3/2}}{2}\beta\theta\nabla f(y)[\Delta x, \Delta x]\ \|y^{-1}\Delta x\| \\ &\le \Delta x^T \Delta s + \frac{3^{3/2}}{2}\beta\theta\left(\frac{\nabla f(y)}{\min(v)^2} + Y^{-2}\right)[\Delta x, \Delta x]\cdot \\ &\qquad \|y^{-1}\Delta x\|\min(v)^2, \end{aligned} \tag{6.36}$$

where the first inequality follows from the β-compatibility and $Y = \mathrm{Diag}(y)$. We apply the relative Lipschitz condition with y and x to the self-concordant mapping

$$\frac{f(x)}{\min(v)^2} - x^{-1}e.$$

Using $\Delta x^T \nabla f(x)\Delta x = \Delta x^T \Delta s$ and the definition of y we have

$$(\theta\lambda\Delta x)^T\left(\frac{\nabla f(x)}{\min(v)^2} + X^{-2}\right)(\theta\lambda\Delta x) \le \theta^2\left(\frac{\Delta x^T \Delta s}{\min(v)^2} + \|x^{-1}\Delta x\|^2\right) \le \frac{1}{2},$$

where the last inequality follows from (6.35). The relative Lipschitz condition gives

$$\begin{aligned} &\left(\frac{\nabla f(y)}{\min(v)^2} + Y^{-2}\right)[\Delta x, \Delta x] \\ &\qquad \le \left(1 + \frac{1}{(1-1/\sqrt{2})^2} - 1\right)\left(\Delta x^T\left(\frac{\nabla f(x)}{\min(v)^2} + X^{-2}\right)\Delta x\right) \\ &\qquad \le 12\left(\frac{\Delta x^T \Delta s}{\min(v)^2} + \|x^{-1}\Delta x\|^2\right). \end{aligned}$$

Furthermore, it holds

$$\|Y^{-1}\Delta x\| \leq \frac{1}{1-\theta\|x^{-1}\Delta x\|}\|x^{-1}\Delta x\| \leq 2\|x^{-1}\Delta x\|.$$

Substituting the latter results in (6.36) gives

$$\begin{aligned} p_x^T p_s(\theta) &\leq \Delta x^T \Delta s + \frac{3^{3/2}}{2}\beta\theta 12\left(\frac{\Delta x^T \Delta s}{\min(v)^2} + \|x^{-1}\Delta x\|^2\right) 2\,\|x^{-1}\Delta x\|\min(v)^2 \\ &\leq \Delta x^T \Delta s + 63\beta\left(\frac{\Delta x^T \Delta s}{\min(v)^2} + \|x^{-1}\Delta x\|^2\right)\min(v)^2, \end{aligned}$$

which completes the proof. □

From the lemma we derive the following corollary in case of applying our primal–dual affine scaling algorithms to MNCPs.

Corollary 6.5.7 *Consider the situation as in Lemma 6.5.6 and let f be a monotone mapping. If Δx and Δs are determined with a primal-dual affine scaling algorithm with parameter $\nu \geq 0$ and $\theta \leq \omega(v)/2$, then*

$$p_x^T p_s(\theta) \leq (1/4 + 79\beta)\max(v)^2.$$

Proof: Analogously to Lemma 5.3.3 and the fact that $\|p_v\| \leq \max(v)$ for any value of $\nu \geq 0$ it follows

$$\Delta x^T \Delta s \leq \frac{1}{4}\max(v)^2.$$

From $\|x^{-1}\Delta x + s^{-1}\Delta s\| \leq 1$ and $\Delta x^T \Delta s \geq 0$ (since f is monotone), it follows

$$\begin{aligned} \|x^{-1}\Delta x\|^2 &= \|v^{-1}d^{-1}\Delta x\|^2 \leq \frac{1}{\min(v)^2}\|d^{-1}\Delta x\|^2 \\ &\leq \frac{1}{\min(v)^2}(\|d^{-1}\Delta x\|^2 + 2\Delta x^T \Delta s + \|d\Delta s\|^2) \\ &= \frac{1}{\min(v)^2}\|d^{-1}\Delta x + d\Delta s\|^2 = \frac{1}{\min(v)^2}\|v(x^{-1}\Delta x + s^{-1}\Delta s)\|^2 \\ &\leq \frac{\max(v)^2}{\min(v)^2}\|x^{-1}\Delta x + s^{-1}\Delta s\|^2 \leq \frac{1}{\omega(v)^2}. \end{aligned}$$

Using also $\theta \leq \omega(v)/2$ it follows that the conditions on θ in (6.35) are satisfied. Consequently,

$$p_x^T p_s(\theta) \leq \frac{1}{4}\max(v)^2 + 63\beta\left(\frac{1}{4}\frac{1}{\omega(v)^2} + \frac{1}{\omega(v)^2}\right)\min(v)^2$$

$$\leq \frac{1}{4}\max(v)^2 + 79\beta\frac{1}{\omega(v)^2}\frac{\min(v)^2}{\max(v)^2}\max(v)^2 = \left(\frac{1}{4} + 79\beta\right)\max(v)^2.$$

This completes the proof. □

The result of this corollary gives an alternative proof for a bound of the type as in Lemma 6.4.2(*iii*). Unfortunately, since our analysis of the primal–dual affine scaling algorithms is based on large neighborhoods, also bounds on $\|p_x p_s(\theta)\|_\infty$ and $\|g(\theta)\|_\infty$ are required (cf. Lemma 6.4.2(*iv*)). We have not been able to derive such a bound using self–concordance and relative Lipschitz. Moreover, it is not clear how self–concordance can be generalized to nonmonotone mappings. However, to have more insights into the relationship between the self–concordance and our condition, we will enforce more strict conditions on the mapping. Recall from (6.12) that Condition 6.2.2 is equivalently written as

$$\|d\,(f(x+\theta\Delta x) - f(x) - \theta\nabla f(x)\Delta x)\,\| \leq \gamma\theta\|d\nabla f(x)\Delta x\|.$$

Using Taylor's expansion we obtain

$$\|d\,(f(x+h) - f(x) - \nabla f(x)h)\,\| = \|d\,(\nabla f(y) - \nabla f(x))\,h\|$$

where $y = x + \lambda h$ for some vector λ satisfying $\lambda_i < 1,\ \forall i$. It is not difficult to derive the following results from these observations.

Lemma 6.5.8 *Let G be a closed convex domain in $\mathbb{R}^n$ with nonempty interior $Q := \mathrm{int}(G)$. Let f be a single–valued monotone operator $f : Q \to \mathbb{R}^n$, and let*

$$M_x := \nabla f(x) \quad and \quad N_{x,y} := \nabla f(y) - \nabla f(x).$$

Then
(i) if $-\left(1/((1-\tau)^2) - 1\right) M_x \preceq N_{x,y} \preceq \left(1/((1-\tau)^2) - 1\right) M_x$ for all $x \in Q$, $y \in Q$ such that $(y-x)^T M_x (y-x) \leq \tau^2$ then f satisfies the relative Lipschitz condition;
(ii) if $N_{x,y}^T D^2 N_{x,y} \preceq \theta^2\gamma^2 M_x^T D^2 M_x$ for all $x \in Q$, $y := x + \theta\Delta x \in Q$ such that $\|x^{-1}(y-x)\| \leq \tau$ for every $\theta \in (0, \Theta]$ and $D = \mathrm{Diag}(d)$ with $d > 0$, then f satisfies Condition 6.2.2 with Θ.

While this lemma seems to be quite trivial, it may be practical since the conditions in the theorem can be more easily checked than the original conditions. In fact, we can find some mappings satisfying the relative Lipschitz condition and/or Condition 6.2.2 using the above theorem.

Example 6.5.9 (LCP) Consider the LCP with $f(x) = Mx + q$, and M a positive semidefinite matrix. Then $\nabla f(x) = M$ and all of the smoothness conditions are satisfied. Specifically, Condition 6.2.2 is satisfied with $\Theta = \infty$ and $\pi = 0$. The last statement also holds true if M is in P_*. For instance, consider the LCP defined by

$$M := \begin{bmatrix} 2 & 1 \\ 4 & 2 \end{bmatrix}, \quad q \in \mathbb{R}^2.$$

Then $\nabla f(x) = M$ and it is easy to see that both of the scaled Lipschitz condition and the relative Lipschitz condition do not hold with $h = \theta(1, -1)^T$ for some θ, while $\kappa = 1/16$. However, Condition 6.2.2 still holds with $\Theta = \infty$ and $\pi = 0$. ◇

Example 6.5.10 (Entropy function) Let $u \in \mathbb{R}^n_{++}$ and let $\phi(x)$ be an entropy function of the form

$$\phi(x) = \sum_{i=1}^{n} x_i \log\left(\frac{x_i}{u_i}\right).$$

Let us define $f(x) = \nabla\phi(x)$, that is $f_i(x) = \log x_i - \log u_i + 1$ for all $i = 1, 2, \ldots, n$. Then f satisfies all of the smoothness conditions (cf. [258, Theorem 4.1]). ◇

Example 6.5.11 (Power function) The mapping $f(x) : \mathbb{R}^n_+ \to \mathbb{R}^n$ defined by

$$f_i(x) = x_i^\alpha$$

satisfies the scaled Lipschitz condition for $\alpha \geq 0$. However, it satisfies Condition 6.2.2 for all $\alpha \in \mathbb{R}$. ◇

6.6 OTHER PRIMAL–DUAL AFFINE SCALING DIRECTIONS

Besides the family of affine scaling directions we introduced in this chapter other authors have developed algorithms that use the affine scaling idea in some way or another.

Tsuchiya [232] proposed a family of algorithms related to those in this chapter. The search–direction is obtained from the system

$$\begin{aligned} A\Delta x &= 0, \\ A^T\Delta y + \Delta s &= 0, \\ H^{-1}\Delta x + H\Delta s &= (XS)^{(\alpha^2+1)/2}e, \end{aligned}$$

where $H = X^{(\alpha+1)/2}S^{(\alpha-1)/2}$ and $\alpha \in [-1, 1]$. Notice, that the difference with our systems (5.10) and (6.17) is that not only the right–hand side in the third equation is scaled, but also the left–hand side. For $\alpha = 0$, Tsuchiya's algorithm gives the primal–dual affine scaling direction, while for $\alpha = 1$ ($\alpha = -1$) the primal (dual) affine scaling direction is obtained.

Sturm and Zhang [224] introduced a *primal–dual cone–affine scaling algorithm*, as an extension of the primal cone–affine scaling direction in Padberg [197] and Goldfarb and Xiao [79]. The main difference with the Dikin–affine scaling algorithm, is that instead of an ellipsoid a cone inscribed in the feasible set is used. The search–direction in Sturm and Zhang [224] is a linear combination of the primal–dual affine scaling and a new centering direction; it also has the property of combining centering and decreasing complementarity. The algorithm only requires $\mathcal{O}(\sqrt{n}\ln 1/\epsilon)$ iterations, however, the iterates are confined to a small neighborhood of the central path.

6.7 CONCLUDING REMARKS

In this chapter we introduced and analyzed a family of primal–dual affine algorithms, which contains the Dikin–affine scaling method as well as the classical one of [182] as special cases. For the analysis of the methods for NCPs we introduced a new smoothness condition, being natural for the analysis of primal–dual large neighborhood methods and, moreover, also applicable to nonmonotone mappings. Hence, we can analyze interior point methods for nonmonotone LCPs and NCPs. We are not aware of earlier complexity results on nonmonotone NCPs.

More research is needed on the applicability of these and other interior point methods to nonconvex NLP problems, and their theoretical properties. As a first extension, quasiconvex programming problems could be considered. Our smoothness condition, being applicable to nonmonotone mappings, might give a theoretical background here. It is very important to study the (local) behavior of Newton's method for nonconvex functions, possibly, by extending the self–concordance condition to this case. This might be done by including a type of trust–region condition and/or trust–region term within the self–concordance condition.

7

COMPUTATIONAL RESULTS WITH AFFINE SCALING METHODS

7.1 INTRODUCTION

In this chapter we report on computational experience with different variants of the primal–dual affine scaling algorithm for some classes of nonlinear problems, as introduced and analyzed in Chapters 5 and 6. The purpose of these experiments is to investigate whether the theoretical properties derived in those chapters show themselves in practice. More specifically, we are interested in the effect of correctors as suggested in Section 5.4, in the performance for various orders of scaling as in Section 6.3, as well as in the influence of the step size strategy.

First, we apply the algorithm to a class of ill–conditioned LCPs, derived from a convex regression problem (e.g., Jongbloed [123], Dykstra [46]). Second, we apply the algorithm to nonlinear entropy optimization problems used for estimating minimally informational distributions given marginal information (e.g., Meeuwissen [167], Agmon et al. [3] and Ben–Tal et al. [21]). Finally, we consider maximum likelihood estimation problems as in Jongbloed [123] and Terlaky and Vial [229].

7.2 IMPLEMENTATIONAL ASPECTS

The research on interior point methods has shown that implementations of exactly the same methods as analyzed theoretically in general will not produce sufficiently good results. This has some important reasons:

- the interior point assumption and the assumption that a strictly feasible point is available: these can be alleviated in several ways;
- to be able to prove polynomial convergence often too pessimistic step sizes are proposed;
- the stopping criteria are too strict for practical use and efficiency.

We will discuss these points in this section.

7.2.1 Starting points and infeasibility

In the algorithmic development in this chapter it is always assumed that an initial interior–feasible point were available. While this may be the case in certain applications, in general the problem of finding an initial point is as hard as solving the complementarity problem itself. We briefly discuss some approaches to handle this facet of the algorithm.

Using a 'big M'

A traditional way to deal with starting points is the use of a 'big M', which essentially means that the error generated by an infeasible solution is penalized with a large number in the objective function. Kojima et al. [136] show that for the LCP the following procedure can be used.

Let the LCP be specified by $f(x) = Qx + p$, for $Q \in \mathbb{R}^{n\times n}$ and $p \in \mathbb{R}^n$. Let $x^{(0)} \in \mathbb{R}^n_{++}$ be arbitrary and define $s^{(0)} = Qx^{(0)}+p$ and $M > e^T x^{(0)}$ sufficiently large. Typically $s^{(0)}$ will contain nonpositive elements. In that case, define

$$\overline{Q} := \begin{pmatrix} Q & e \\ -e^T & 0 \end{pmatrix}, \quad \overline{p} := \begin{pmatrix} p \\ M \end{pmatrix}, \quad \overline{x} := \begin{pmatrix} x^{(0)} \\ 1 - \min(s^{(0)}) \end{pmatrix}.$$

Consider the LCP with $\overline{f}(\xi) = \overline{Q}\xi + \overline{p}$; it is easy to see that $\xi = \overline{x}$ is interior–feasible, since $\overline{x} > 0$ and $\overline{f}(\overline{x}) > 0$. Comparing the conditions for complementarity in both LCPs it follows that the auxiliary problem will give a solution to the original LCP as long as M is large enough (theoretically $M = 2^{\mathcal{O}(L)}$). Unfortunately, a practically acceptable value is often hard to compute.

Using Newton's method

A second way to overcome infeasible starting points is motivated as follows: whenever Newton's method is efficient in solving the nonlinear constraints in the KKT–systems evolving in interior point methods, then it is likely that Newton's method will also be able to handle infeasibility in the linear equations in those systems. For LP the resulting infeasible interior point methods were analyzed by Kojima et al. [134], Zhang [256] and Mizuno [177], for LCPs by Potra [201] and Wright [246], among others. Extensive and encouraging computational experience by Lustig et al. [156, 157] and Mehrotra [170] motivated this theoretical work. A similar procedure for complementarity problems was outlined in Kojima et al. [136]. Kojima et al. [140] studied global convergence of this type of approach in a general framework.

We describe the resulting procedure for NCPs with mapping $f(x)$. Let an arbitrary point $(x, s) \in \mathbb{R}^{2n}_{++}$ be given. The search–mappings Δx and Δs as used in Chapter 6 (see (6.1)–(6.4)) are taken as combinations of two directions as follows:

$$\begin{aligned} (\Delta x, \Delta s) &= (\Delta x^o + \lambda \Delta x^f, \Delta s^o + \lambda \Delta s^f), \\ s\Delta x^o + x\Delta s^o &= h, \\ -\nabla f(x)\Delta x^o + \Delta s^o &= 0, \\ s\Delta x^f + x\Delta s^f &= 0, \\ -\nabla f(x)\Delta x^f + \Delta s^f &= -(s - f(x)). \end{aligned} \tag{7.1}$$

where $\lambda \in \mathbb{R}_+$ and h depends on the specific algorithm used. Using $x(\theta)$ and $s(\theta)$ as in (6.1) and (6.4) we have

$$\begin{aligned} s(\theta) - f(x(\theta)) &= s + \theta\Delta s + f(x + \theta\Delta x) - f(x) - \theta\nabla f(x)\Delta x - f(x + \theta\Delta x) \\ &= s - f(x) + \theta(-\nabla f(x)\Delta x + \Delta s) = (1 - \theta\lambda)(s - f(x)) \end{aligned}$$

for every $\theta \geq 0$, i.e., if $s - f(x) \neq 0$ then the infeasibility is decreased with the factor $(1 - \theta\lambda)$ and otherwise feasibility is maintained for every θ and λ.

Using an auxiliary problem

Recently, Andersen and Ye [7] proposed for solving MNCPs a homogeneous auxiliary problem, as an extension of similar problems for LP in Ye et al. [254] and Jansen et al. [114] for LP (Section 2.3) and in Ye [251] for LCPs. The basic idea is to consider the monotone mapping

$$\psi(x, \tau) := \begin{pmatrix} \tau f(x/\tau) \\ -x^T f(x/\tau) \end{pmatrix},$$

and to compute a solution to the NCP connected to this mapping using straightforward interior point techniques. Given $x^{(0)} = s^{(0)} = e$, $\tau_0 = \kappa_0 = 1$ initial residuals are defined by

$$r^{(0)} := s^{(0)} - \tau_0 f(x^{(0)}/\tau_0), \quad z^{(0)} := \kappa_0 + (x^{(0)})^T f(x^{(0)}/\tau_0).$$

In [7] it is shown that the system

$$\begin{aligned} s - \tau f(x/\tau) &= \mu r^{(0)}, \\ \kappa + x^T f(x/\tau) &= \mu z^{(0)}, \\ xs &= \mu e, \\ \tau\kappa &= \mu, \end{aligned}$$

has a unique solution for all $0 < \mu \leq 1$. Furthermore, a limiting solution $(x^*, s^*, \tau^*, \kappa^*)$ as μ goes to zero renders a solution to the original MNCP (if $\tau^* > 0$) or shows that it is strongly infeasible ($\kappa^* > 0$), which means that there is no sequence of iterates for which the infeasibility converges to zero. An interior point method based on this system was shown to run in polynomial time. Initial computational experiments reported in [7] are promising, although fairly easy test problems are used. Moreover, in case $\kappa^* = \tau^* = 0$ it remains undecided whether the problem is infeasible (i.e., there is no $(x, s) \geq 0$ such that $s = f(x)$) or is feasible but does not have a solution (i.e., for all feasible solutions $x^T s \neq 0$). Recall that this is a similar problem as with the embedding presented for LP in Section 2.3.

7.2.2 Long–step algorithms

In the theoretical analysis of primal–dual affine scaling algorithms fairly short steps are taken to be able to prove polynomiality. Unfortunately, such strategies are rather unpractical. Various long–step strategies have been proposed in the literature, of which we mention a few. First, we recall that the theoretically guaranteed (worse case) step size can be (much) smaller than what could be obtained in practice; in that case it is important whether large or small neighborhoods are used. Mehrotra [170] and Lustig et al. [156, 157] advocate the use of predictor–corrector algorithms. The idea is to take a long affine step to compute a 'target–point' on the central path and to use a corrector step to try to approximate this target sufficiently. A drawback of this approach is that the target on the path is chosen to depend in a specific way on the affine step; the actual way this is done is motivated by extensive computational experiments.

Gonzaga [88] proposes a theoretical way as follows. Given a proximity measure for being close to the central path and a threshold value for this measure, take the step size such that the value of the proximity for each iterate exactly equals the threshold. Gonzaga has shown this method to be quadratically convergent. Nesterov [191] analyzes similar techniques using potential functions rather than proximity measures. In the case of neighborhood $\mathcal{N}_\omega(\rho)$, see (5.12), this approach would mean to compute the step size such that $\omega(v) = \rho$ for all iterates.

Roos and Vial [209] and Den Hertog [100] analyze yet a different approach for central path–following algorithms. Instead of updating the centering parameter μ with a small amount, they divide it by a constant; then several damped Newton steps are needed to reach the vicinity of this new target (cf. Section 8.3).

In the analysis of primal affine scaling methods (see e.g., Vanderbei et al. [238], Tsuchiya et al. [233]) it is proposed to compute the step size that leads to the boundary and to take a fixed proportion of this maximal step. Kojima et al. [135] analyze such a long–step variant for a primal–dual method. Similar to the analysis in [135] it can be shown that such long–step variants of the primal–dual Dikin–affine scaling method are globally convergent and polynomial under certain conditions. In the computational experiments we use this type of long–step algorithms.

7.2.3 Computing environment and stopping criteria

The family of algorithms has been implemented in MATLAB^{TM} (version 4.2c) with efficient sparse matrix handling facility [165]. For the computations we used an HP9000/720 workstation; the memory requirement was less than 16 MB. We use Newton's method to handle infeasibility as in (7.1) with $\lambda = 1$; the starting point used is a multiple of the vector of all ones for both x and s. Corrector steps are used as outlined in Section 5.4. Systems of equations are solved using Cholesky or LU–decomposition. The maximal step size such that $x(\theta) \geq 0$ and $s(\theta) \geq 0$ is computed with bisection; for the actual step size we take in principal 2/3 of the maximal step. As stopping criteria we use similar ones as in Terlaky and Vial [229] and Shanno [215]. For the infeasibility we require

$$\frac{||s - f(x)||}{1 + ||f(x)||} < \epsilon_1;$$

for the complementarity we require

$$\frac{x^T s}{1 + \|x\|} < \epsilon_2. \tag{7.2}$$

Since the type of implementation is comparable with the one reported in [229] for the primal and primal–dual logarithmic barrier method, we will be able to compare our algorithms with these on the maximum likelihood problems.

7.3 CONVEX REGRESSION PROBLEMS

The convex regression problem is an estimation problem from statistics. Special algorithms have been constructed by Dykstra [46] and Jongbloed [123]. The problem is a convex quadratic programming (CQP) problem, however, in general very ill–conditioned.

The problem is stated as follows. We have been given two vectors y and c in $\mathbb{R}^n$. The values y_i are sample points and the values c_i are observed values. The problem is to find a convex (regression) function with function values ξ_i in the sample points y_i, such that the 'distance' between the observations c and function values ξ is minimal. Assuming $y_1 < y_2 < \cdots < y_n$, we formally have the following CQP problem

$$\begin{array}{lll} \min\limits_{\xi} & \displaystyle\sum_{i=1}^{n} (\xi_i - c_i)^2 & \\ \text{s.t.} & \dfrac{\xi_{i+1} - \xi_i}{y_{i+1} - y_i} \geq \dfrac{\xi_i - \xi_{i-1}}{y_i - y_{i-1}} & \text{for all} \quad i = 2, 3, ..., n-1. \end{array}$$

Obviously, the optimal values of ξ will be the values of a convex piecewise linear function with breakpoints at y_i. Rewriting the problem as an LCP was done as in Section 5.3.2.

To analyze variants of the algorithms random problems have been generated. In the test set the values y_i are taken randomly from a uniform distribution on the interval $[-1, 1]$. We define $c_i = (y_i - 0.5)^2 + \zeta_i$, where ζ_i is randomly generated from the normal distribution with zero mean and variance 0.1. As n increases the resulting problem will be more difficult to solve. An important reason for this is the following. The average value of $y_{i+1} - y_i$ over all i will be $2/(n+1)$. However, the probability that $\min_i(y_{i+1} - y_i)$ is smaller than $2/(n^2)$ is equal to $1 - (1 - 1/n)^n$, i.e., converging to $1 - 1/e$ for large n. Consequently, it is very likely that one of the distances $y_{i+1} - y_i$ is very small. This implies that the matrix involved in the LCP will contain very large as well as small numbers and will be ill-conditioned.

In Table 7.1 the average number of iterations needed to reach a solution satisfying the convergence criteria with $\epsilon_1 = 10^{-7}$ and $\epsilon_2 = 10^{-6}$ is presented. Table 7.2 gives the minimal and maximal number of iterations needed over the problems involved. The numbers of sample points n are in the range 100–500, the scaling parameter ν varies from 0 to 10 and the parameter r is taken to be 1 (i.e., no corrector) to 4. For each setting of the parameters 10 problems have been solved. In a few cases (indicated by superscripts in Table 7.1) the method failed to attain the required precision in complementarity. The reason for focusing on the number of iterations is that in interior point methods the major amount of work is performing one Cholesky factorization per iteration, while in our case the work per iteration does not depend on the scaling. Moreover, since the use of corrector steps increases the amount of work per iteration a drastic decrease in the number of iterations is necessary for corrector steps to be computationally efficient.

From the experiments we conclude the following.

- When using no correctors ($r = 1$):
  - the number of iterations increases approximately linearly in n;
  - the number of iterations is smallest for $\nu \in \{0.5, 1, 2\}$;
  - for larger values of ν and n the number of failures increases, due to the huge numbers arising in the computations.
- When using correctors ($r \geq 2$)
  - the number of iterations is (much) smaller (up to 60%) than without correctors;
  - the number of iterations does not benefit significantly from using more than one corrector;
  - the influence of the scaling ν is much smaller than without using correctors.

Besides the number of iterations the computational effort per iteration needs to be taken into account. Not using corrector steps has two advantages: (*i*) the corrector needs not be computed, and (*ii*) the maximal step size can be

n	ν / r	0	0.01	0.1	0.5	1	2	5	10
100	1	47.1	47.2	46.0	42.6	42.8	42.5	54.2	72.0
	2	34.2	35.1	33.8	34.2	33.0	33.2	44.9	68.8
	3	35.1	34.8	34.2	34.9	33.3	34.2	45.3	69.2
	4	33.4	35.4	32.6	31.2	35.0	35.2	44.9	69.9
200	1	55.0	54.0	53.5	49.8	46.8	47.9	59.0	78.3
	2	36.5	36.8	37.3	35.9	35.3	38.2	47.6	72.7
	3	36.4	37.2	36.3	35.1	38.0	39.6	47.3	74.0
	4	36.3	37.3	37.6	36.7	38.2	37.7	47.6	73.0
300	1	59.6	60.1	60.5^{1}	58.5	54.2	53.7	63.4	83.1
	2	40.0	39.0	38.2	37.8	39.1	40.1	48.6	75.3
	3	38.2	38.1	39.0	38.4	39.3	39.3	49.5	73.7
	4	39.7	38.8	39.0	38.0	37.7	39.2	50.0	73.8
400	1	66.7	68.5	68.4	60.0	57.9	58.1	64.8	87.5
	2	40.6	41.3	40.3	40.3	39.0	42.8	51.2	75.3
	3	40.0	41.9	41.3	40.6	40.2	42.6	51.3	76.7
	4	40.2	41.3	41.8	40.0	40.4	41.3	50.5	77.8
500	1	70.2	73.3	68.4	66.2	65.2	67.4	80.6^{1}	90.4^{2}
	2	44.6	42.0	43.9	44.5	44.4	55.8	76.1^{1}	94.4^{2}
	3	43.3	43.2	44.9	43.6	47.6	65.4	68.4^{1}	94.0^{3}
	4	44.4	42.3	44.1	44.0	50.1	50.3^{1}	74.9^{2}	80.8^{4}

Table 7.1 Average iteration numbers and number of failures for various values of n and r for primal–dual affine scaling algorithms with scaling ν.

n	ν / r	0	0.01	0.1	0.5	1	2	5	10
100	1	43/52	41/52	40/55	37/49	39/47	37/51	49/60	66/80
	2	30/38	29/42	26/41	30/39	29/37	28/36	42/49	67/71
	3	29/40	29/39	30/40	30/39	28/42	31/37	42/50	67/71
	4	29/37	30/40	30/35	26/41	31/40	32/39	44/47	69/72
200	1	49/62	45/71	46/63	43/58	42/53	42/52	53/68	71/83
	2	28/42	32/41	33/41	34/38	30/38	34/46	44/53	68/79
	3	31/41	34/41	32/42	31/38	34/44	35/46	46/53	69/92
	4	31/41	32/41	33/43	32/42	34/43	32/48	46/50	69/80
300	1	51/67	52/67	54/66	53/76	47/59	50/60	53/74	74/90
	2	33/47	35/48	33/41	33/43	36/42	36/46	46/54	72/92
	3	33/41	33/48	36/44	32/44	34/43	35/50	46/54	70/82
	4	33/45	34/42	30/44	34/43	32/46	34/52	46/67	70/83
400	1	57/72	60/75	60/81	52/66	48/70	49/74	58/72	78/96
	2	35/48	37/47	37/46	35/47	34/44	37/48	49/54	73/80
	3	35/46	38/46	37/47	35/50	35/46	38/46	50/54	75/81
	4	35/44	36/46	34/46	31/48	37/44	36/49	48/54	72/91
500	1	55/84	57/90	59/78	49/83	51/104	50/130	65/139	79/119
	2	35/57	39/50	35/55	34/60	34/57	39/146	50/176	73/171
	3	36/54	35/54	34/56	34/61	34/67	39/177	49/126	75/180
	4	35/63	32/63	34/62	37/67	37/103	38/75	49/194	76/95

Table 7.2 Minimal and maximal iteration numbers for various values of n and r for primal–dual affine scaling algorithms with scaling ν.

computed with a ratiotest. In Table 7.3 we give the average number of floating point operations (measured by the 'flops'-command in MATLAB^{TM}) required by the Dikin–affine scaling algorithm ($\nu = 1$). We conclude that the algorithm with one corrector performs best for all problem sizes. A similar behavior was found for the other scaling factors.

$r \backslash n$	100	200	300	400	500
1	2.10	4.78	8.50	11.77	16.55
2	1.96	4.21	7.39	9.56	13.98
3	2.47	5.87	9.82	12.74	19.35
4	3.21	7.13	10.50	15.81	24.36

Table 7.3 Average number of megaflops for various values of n and r for the primal–dual Dikin–affine scaling algorithm (ν=1).

The other problems used in our testing are computationally more easily than the convex regression problem. Henceforth, we mainly use the primal–dual Dikin–affine method.

7.4 MAXIMAL ENTROPY DISTRIBUTIONS

Let be given two random variables X and Y on $A = [-1/2, 1/2] \times [-1/2, 1/2]$, which are known to have a bivariate distribution with uniform marginals (having mean 0 and standard–deviation $1/\sqrt{12}$) and correlation ρ. The problem is to find a distribution that adds minimal information (in entropy sense) of all distributions with the given marginals and correlation. Although the problem is actually a continuous optimization problem, it is common to use grids. Let n be a given grid–size. Define

$$x_i = \frac{2i-1-n}{2n}, \quad y_i = x_i, \quad i = 1, \ldots, n,$$

and variables p_{ij} for $i, j = 1, \ldots, n$. The variables represent values of the density on subsquares of the square A. The mathematical problem is

$$\min_{p \geq 0} \quad \sum_{i,j=1}^{n} p_{ij} \ln p_{ij}$$

$$\text{s.t.} \quad \sum_{i=1}^{n} p_{ij} = \frac{1}{n}, \quad j = 1, \ldots, n,$$
$$\sum_{j=1}^{n} p_{ij} = \frac{1}{n}, \quad i = 1, \ldots, n,$$
$$\sum_{i,j=1}^{n} x_i y_j p_{ij} = \frac{\rho}{12}.$$

The problem described here is taken from Meeuwissen [167]; similar problems can be found in e.g., Ben–Tal et al. [21]. Since the probabilities sum to one, the problem can also be viewed as a dual geometric programming problem.

This problem is written as an NCP as follows. First notice, that we may change the first two sets of equality constraints in inequality constraints, since they will certainly be binding in any optimal solution. The Lagrangian is given by

$$\begin{aligned} L(p; \lambda, \xi, \zeta, \zeta') = & \sum_{i,j=1}^{n} p_{ij} \ln p_{ij} + \sum_{j=1}^{n} \lambda_j \left(\sum_{i=1}^{n} p_{ij} - \frac{1}{n} \right) \\ & + \sum_{i=1}^{n} \xi_i \left(\sum_{j=1}^{n} p_{ij} - \frac{1}{n} \right) + (\zeta' - \zeta) \left(\sum_{i,j=1}^{n} x_i y_j p_{ij} - \frac{\rho}{12} \right). \end{aligned}$$

Then the mapping $f : \mathbb{R}_+^{n^2+2n+2} \to \mathbb{R}^{n^2+2n+2}$ given by

$$\begin{aligned} f_k(\cdot) &= \ln p_{ij} + 1 + \lambda_j + \xi_i + (\zeta' - \zeta) x_i y_j, & k = (i-1)n + j, \\ f_{n^2+j}(\cdot) &= \tfrac{1}{n} - \textstyle\sum_{i=1}^{n} p_{ij}, & j = 1, \ldots, n, \\ f_{n^2+n+i}(\cdot) &= \tfrac{1}{n} - \textstyle\sum_{j=1}^{n} p_{ij} & i = 1, \ldots, n, \\ f_{n^2+2n+1}(\cdot) &= \textstyle\sum_{i,j=1}^{n} x_i y_j p_{ij} - \tfrac{\rho}{12}, & \\ f_{n^2+2n+2}(\cdot) &= \tfrac{\rho}{12} - \textstyle\sum_{i,j=1}^{n} x_i y_j p_{ij}, & \end{aligned}$$

defines the NCP. The definition of $f(\cdot)$ combines the primal feasibility with the dual feasibility constraints; duality theory for entropy problems is contained in Kas and Klafszky [128], among others. The size of the Jacobian of f is n^2+2n+2, which means that factorizing the Jacobian is computationally prohibitive. Fortunately, the search–direction can be computed by solving systems of size $2n+2$ only by first solving explicitly for the displacement in the p–variables. The computational effort per iteration is therefore dominated by the number of times vectors of size n^2 have to be manipulated with, i.e., in the computation of the displacement in p and in the line search via function evaluations. The parameters in the stopping criteria are $\epsilon_1 = \epsilon_2 = 10^{-8}$. The starting point is taken to be $x^{(0)} = s^{(0)} = \sqrt{n}\, e$.

In Table 7.4 we give iteration numbers of the primal–dual Dikin–affine scaling algorithm without corrector, for various sizes n of the grid and values ρ of the correlation. The last column contains the number of variables x in the NCP to be solved. Two different step sizes have been used, namely 2/3 and 0.95. For $n = 25$ the latter appeared to be too optimistic, since some of the elements in s converged to zero too fast. Observe that the number of iterations grows very slowly with the gridsize, and is still small for the NCPs with over 15000 variables. When the algorithm stops the infeasibility and complementarity are of the same order of magnitude. The number of iterations increases with ρ because the condition of the problem changes with ρ: for $\rho = 0$ all variables are equal in the optimum, while for larger ρ there is a large difference in the optimal variable values.

Experimentation with the use of correctors showed that a small decrease in the required number of iterations can be achieved, however, at the cost of increasing total computational effort measured in megaflops used. This is caused by the fact that every corrector requires manipulations with an extra vector of $\mathcal{O}(n^2)$ elements, which is highly unattractive in MATLAB^{TM}.

We also experimented with the use of a linear update as in [7]; then the computation of the maximal step size is very cheap. We found that for the small problems (and using 2/3 of the maximal step) the number of iterations increased with 10-20%, while the number of megaflops decreased. For larger problems, however, the method failed to obtain the required precision in the infeasibility.

Solving these problems with the primal–dual affine scaling algorithm ($\nu = 0$) it appeared that the number of iterations for the 'easier' problems ($\rho \leq 0.3$) can be drastically reduced (up to 40%). However, this algorithm halted for the more difficult ones due to numerical difficulties. It was also tried to solve these entropy optimization problems with AIMMS-CONOPT [23] and GAMS-MINOS [31]; however, both packages already failed to solve them for values of n larger than 20.

7.5 MAXIMUM LIKELIHOOD ESTIMATION

In Terlaky and Vial [229] two algorithms are tested on various maximum likelihood estimation problems having a convex nonlinear objective and linear constraints. The algorithms are the primal logarithmic barrier method and

$n \setminus \rho$	0	0.1	0.3	0.5	0.7	0.9	# vars.
25	24/19	25/19	26/20	26/21	26/21	26/24	677
50	26/17	26/19	27/20	27/20	28/21	28/22	2602
75	27/16	27/20	28/20	28/21	29/21	29/20	5777
100	27/17	28/20	28/20	29/20	30/21	30/20	10202
125	28/18	28/20	29/20	29/20	31/21	31/21	15877
150	29/18	29/20	30/21	30/21	32/21	31/21	22802

Table 7.4 Iteration numbers for various values of grid–size n and correlation ρ. The left number in each cell is obtained with step size 2/3, the right with 0.9 for $n = 25$ and 0.95 for $n \geq 50$.

the primal–dual method developed by Vial [240]. We have tested the primal–dual Dikin–affine method on the same set of problems. The statistical problem is as follows.

Let Y be a real–valued random variable with (unknown) convex density function $g : \mathbb{R}_+ \to \mathbb{R}_+$. Let $\{y_1, \cdots, y_n\}$ be an ordered sample of n outcomes of Y with $0 < y_1 < \cdots < y_n$. Let us define $y_0 = y_{-1} = 0$. The problem is to compute the maximum likelihood estimator of the sample, i.e., a convex function having values ξ_i in the sample points y_i, $i = 0, \ldots, n-1$ such that the log–likelihood function

$$L(\xi) = \sum_{i=0}^{n-1} \ln \xi_i$$

is maximized. As in the convex regression problem, the estimator of g is a piecewise linear function, where we assume $\xi_n = 0$ for decreasing densities. The optimization problem can now be stated as follows.

$$\begin{array}{lll} \min\limits_{\xi} & -\sum\limits_{i=0}^{n-1} \ln \xi_i & \\ \text{s.t.} & \dfrac{\xi_{i+1} - \xi_i}{y_{i+1} - y_i} \geq \dfrac{\xi_i - \xi_{i-1}}{y_i - y_{i-1}} & i = 1, 2, \ldots, n-1, \\ & \dfrac{1}{2} \sum\limits_{i=1}^{n-1} (y_{i+1} - y_{i-1}) \xi_i = 1, & \\ & \xi_i \geq 0, \quad i = 1, 2, \ldots, n-1, \quad \xi_n = 0. & \end{array}$$

The first set of constraints models the convexity requirement. The second constraint models the condition that the estimator must be a density function.

The constraint $\xi_n = 0$ ensures that the total probability below the computed function will be one. Using the optimality conditions it is easy to write the problem as a monotone complementarity problem with mapping $f(x)$. The set of test problems is drawn from the distributions in Table 7.5. Since the arcsine law has a nondecreasing density, the constraint $\xi_n = 0$ is dropped for problems with this law, as well as the last of the convexity constraints.

law	distribution	density	domain
exponential	$1 - e^{-y}$	e^{-y}	$y \geq 0$
arcsine	$2\ \text{arcsine}(\sqrt{y})/\pi$	$1/(\pi\sqrt{y(1-y)})$	$0 < y < 1$
quadratic	$y - y^2/4$	$1 - y/2$	$0 \leq y \leq 2$
inverse	$1 - 1/y$	$1/y^2$	$y \geq 1$

Table 7.5 Distributions used in testing.

For each law we randomly generated 30 problems, 10 per number of observations being 500, 1000 and 2000 each. The algorithm used is the primal–dual Dikin–affine scaling method ($\nu = 1$) with one corrector.

The stopping criteria for the complementarity is as in (7.2) with $\epsilon_2 = 10^{-6}$. For the infeasibility we use the condition

$$\min_i f_i(x) > -\epsilon_1 := -10^{-8}.$$

In all problems the starting point was taken to be $x = s = \sqrt{2n}\, e$. It appeared to be hard (and often impossible) to solve the problems using the nonlinear update (6.4) in s, used before in the theory and computational experiments. Instead, we used a simple update

$$s^+ := s + \alpha\Delta s^{(1)} + \alpha^2 \Delta s^{(2)},$$

where α is the step size. A similar strategy is used in Andersen and Ye [7]. For the step size we first used a fraction 2/3 of the maximal step to the boundary. However, experimentation showed the following procedure to perform significantly better on these problems. It is inspired by the desire to stay in a large neighborhood of the central path. Starting from $\alpha = 0.95$, check whether

$$n\frac{\min_i (x_i(\alpha)s_i(\alpha))}{x(\alpha)^T s(\alpha)} \geq 0.05.$$

If this condition is not yet satisfied we multiply α by 0.95 and check again, which is repeated until the condition is met.

Terlaky and Vial [229] report computational difficulties for increasing sample sizes when the distance between observations is very small (see also the discussion in Section 7.3). To circumvent this they cluster observations that are closer to each other than a threshold, taken to be 10^{-5}.

In Table 7.6 we report the results obtained for the set of problems. For the larger problems, it sometimes appeared to be impossible to attain the required accuracy. Using the clustering scheme all problems were solved without numerical difficulties, see Table 7.7. Note that the clustering improves the efficiency of the algorithm for these problems as well. There appeared to be no need to incorporate an iterative refinement scheme (Golub and Van Loan [81]) for the computation of the search–direction as needs to be done in [229].

Comparing the performance of our algorithm with those in [229] we conclude that our algorithm shows the same behavior with respect to the order of difficulty of the various distribution laws and problem sizes. With respect to number of iterations our algorithm performs much better than the primal logarithmic barrier method, and slightly worse than Vial's [240] method. With respect to megaflops used our algorithm is much worse; this is due to the fact that we have to compute corrector steps, and as a consequence, the line–search is more expensive. Since we use no explicit targets (on the central path) in our algorithm the use of correctors is necessary to stay sufficiently away from the boundary.

7.6 CONCLUDING REMARKS

To investigate the primal–dual affine scaling algorithms described in Chapters 5 and 6 a practical version of the algorithms was implemented and used to solve various optimization problems from statistics.

On all these problems the primal–dual Dikin–affine scaling method outperforms the classical primal–dual affine scaling method by Monteiro et al. [182] (it is faster and more stable and robust). The main reason is its property of simultaneously driving the iterates to optimality as well as to the central path.

The algorithms are attractively simple to understand and implement, and require little parameter setting or inclusion of safeguards. The primal–dual

n		gap	inf	iter	mflops
		exponential			
	min	7.6E-10	0	29	7.3
500	average	2.6E-7	1.7E-9	31.1	8.3
	max	6.7E-7	8.8E-9	36	10.2
	min	4.8E-12	0	32	16.9
1000	average	8.7E-8	1.1E-9	34.7	18.7
	max	2.3E-7	9.0E-9	46	26.4
	min	1.4E-10	0	35	36.5
2000	average	1.6E-8	2.8E-9	40.0	44.3
	max	4.0E-8	8.1E-9	49	58.6
		arcsine			
	min	3.0E-11	0	29	7.4
500	average	1.0E-7	2.6E-9	33.9	9.2
	max	3.8E-7	1.0E-8	42	12.4
	min	7.2E-10	0	33	17.7
1000	average	1.4E-7	1.1E-9	37.5	20.8
	max	6.1E-7	7.7E-9	42	23.7
	min	7.3E-10	0	38	40.5
2000	average	5.2E-8	3.8E-9	41.8	46.8
	max	2.0E-7	9.5E-9	47	54.1
		quadratic[a]			
	min	5.5E-10	0	24	6.1
500	average	1.7E-7	3.8E-10	27.6	7.2
	max	8.9E-7	3.5E-9	31	8.6
	min	4.4E-14	0	26	13.1
1000	average	1.3E-8	1.8E-9	34.9	18.9
	max	7.0E-8	8.0E-9	56	34.6
	min	4.1E-15	0	31	31.1
2000	average	4.3E-10	2.5E-9	41.4	46.0
	max	2.2E-9	9.3E-9	56	74.7
		inverse			
	min	1.2E-14	0	30	8.4
500	average	8.9E-8	6.2E-10	35.8	10.2
	max	8.4E-7	4.2E-9	42	13.0
	min	1.9E-14	0	31	17.7
1000	average	6.4E-9	5.1E-10	40.2	23.8
	max	5.1E-8	2.4E-9	50	31.2
	min	1.2E-13	4.6E-11	36	41.8
2000	average	3.0E-9	1.8E-9	47.0	58.9
	max	1.9E-8	4.8E-9	75	110.0

[a]For $n = 2000$ one problem could not be solved with the required accuracy for the quadratic law, and three were not solved for the inverse law.

Table 7.6 Results for maximum likelihood problems, without clustering. For each size and law 10 problems were solved.

n		clusters	gap	inf	iter	mflops
			exponential			
500	min	498	1.3E-11	0	28	7.2
	average	499.2	2.6E-7	1.3E-9	30.6	7.9
	max	500	5.8E-7	8.8E-8	35	8.7
1000	min	992	3.2E-8	0	28	14.3
	average	995.7	4.2E-7	1.7E-9	31.3	16.2
	max	999	7.8E-7	8.1E-8	34	18.3
2000	min	1974	2.4E-8	0	31	31.8
	average	1981.0	1.4E-7	1.8E-9	34.1	35.2
	max	1987	3.0E-7	7.7E-8	38	40.1
			arcsine			
500	min	489	6.0E-12	0	29	7.9
	average	493.1	3.4E-8	0	31.9	8.9
	max	497	1.7E-7	0	35	9.6
1000	min	967	3.8E-10	0	32	17.5
	average	976.5	5.5E-9	0	34.4	19.7
	max	984	1.7E-8	0	40	23.2
2000	min	1893	8.1E-12	0	29	33.1
	average	1917.2	5.4E-8	9.4E-10	38.5	46.1
	max	1932	4.1E-7	9.3E-9	48	60.1
			quadratic			
500	min	495	1.0E-10	0	24	6.0
	average	498.3	2.4E-7	2.5E-9	26.9	6.8
	max	500	9.6E-7	9.8E-9	33	8.4
1000	min	988	1.5E-9	0	25	12.4
	average	992.8	9.7E-8	1.1E-9	28.3	14.4
	max	996	3.8E-7	6.3E-9	31	16.6
2000	min	1962	3.2E-10	0	26	25.8
	average	1973.2	5.0E-8	2.1E-9	31.4	31.9
	max	1981	1.9E-7	8.0E-9	38	40.3
			inverse			
500	min	497	6.0E-17	0	28	7.5
	average	499.1	1.1E-7	9.9E-10	34.1	9.6
	max	500	8.4E-7	4.7E-9	49	13.2
1000	min	990	1.0E-9	0	27	15.3
	average	996.0	2.0E-7	0	32.9	19.1
	max	998	9.8E-7	0	38	23.4
2000	min	1976	2.3E-11	0	30	34.3
	average	1985.7	1.6E-7	1.3E-10	38.7	46.9
	max	1993	7.4E-7	9.1E-10	51	69.1

Table 7.7 Results for maximum likelihood problems, with clustering. For each size and law 10 problems were solved.

Dikin–affine method performs comparable to the primal logarithmic barrier method on a set of difficult maximum likelihood problems, but cannot compete with the more involved method of Vial [240] on the same set of problems. In the current implementation, the expensive line–search is the main bottle–neck. The computational results confirm the theoretical ordering in efficiency of the family of primal–dual affine, with and without corrector steps. The use of one corrector step saves computational effort, and is sometimes even necessary to derive practical convergence. Using more correctors typically is unattractive.

An extensive comparison of interior point methods for general NLP problems versus commercially available software as MINOS and CONOPT is still lacking; to date, only isolated comparisons on specific problem classes have been performed.

8

TARGET–FOLLOWING FOR LINEAR PROGRAMMING

8.1 INTRODUCTION

In Chapter 5 we analyzed a family of primal–dual affine scaling algorithms, using a direction having the property of combining centering and moving towards optimality. As discussed in Section 5.2 the main difference between affine scaling algorithms and path–following algorithms is that in the former the search–direction only depends on the given iterate, while the latter use reference–points (*target–points*) in the v–space (see Section 2.1). Not using targets may cause the step size to become extremely small if for some reason an iterate comes close to the boundary of the feasible region. This behavior can specifically be observed in the primal–dual affine scaling algorithm of Monteiro et al. [182] that does not contain a centering effect. In efficient primal–dual methods for linear programming (LP), developed by Monteiro and Adler [180], Kojima et al. [139], Lustig et al. [156, 157] and Mehrotra [170] among others, the central path is therefore used to keep the iterates sufficiently away from the boundary. So–called weighted paths have been used with the same objective in e.g., Den Hertog et al. [104], Ding and Li [45] and Mizuno [174]. In this chapter we propose the use of a different path, that, unlike the weighted paths improves the centering along the path. More specifically, such a path may start in any non–central point but is tangential to the central path in the limit. The path can be viewed as a continuous extension of the primal–dual Dikin–affine scaling direction. A path–following algorithm is developed that uses such a path as a guideline to optimality; this will be called the *Dikin–path–following algorithm.* We stress that centering is very important in interior point methods. A sequence of iterates that approximate the central path (in the limit) will generate points converging to the analytic center of the optimal face (Güler and Ye [97]). It is well known that this center is a strictly complementary solution, thereby defining the optimal partition of the problem, which is very useful in sensitivity analysis (see Chapter 2). Also, the asymptotic analysis of certain interior point methods uses the centering to prove superlinear or quadratic convergence of algorithms, e.g., [88, 89].

The analysis of the Dikin–path–following method offers a general framework for the convergence analysis of primal–dual interior point methods. This framework is general enough to apply to very diverse existing methods and still yield simple convergence proofs. The methods being analyzable in this context are called *target-following*. These methods appeared to be closely related to the methods using *α–sequences* developed by Mizuno [174, 176] for LCPs. The basic observation for the development and analysis of target–following methods is contained in Theorem 2.1.5. To be more specific we use the LP problem in standard form

$$\text{(P)} \quad \min_{x} \left\{ c^T x \; : \; Ax = b, \; x \geq 0 \right\},$$

and its dual

$$\text{(D)} \quad \max_{y,s} \left\{ b^T y \; : \; A^T y + s = c, \; s \geq 0 \right\}.$$

Under the assumption of the existence of a positive primal–dual pair for (P) and (D) the system

$$\begin{aligned} Ax &= b, \; x > 0 \\ A^T y + s &= c, \; s > 0 \\ xs &= \overline{v}^2, \end{aligned} \tag{8.1}$$

has a unique solution for any $\overline{v} \in \mathbb{R}^n_{++}$, see Theorem 2.1.5. The existence of the solution follows from the observation (cf. Theorem 2.2.6) that the given system is the KKT–system for minimizing the *weighted logarithmic barrier function*

$$f(x, s; \overline{v}) = x^T s - \sum_{i=1}^{n} \overline{v}_i^2 \ln x_i s_i.$$

Recall from Section 2.1 that the *v–space* of a given LP problem is defined as the space of the square roots of the complementary products of positive primal–dual pairs:

$$\mathcal{V} = \{ \, v \in \mathbb{R}^n \; : \; v_i = \sqrt{x_i s_i}, \; Ax = b, \; A^T y + s = c, \; x > 0, \; s > 0 \, \}.$$

Note that if $v = \sqrt{xs}$ then $\|v\|^2 = x^T s$, so in the v–space the points with constant norm represent all positive primal–dual pairs with a fixed duality gap. Observe that all optimal pairs (x, s) correspond to the vector $v = 0$. The image of the central path in the v–space is the main diagonal; also the image of the weighted path that passes through an initial point $(x^{(0)}, s^{(0)})$ is the positive ray passing through $v^{(0)} = \sqrt{x^{(0)}s^{(0)}}$. Atkinson and Vaidya [10] discuss how the efficiency of Newton's method is affected by differences in the elements of a weight-vector. They give a simple example demonstrating that when the ratio between the smallest and the largest weight decreases, the region where Newton's method converges gets smaller. Hence, a natural way of measuring the closeness of a point to the central path appears to be this ratio, which is denoted as

$$\omega(\overline{v}) := \frac{\min(\overline{v})}{\max(\overline{v})}. \tag{8.2}$$

Note that $0 < \omega(\overline{v}) \leq 1$, with equality if and only if $\overline{v}$ is on the central path. To combine centering and improving complementarity we will be interested in trajectories of which the image in the v-space passes through $v^{(0)}$ and is tangent to the main diagonal at the origin of the positive orthant.

To analyze primal-dual algorithms we focus on a few general concepts. The basic algorithmic step in path-following primal-dual interior point methods is a Newton step in the (x, s)-space. This step is defined with respect to some *target(-point)* $\overline{v}$ in the v-space. The fundamental property in interior point methods is that the step is feasible (i.e., preserves the interior point property) if the current iterate (x, s) is close enough to the target $\overline{v}$, where closeness is defined with some appropriate measure of proximity. With this in mind, we can define the concept of a *target-sequence*, by which we mean any sequence of vectors in the v-space. A *traceable* target-sequence is a target-sequence with the property that: (*i*) it can be approximated, in the sense of the above mentioned proximity measure, by a sequence of points in the (x, s)-space, such that (*ii*) successive points in the (x, s)-space are obtained by some 'easy' computations such as one or a few Newton steps. If the target-sequence converges to some point, then we may enforce convergence of the associated (x, s)-sequence to the target limit. We now define a *target-following algorithm* as an algorithm that generates iterates $(x^{(k)}, s^{(k)})$ which are close to their corresponding targets $\overline{v}^{(k)}$. In the standard (central) path-following methods the targets are points on the central path. Then the (traceable) target-sequence is determined by

$$\overline{v}^{(0)} = \mu_0 e, \quad \overline{v}^{(k+1)} = \sqrt{1-\theta_k}\, \overline{v}^{(k)},$$

for certain values $\mu_0 > 0$ and $0 \leq \theta_k \leq 1$, where k is the iteration number. A weighted-path following algorithm has a given $\overline{v}^{(0)} > 0$ and sets $\overline{v}^{(k+1)} = \sqrt{1-\theta_k}\, \overline{v}^{(k)}$. The one-to-one correspondence between points in the v-space and positive primal-dual pairs (x, s) suggests that, to solve the LP problem, we can follow any sequence of targets $\{\overline{v}^{(k)}\}$ in the v-space for which $e^T(\overline{v}^{(k)})^2$ tends to zero, hence leads to optimality. The same methodology can be used to solve other problems, like computing weighted centers. Note that a target–sequence may consist of an infinite as well as a finite number of targets; a target–sequence can be predetermined, but also adaptively constructed during the algorithm.

The striking feature of the convergence analysis we propose is that it is essentially performed in the v–space. We express a simple condition on the target–sequence to be traceable by a sequence of primal–dual pairs (x, s). By verifying that a given target–sequence satisfies the condition, we have a simple methodology to derive complexity bounds. The general results are developed in Section 8.2. In this way we are able to analyze and prove convergence of a great variety of algorithms (see Sections 8.2.3 and 8.2.4) such as two variants of the Dikin–path–following method (cf. Chapter 5), the standard path–following method [139, 180], the weighted path–following method [45], a variant of the cone–affine scaling algorithm [224], a variant of Freund's shifted barrier method [59], algorithms for computing analytic centers [100, 174] and algorithms for computing weighted centers [10, 176]. The convergence proofs are short and similar, thereby demonstrating the unifying value of an analysis focusing on the v–space.

Whereas the applications considered so far are all short–step methods, we show how to transfer the target–following methodology to long–step algorithms (cf. Roos and Vial [209], Gonzaga [85], Den Hertog [100] and Jansen et al. [116]). In the analysis in Section 8.3 we introduce a weighted logarithmic barrier function which serves the role of a proximity measure between primal–dual pairs and targets in the v–space. One of the striking outcomes will be that the complexity of long–step methods is negatively influenced by the use of weights.

8.2 SHORT–STEP PRIMAL–DUAL ALGORITHMS FOR LP

8.2.1 Directions in v–space and (x, s)–space

In this section we analyze the (iteration) complexity of primal–dual methods for LP that follow a traceable target–sequence. Methods of this type have an iterative nature, meaning that in every iteration a direction is computed that leads from the current iterate to the next. Let (x, s) be a pair of primal–dual interior–feasible solutions, and let v be the corresponding point in the v–space, i.e., $v = \sqrt{xs}$. Furthermore, let $\overline{v}$ be the current target–point in the v–space. The aim is to find an approximate solution of the system of equations (8.1), or stated otherwise, we seek directions $(\Delta x, \Delta y, \Delta s)$ such that

$$\begin{aligned} A(x + \Delta x) &= b, \\ A^T(y + \Delta y) + s + \Delta s &= c, \\ (x + \Delta x)(s + \Delta s) &= \overline{v}^2. \end{aligned}$$

Applying Newton's method to this system we remove the nonlinear term in the last equation and obtain the following relations for the displacements:

$$\begin{aligned} A\Delta x &= 0, \\ A^T\Delta y + \Delta s &= 0, \\ x\Delta s + s\Delta x &= \overline{v}^2 - v^2. \end{aligned} \tag{8.3}$$

For the analysis it is convenient to work in scaled space as has become more or less standard in the literature on primal–dual methods for LP (see Gonzaga [87]). To this end we introduce the vector

$$d := \sqrt{xs^{-1}}.$$

Using d we can rescale both x and s to the same vector, namely v:

$$d^{-1}x = ds = v.$$

The main property of the scaling is that it maps both x and s to the vector v; this property is extended to a nonlinear setting by Nesterov and Todd [194]. We also use d to rescale Δx and Δs:

$$p_x := d^{-1}\Delta x, \qquad p_s := d\Delta s.$$

Note that the orthogonality of Δx and Δs implies that p_x and p_s are orthogonal as well. In scaled space, the search–directions p_x and p_s are orthogonal components of a vector. Indeed, we may write

$$x\Delta s + s\Delta x = xd^{-1}d\Delta s + sdd^{-1}\Delta x = v(p_x + p_s).$$

Obviously Δy should not be scaled, hence we define $p_y = \Delta y$. So, Newton's direction is determined by the following linear system:

$$\begin{aligned} ADp_x &= 0 \\ DA^Tp_y + p_s &= 0 \\ p_x + p_s &= v^{-1}\left(\overline{v}^2 - v^2\right). \end{aligned}$$

Denoting

$$p_v := v^{-1}\left(\overline{v}^2 - v^2\right), \tag{8.4}$$

we have $p_x + p_s = p_v$, and p_x and p_s are simply the orthogonal decomposition of p_v in the nullspace of AD and the row space of AD respectively. Note that this is established by the scaling with d. We mention here that this is the last time that the data A, b, c explicitly appear in this section, and that the data only come in via an initial starting point. This has the advantage that we work completely in the v–space from now on.

8.2.2 Analysis of the Newton step

Since we will use Newton's method for following a traceable target–sequence we need to analyze its behavior. Let us define the vector q_v as follows:

$$q_v := p_x - p_s.$$

Note that the orthogonality of p_x and p_s implies that $\|q_v\| = \|p_v\|$. We also have

$$\begin{aligned} p_x &= \tfrac{1}{2}(p_v + q_v), \\ p_s &= \tfrac{1}{2}(p_v - q_v), \end{aligned}$$

whence

$$p_x p_s = \tfrac{1}{4}(p_v^2 - q_v^2). \tag{8.5}$$

The product $p_x p_s$ plays an important role in the analysis. It represents the second order effect in the Newton step, which needs to be small to prove efficiency of Newton's method. Indeed,

$$(x + \Delta x)(s + \Delta s) = xs + x\Delta s + s\Delta x + \Delta x\Delta s = v^2 + vp_v + p_x p_s = \overline{v}^2 + p_x p_s.$$

Unless the nonlinear term $\Delta x \Delta s$ (that was left out in (8.3) to obtain a linear system) is zero, the vector of complementarity products after the step will not be exactly $\overline{v}^2$. We relate the euclidean and the infinity norms of this product to the norm of p_v as follows (cf. Lemma 5.3.3; a similar lemma for the case $\overline{v}$ is on the central path is proved by Mizuno et al. [179]).

Lemma 8.2.1 *It holds $\|p_x p_s\|_\infty \le \|p_v\|^2/4$ and $\|p_x p_s\| \le \|p_v\|^2/(2\sqrt{2})$.*

Proof: Using (8.5) we may write

$$\|p_x p_s\|_\infty \le \tfrac{1}{4}\max\left(\|p_v\|_\infty^2, \|q_v\|_\infty^2\right) \le \tfrac{1}{4}\max\left(\|p_v\|^2, \|q_v\|^2\right) = \tfrac{1}{4}\|p_v\|^2.$$

Using (8.5) once more we obtain

$$\begin{aligned} \|p_x p_s\|^2 &= e^T (p_x p_s)^2 = \tfrac{1}{16} e^T (p_v^2 - q_v^2)^2 = \tfrac{1}{16} \left\|p_v^2 - q_v^2\right\|^2 \\ &\le \tfrac{1}{16} \left(\left\|p_v^2\right\|^2 + \left\|q_v^2\right\|^2\right) \le \tfrac{1}{16} \left(\|p_v\|^4 + \|q_v\|^4\right) = \tfrac{1}{8} \|p_v\|^4 . \end{aligned}$$

This proves the lemma. □

In the analysis of target–following algorithms we need a measure for the proximity of the current iterate v to the current target $\overline{v}$. For this purpose we introduce the following proximity measure:

$$\delta(v; \overline{v}) := \frac{1}{2\min(\overline{v})} \|p_v\| = \frac{1}{2\min(\overline{v})} \left\| \frac{\overline{v}^2 - v^2}{v} \right\| . \tag{8.6}$$

We point out that this proximity measure is in the spirit of the Roos–Vial measure [210], and the primal–dual measures discussed in Jansen et al. [116]; note that it is not symmetric in the iterate v and the target $\overline{v}$. Defining

$$u := \frac{\overline{v}}{v}, \tag{8.7}$$

the measure can be rewritten as

$$\delta(v; \overline{v}) = \frac{1}{2\min(\overline{v})} \left\|v^{-1}\left(\overline{v}^2 - v^2\right)\right\| = \frac{1}{2\min(\overline{v})} \left\|\overline{v}\left(u - u^{-1}\right)\right\| . \tag{8.8}$$

If $\overline{v}^2 = \mu e$ for some positive μ then the measure is

$$\delta(v; \overline{v}) = \frac{1}{2} \left\|u - u^{-1}\right\| ,$$

which is up to the factor 1/2 equal to the proximity measure used in [116]. A similar measure, namely

$$\frac{1}{2\min(\overline{v})} \left\| \frac{\overline{v}^2 - v^2}{\overline{v}} \right\| ,$$

was used by Mizuno [174, 176]. This proximity measure differs from ours by a factor involving

$$\|u\|_\infty = \left\| \frac{\overline{v}}{v} \right\|_\infty \quad \text{and} \quad \|u^{-1}\|_\infty = \left\| \frac{v}{\overline{v}} \right\|_\infty .$$

The next lemma is concerned with bounding these quantities. Moreover, the analysis will show that these quantities are very important for the proximity in the v–space.

Lemma 8.2.2 *Let $\delta := \delta(v;\overline{v})$ and u as defined in (8.6) and (8.7). Then it holds*

$$\frac{1}{\rho(\delta)} \leq u_i \leq \rho(\delta), \quad i = 1, \ldots n,$$

where

$$\rho(\delta) := \delta + \sqrt{1+\delta^2}. \tag{8.9}$$

Proof: Observe that

$$\delta = \frac{1}{2\min(\overline{v})} \left\|\overline{v}\left(u - u^{-1}\right)\right\| \geq \frac{1}{2\min(\overline{v})}\min(\overline{v}) \left\|u - u^{-1}\right\| = \frac{1}{2}\left\|u - u^{-1}\right\|.$$

So, for each $i, 1 \leq i \leq n$,

$$-2\delta \leq u_i^{-1} - u_i \leq 2\delta.$$

Since u_i is positive, this is equivalent to

$$-2u_i\delta \leq 1 - u_i^2 \leq 2u_i\delta,$$

or

$$u_i^2 - 2u_i\delta - 1 \leq 0 \leq u_i^2 + 2u_i\delta - 1.$$

One easily verifies that this is equivalent to $\rho(\delta)^{-1} \leq u_i \leq \rho(\delta)$. This proves the lemma. □

We proceed by investigating when the (full) Newton step to the target–point $\overline{v}$ can be made without becoming infeasible, i.e., under which conditions the new iterates $x^+ := x + \Delta x$ and $s^+ := s + \Delta s$ are positive.

Lemma 8.2.3 *The Newton step is feasible if $\|\overline{v}^{-2}p_x p_s\|_\infty < 1$. This condition is satisfied if $\delta := \delta(v;\overline{v}) < 1$.*

Proof: Let $0 \leq \alpha \leq 1$ be a step length along the Newton direction. Define $x(\alpha) = x + \alpha\Delta x$ and $s(\alpha) = s + \alpha\Delta s$. Then we have

$$\begin{aligned} x(\alpha)s(\alpha) &= (v + \alpha p_x)(v + \alpha p_s) = v^2 + \alpha v(p_x + p_s) + \alpha^2 p_x p_s \\ &= v^2 + \alpha(\overline{v}^2 - v^2) + \alpha^2 p_x p_s = v^2(1-\alpha) + \alpha\overline{v}^2\left(e + \alpha\overline{v}^{-2}p_x p_s\right). \end{aligned} \tag{8.10}$$

We obtain that $x(\alpha)s(\alpha) > 0$ if $\|\overline{v}^{-2}p_x p_s\|_\infty < 1$ and $\alpha \leq 1$, which proves the first statement. The condition on δ follows from the observation

$$\left\|\frac{p_x p_s}{\overline{v}^2}\right\|_\infty \leq \frac{\|p_x p_s\|_\infty}{\min(\overline{v})^2} \leq \frac{\|p_v\|^2}{4\min(\overline{v})^2} = \delta^2.$$

where the last inequality follows from Lemma 8.2.1. □

Letting $\alpha = 1$ in (8.10) and denoting $(v^+)^2 = x^+s^+$ we get the useful relation

$$(v^+)^2 = \overline{v}^2 + p_x p_s. \tag{8.11}$$

The following lemma shows that if the current iterate v is close enough to the target $\overline{v}$, then the Newton step ensures quadratic convergence of the proximity measure.

Lemma 8.2.4 *Assume that $\delta := \delta(v;\overline{v}) < 1$ and let v^+ result from a Newton step at v with respect to $\overline{v}$. Then it holds*

$$\delta(v^+;\overline{v})^2 \leq \frac{\delta^4}{2\,(1-\delta^2)}.$$

Proof: Lemma 8.2.3 implies that x^+ and s^+ are feasible. For the calculation of $\delta(v^+;\overline{v})$ we need v^+. From (8.11) and Lemma 8.2.1 we get

$$\min(v^+)^2 \geq \min(\overline{v})^2 - \|p_x p_s\|_\infty \geq \min(\overline{v})^2 - \frac{1}{4}\|p_v\|^2 = \min(\overline{v})^2(1-\delta^2). \tag{8.12}$$

Using this relation, (8.8) and (8.11) we derive

$$\begin{aligned} \delta(v^+;\overline{v})^2 &= \frac{1}{4\min(\overline{v})^2}\left\|(v^+)^{-1}\left(\overline{v}^2 - (v^+)^2\right)\right\|^2 = \frac{1}{4\min(\overline{v})^2}\left\|(v^+)^{-1}p_x p_s\right\|^2 \\ &\leq \frac{1}{4\min(\overline{v})^2}\frac{\|p_x p_s\|^2}{\min(v^+)^2}. \end{aligned}$$

Substitution of the bounds in Lemma 8.2.1 and (8.12) yields

$$\delta(v^+;\overline{v})^2 \leq \frac{1}{32\min(\overline{v})^2}\frac{\|p_v\|^4}{\min(\overline{v})^2(1-\delta^2)}.$$

Performing the substitution $\|p_v\| = 2\min(\overline{v})\delta$ gives

$$\delta(v^+;\overline{v})^2 \leq \frac{\delta^4}{2\,(1-\delta^2)},$$

which proves the lemma. □

For $\delta := \delta(v;\overline{v}) < \sqrt{2/3}$ it holds $\delta(v^+;\overline{v}) < \delta$, implying convergence of the sequence of Newton steps, while for $\delta < 1/\sqrt{2}$ it holds $\delta(v^+;\overline{v}) < \delta^2$, guaranteeing quadratic convergence. The Newton step has another important property, namely that the duality gap after the step has the same value as the gap in the target $\overline{v}$.

Lemma 8.2.5 *Let the primal–dual feasible pair (x^+, s^+) be obtained from a full Newton step with respect to $\overline{v}$. The corresponding duality gap achieves its target value: $(x^+)^T s^+ = \|\overline{v}\|^2$.*

Proof: Using (8.11) and orthogonality of p_x and p_s we have

$$(x^+)^T s^+ = e^T (v^+)^2 = e^T \overline{v}^2 + p_x^T p_s = e^T \overline{v}^2 = \|\overline{v}\|^2 .$$

□

This lemma has two important implications. First, if subsequent Newton steps would be taken with $\overline{v}$ fixed then the duality gap would remain constant. Furthermore, if we take only full Newton steps in an algorithm (as is typically done in short–step methods) we do not have to bother about the duality gap in the iterates themselves, it suffices to consider the duality gap in targets.

To complete the general results we analyze the effect on the proximity measure of a Newton step followed by an update in the target. This is technically a bit more easy than analyzing the effect of an update in the target followed by a Newton step, since now we can just use p_v as defined before. Although the latter might seem more natural both approaches are of course equivalent. We perform the analysis in a general setting, such that in the sequel it will be an easy task to apply this theorem and derive polynomial complexity bounds for various applications.

Theorem 8.2.6 *Let v and $\overline{v}$ be such that $\delta := \delta(v; \overline{v}) \leq 1/2$. Let v^+ be obtained from v by a full Newton step with respect to $\overline{v}$ and let $\overline{v}^+ \in \mathrm{IR}^n_{++}$ be arbitrary. Then*

$$\delta(v^+; \overline{v}^+) \leq \frac{\sqrt{6}}{2} \delta(\overline{v}; \overline{v}^+) + \frac{1}{2\sqrt{6}} \frac{\min(\overline{v})}{\min(\overline{v}^+)}.$$

Proof: From Lemma 8.2.3 it follows that v^+ is well–defined. By definition we have

$$\delta(v^+; \overline{v}^+) = \frac{1}{2\min(\overline{v}^+)} \left\| \frac{(\overline{v}^+)^2 - (v^+)^2}{v^+} \right\| .$$

Recall from (8.11) that $(v^+)^2 = \overline{v}^2 + p_x p_s$ and from (8.12) that

$$\min(v^+)^2 \geq \min(\overline{v})^2 (1 - \delta^2). \tag{8.13}$$

Using these and Lemmas 8.2.1 and 8.2.2 gives

$$\begin{aligned}
\delta(v^+;\overline{v}^+) &\leq \frac{1}{2\min(\overline{v}^+)}\left\|\frac{(\overline{v}^+)^2-\overline{v}^2}{\overline{v}}\frac{\overline{v}}{v^+}\right\| + \frac{1}{2\min(\overline{v}^+)}\left\|\frac{p_x p_s}{v^+}\right\| \\
&\leq \delta(\overline{v};\overline{v}^+)\left\|\frac{\overline{v}}{v^+}\right\|_\infty + \frac{1}{2\min(\overline{v}^+)\min(v^+)}\frac{1}{2\sqrt{2}}\|p_v\|^2 \\
&\leq \delta(\overline{v};\overline{v}^+)\rho(\delta(v^+;\overline{v})) + \frac{\min(\overline{v})^2}{\sqrt{2}\min(\overline{v}^+)\min(v^+)}\delta^2 \\
&\leq \delta(\overline{v};\overline{v}^+)\rho(\delta(v^+;\overline{v})) + \frac{\min(\overline{v})}{\min(\overline{v}^+)}\frac{\delta^2}{\sqrt{2(1-\delta^2)}},
\end{aligned}$$

where the last inequality follows from (8.13). Finally, from Lemma 8.2.4 we obtain

$$\delta(v^+;\overline{v}) \leq \frac{\delta^2}{\sqrt{2(1-\delta^2)}}.$$

Substituting $\delta \leq 1/2$ yields $\delta^2/\sqrt{2(1-\delta^2)} \leq 1/(2\sqrt{6})$ and $\rho(\delta(v^+;\overline{v})) \leq \sqrt{6}/2$. This gives the required bound. □

We will later apply this theorem several times in the following way. Given v close to a target $\overline{v}$ such that $\delta(v;\overline{v}) < 1/2$, we need to determine a condition on the new target $\overline{v}^+$ such that v^+ will be in the region of quadratic convergence around $\overline{v}^+$, in other words, such that $\delta(v^+;\overline{v}^+) < 1/2$. The theorem implies that this can be done measuring the proximity $\delta(\overline{v};\overline{v}^+)$ between the targets, and the ratio $\min(\overline{v})/\min(\overline{v}^+)$.

8.2.3 Algorithms using Dikin–affine steps

Motivation

In Section 5.3 we introduced the primal–dual Dikin–affine scaling direction at $\overline{v}$ using the solution of the subproblem

$$\min_{\Delta v}\{\, \overline{v}^T\Delta v \;:\; \|\overline{v}^{-1}\Delta v\| \leq 1 \,\},$$

defined in the v–space. This problem can be interpreted as finding the direction in the v–space that aims at a maximum decrease in the duality gap within the Dikin–ellipsoid in the v–space. The solution Δv is given by $-\overline{v}^3/\|\overline{v}^2\|$. Let us now use the vector field of the primal–dual Dikin direction and its associated set of trajectories. The equation of the trajectory passing through $\overline{v} \in \mathbb{R}^n_{++}$ and tangent to the vector field is given by

$$\Phi(t;\overline{v}) = \frac{\overline{v}}{\sqrt{\overline{v}^2 t + e}}, \qquad t \geq 0. \tag{8.14}$$

It holds $\Phi(0;\overline{v}) = \overline{v}$ and, for $t \to \infty$, $\Phi(t;\overline{v})$ tends to zero tangentially to the vector e. We first show that $\Phi(t;\overline{v})$ defines a path in the v–space, henceforth called the *Dikin-path starting at* $\overline{v}$, and derive some interesting properties.

Lemma 8.2.7 *Let $\Phi(t;\overline{v})$ be as defined in (8.14).*
(i) For any $t_1, t_2 \geq 0$ it holds

$$\Phi(t_1 + t_2;\overline{v}) = \Phi(t_2;\Phi(t_1;\overline{v}));$$

(ii) For any $t \geq 0$ it holds that if $\overline{v}_i \leq \overline{v}_j$ then $\Phi_i(t;\overline{v}) \leq \Phi_j(t;\overline{v})$;
(iii) For any $t \geq 0$ it holds $\omega(\Phi(t;\overline{v})) \geq \omega(\overline{v})$, where $\omega(\cdot)$ is defined in (8.2).

Proof: (i) It holds

$$\Phi(t_1 + t_2;\overline{v})^2 = \frac{\overline{v}^2}{\overline{v}^2(t_1+t_2)+e} = \frac{\frac{\overline{v}^2}{\overline{v}^2 t_1 + e}}{\frac{\overline{v}^2}{\overline{v}^2 t_1 + e} t_2 + e} = \Phi(t_2;\Phi(t_1;\overline{v}))^2.$$

(ii) If $\overline{v}_i \leq \overline{v}_j$ then it also holds $\overline{v}_i^2(\overline{v}_j^2 t + 1) \leq \overline{v}_j^2(\overline{v}_i^2 t + 1)$, from which the statement follows.
(iii) Using the fact from (ii) that the ordering of the coordinates of $\overline{v}$ is the same along the path we have

$$\begin{aligned} \omega(\Phi(t;\overline{v})) &= \frac{\min(\Phi(t;\overline{v}))}{\max(\Phi(t;\overline{v}))} = \frac{\min(\overline{v})}{\max(\overline{v})}\sqrt{\frac{\max(\overline{v})^2 t + 1}{\min(\overline{v})^2 t + 1}} \\ &= \omega(\overline{v})\sqrt{\frac{\max(\overline{v})^2 t + 1}{\min(\overline{v})^2 t + 1}} \geq \omega(\overline{v}). \end{aligned}$$

□

We consider two algorithms. The first is called a *Dikin-path-following method*. Given an initial target $\overline{v}^{(0)}$, the other targets will all be at the Dikin–path starting at $\overline{v}^{(0)}$. The second algorithm we consider uses the tangent at $\overline{v}^{(0)}$ and moves the target with a certain step size in this direction. This brings the new target to a different Dikin–path, from which the algorithm proceeds, see Figure 8.1. We will show that from a complexity point of view both algorithms behave similarly. Observe, that in the case of a weighted path–following method both approaches are equivalent.

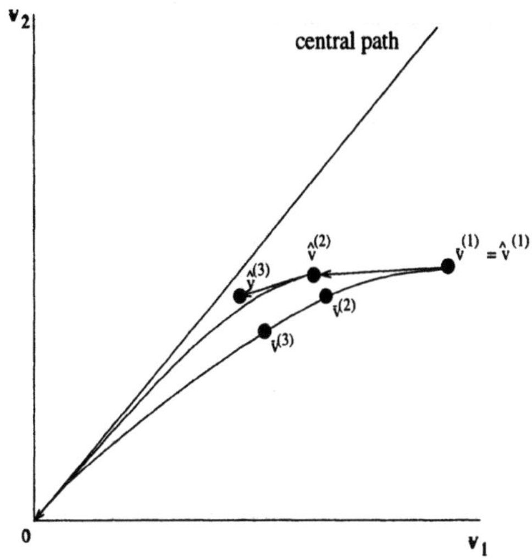

Figure 8.1 The Dikin–path–following in the v–space uses targets $\overline{v}^{(k)}$; the algorithm using Dikin steps has targets $\hat{v}^{(k)}$ on different paths.

Algorithm 1, properties and complexity

Let the initial target be denoted by $\overline{v}^{(0)}$ and let $(x^{(0)}, s^{(0)})$ be such that for $v^{(0)} := \sqrt{x^{(0)}s^{(0)}}$ we have $\delta(v^{(0)}; \overline{v}^{(0)}) \leq 1/2$. The target–sequence is determined by values $t_k > 0$ and the targets are defined by

$$\overline{v}^{(k)} := \Phi(t_k; \overline{v}^{(k-1)}).$$

In view of Lemma 8.2.7(ii) we assume

$$\min(\overline{v}^{(0)}) = \overline{v}_1^{(0)}, \quad \max(\overline{v}^{(0)}) = \overline{v}_n^{(0)},$$

and know that the ordering of the elements of $\overline{v}$ is the same for all targets. Since we are interested in the behavior of Newton's method per iteration we just denote $\overline{v} := \overline{v}^{(k-1)}$, $\overline{v}^+ := \overline{v}^{(k)}$ and $t := t_k$. We also use $\overline{\omega} := \omega(\overline{v})$ (recall the definition from (8.2)). Taking for Δx and Δs the displacements according to a full Newton step with respect to the target–point $\overline{v}$, we can now formally state the algorithm as in Figure 8.2.

From Section 8.2.2 it is clear that the only thing remaining to analyze a target-following method, is to guarantee that a sufficiently large step size in the v–space can be taken, and to use this to compute the number of steps needed by

---

Input

$(x^{(0)}, s^{(0)})$: the initial pair of interior–feasible solutions;

Parameters

ϵ is the accuracy parameter;

t is the step size (default value $\overline{\omega}/(3\sqrt{n}\overline{v}_n^2)$);

begin

$x := x^{(0)}$; $s := s^{(0)}$; $\overline{v} := \sqrt{xs}$;

while $x^T s > \epsilon$ **do**

$\overline{v} := \overline{v}/\sqrt{\overline{v}^2 t + e}$;

compute $(\Delta x, \Delta s)$ from (8.3);

$x := x + \Delta x$;

$s := s + \Delta s$;

end

end.

Figure 8.2 Dikin–path–following algorithm.

the algorithm. Specifically, we should check for which value of t the conditions of Theorem 8.2.6 hold.

Lemma 8.2.8 *Let $\overline{v}^+$ result from a step along the Dikin–path with step size $t := \overline{\omega}/(3\sqrt{n}\overline{v}_n^2)$. Then*

$$\frac{\min(\overline{v})}{\min(\overline{v}^+)} \le \frac{9}{8} \quad \textit{and} \quad \delta(\overline{v}; \overline{v}^+) \le \frac{1}{6}.$$

Proof: Using $\min(\overline{v}) = \overline{v}_1$ and $\min(\overline{v}^+) = \overline{v}_1^+$, the first bound follows from

$$\frac{\overline{v}_1}{\overline{v}_1^+} = \sqrt{\overline{v}_1^2 t + 1} = \sqrt{\overline{\omega}^3/(3\sqrt{n}) + 1} \le \sqrt{\frac{1}{3\sqrt{2}} + 1} < \frac{9}{8}.$$

Furthermore

$$\begin{aligned} \delta(\overline{v}; \overline{v}^+) &= \frac{1}{2\overline{v}_1^+} \left\| \overline{v}^{-1}\left((\overline{v}^+)^2 - \overline{v}^2\right) \right\| = \frac{\sqrt{\overline{v}_1^2 t + 1}}{2\overline{v}_1} \left\| t \frac{\overline{v}^3}{\overline{v}^2 t + e} \right\| \\ &\le \frac{\sqrt{\overline{\omega}^3/(3\sqrt{n}) + 1}}{2\overline{v}_1} \frac{\overline{\omega}}{3\sqrt{n}\overline{v}_n^2} \frac{\overline{v}_n^3}{\overline{\omega}^3/(3\sqrt{n}) + 1} \sqrt{n} \end{aligned}$$

$$= \frac{1}{6\sqrt{\overline{\omega}^3/(3\sqrt{n})}+1} \leq \frac{1}{6}.$$

This completes the proof. □

Assuming $\delta(v;\overline{v}) < 1/2$, combining Theorem 8.2.6 with Lemma 8.2.8 shows that we can compute v^+ in one Newton step such that $\delta(v^+;\overline{v}^+) < 1/2$. We proceed by considering the reduction of the duality gap in the algorithm. Recall from Lemma 8.2.5 that after a full Newton step the duality gap attains its target value, so we only need to consider the duality gaps $e^T\overline{v}^2$ resulting from successive target values. Using this, we prove the following theorem.

Theorem 8.2.9 *Let $(x^{(0)}, s^{(0)})$ be a given initial point and let*

$$\overline{v}^{(0)} := \sqrt{x^{(0)}s^{(0)}} \quad \textit{and} \quad \overline{\omega}_0 := \omega(\overline{v}^{(0)}).$$

If the step size t has in every iteration the value $\overline{\omega}/(3\sqrt{n}\overline{v}_n^2)$ then after at most

$$\mathcal{O}\left(\frac{\sqrt{n}}{\overline{\omega}_0^3}\ln\frac{(x^{(0)})^Ts^{(0)}}{\epsilon}\right)$$

iterations the Dikin-path-following algorithm stops with a positive primal-dual pair (x^*, s^*) satisfying $(x^*)^Ts^* \leq \epsilon$.

Proof: At the start of the algorithm the duality gap is given by

$$(x^{(0)})^Ts^{(0)} = \|\overline{v}^{(0)}\|^2.$$

If, as before, the target-point at the beginning of some iteration is denoted as $\overline{v}$ and at the end of the same iteration as $\overline{v}^+$, then we have

$$e^T(\overline{v}^+)^2 = e^T\left(\frac{\overline{v}^2}{\overline{v}^2t+e}\right) \leq \frac{e^T\overline{v}^2}{v_1^2t+1} = \frac{e^T\overline{v}^2}{\overline{\omega}^3/(3\sqrt{n})+1},$$

where $\overline{\omega} := \omega(\overline{v})$. Since $\overline{\omega} \geq \overline{\omega}_0$ by Lemma 8.2.7(*iii*), at each iteration the duality gap is reduced by at least the factor

$$\frac{1}{1+\overline{\omega}_0^3/(3\sqrt{n})}.$$

From this the theorem follows. □

The theorem implies that the target-following algorithm runs in $\mathcal{O}(\sqrt{n}\ln 1/\epsilon)$ iterations whenever $(x^{(0)})^Ts^{(0)} = \mathcal{O}(1)$ and $\overline{\omega}_0 = \Omega(1)$. Unfortunately, whenever $\overline{\omega}_0$ is smaller than $\mathcal{O}(1)$ the complexity bound is heavily negatively influenced. We will later show how the bound can be improved by adjusting the analysis and using the fact that the proximity $\overline{\omega}$ increases along the Dikin-path.

Algorithm 2, properties and complexity

The second algorithm we consider determines the target–sequence by moving from one target to the other using tangents to successive Dikin–paths. Specifically, given a current target $\overline{v}$ let us define the next target by the *Dikin step*

$$\overline{v}^+ := \overline{v} - \alpha \frac{\overline{v}^3}{\|\overline{v}^2\|} = \overline{v}\left(e - \alpha \frac{\overline{v}^2}{\|\overline{v}^2\|}\right),$$

for some positive number α. Since we require $\overline{v}^+$ to be positive, it is well defined only if

$$\alpha < \alpha_{\max} := \frac{\|\overline{v}^2\|}{\max(\overline{v})^2}.$$

Defining the step size θ by $\theta := \alpha/\alpha_{\max}$ we have $0 < \theta < 1$ and

$$\overline{v}^+ := \overline{v} - \theta \frac{\overline{v}^3}{\max(\overline{v})^2} = \overline{v}\left(e - \theta \frac{\overline{v}^2}{\max(\overline{v})^2}\right). \tag{8.15}$$

Note that each element of $\overline{v}^+$ is smaller than the corresponding element of $\overline{v}$. This property is important, since the Newton process in the (x, s)–space forces equality between the duality gap and $e^T(\overline{v}^+)^2$, see Lemma 8.2.5. So the duality gap will be decreasing and is bounded by

$$\|\overline{v}\|(1-\theta) \le \|\overline{v}^+\| \le \|\overline{v}\|\left(1 - \theta \frac{\min(\overline{v})^2}{\max(\overline{v})^2}\right) = \|\overline{v}\|\left(1 - \theta\overline{\omega}^2\right), \tag{8.16}$$

where $\overline{\omega} := \omega(\overline{v})$. If we choose $\theta \le 1/3$ then the Dikin step has two interesting properties, which are similar to the ones in Lemma 8.2.7: it preserves the ranking of the coordinates of $\overline{v}$, and it causes the ratio $\overline{\omega}$ to increase monotonically.

Lemma 8.2.10 *Assume that* $0 < \overline{v}_1 \le \overline{v}_2 \le \cdots \le \overline{v}_n$ *and let* $\theta \le 1/3$. *Then*

$$0 < \overline{v}_1^+ \le \overline{v}_2^+ \le \cdots \le \overline{v}_n^+.$$

Proof: Let $i < j$. We have

$$\begin{aligned} \overline{v}_j^+ - \overline{v}_i^+ &= \overline{v}_j - \overline{v}_i - \frac{\theta}{\overline{v}_n^2}\left(\overline{v}_j^3 - \overline{v}_i^3\right) = (\overline{v}_j - \overline{v}_i)\left(1 - \frac{\theta}{\overline{v}_n^2}\left(\overline{v}_j^2 + \overline{v}_i\overline{v}_j + \overline{v}_i^2\right)\right) \\ &\ge (\overline{v}_j - \overline{v}_i)(1 - 3\theta) \ge 0. \end{aligned}$$

Thus it follows that $\overline{v}_i^+ \le \overline{v}_j^+$ with equality if and only if $\overline{v}_j = \overline{v}_i$. □

Remark 8.2.11 *An alternative proof of Lemma 8.2.10 can be given using the function $\phi(t) = t(1-\theta t^2)/(1-\theta)$. Assuming $\overline{v}_n = 1$, after the Dikin step it holds $\overline{v}_i^+ = \phi(\overline{v}_i)$ if the maximal component of $\overline{v}^+$ is rescaled to 1. The function $\phi(t)$ is monotonically increasing and concave for $\theta \leq 1/3$.*

In the sequel we use $\theta \leq 1/3$, hence we may assume that the coordinates of $\overline{v}$ are ranked as in Lemma 8.2.10. So $\overline{v}_1$ is the smallest and $\overline{v}_n$ the largest element of $\overline{v}$ and $\overline{\omega} = \overline{v}_1/\overline{v}_n$.

Lemma 8.2.12 *Assume that $\theta \leq 1/3$ and let $\overline{\omega}^+ := \omega(\overline{v}^+)$. Then*

$$\overline{\omega}^+ = \left(\frac{1-\theta\overline{\omega}^2}{1-\theta}\right)\overline{\omega} \geq \overline{\omega}, \tag{8.17}$$

and

$$1-\overline{\omega}^+ \leq \left(1-\frac{\theta\overline{\omega}}{1-\theta}\right)(1-\overline{\omega}). \tag{8.18}$$

Proof: Since $\theta \leq 1/3$ Lemma 8.2.10 implies that $\overline{\omega}^+ = \overline{v}_1^+/\overline{v}_n^+$. Hence, from the definition of $\overline{v}_1^+$ and $\overline{v}_n^+$ we get

$$\overline{\omega}^+ = \frac{\overline{v}_1^+}{\overline{v}_n^+} = \frac{\overline{v}_1}{\overline{v}_n}\frac{1-\theta\overline{\omega}^2}{1-\theta} = \left(\frac{1-\theta\overline{\omega}^2}{1-\theta}\right)\overline{\omega} \geq \overline{\omega}.$$

For (8.18), note that

$$\begin{aligned} 1-\overline{\omega}^+ &= 1-\frac{1-\theta\overline{\omega}^2}{1-\theta}\overline{\omega} = \frac{1-\theta-\overline{\omega}+\theta\overline{\omega}^3}{1-\theta} = \frac{1-\theta(1+\overline{\omega}+\overline{\omega}^2)}{1-\theta}(1-\overline{\omega}) \\ &= \left(1-\frac{\theta(\overline{\omega}+\overline{\omega}^2)}{1-\theta}\right)(1-\overline{\omega}) \leq \left(1-\frac{\theta\omega}{1-\theta}\right)(1-\overline{\omega}). \end{aligned}$$

This proves the lemma. □

Remark 8.2.13 *If we use a value $\theta > 1/3$, the ranking of $\overline{v}$ may not be preserved and the proof of Lemma 8.2.12 does not go through. However, it is still possible to prove the monotonicity of $\overline{\omega}$ for $\theta \leq 1/2$. We omit the proof since this property will not be used in the analysis.*

Again it is important to analyze the influence of a target update on the proximity measure by applying Theorem 8.2.6.

Lemma 8.2.14 *Let $\overline{v}^+$ result from a Dikin step at $\overline{v}$ with step size $\theta \leq 1/3$ using (8.15). Then*

$$\frac{\min(\overline{v})}{\min(\overline{v}^+)} \leq \frac{1}{1-\theta} \quad \textit{and} \quad \delta(\overline{v};\overline{v}^+) \leq \frac{1}{1-\theta}\frac{\theta\sqrt{n}}{\overline{\omega}}.$$

Proof: Lemma 8.2.10 guarantees the same ordering for $\overline{v}^+$ as for $\overline{v}$. So

$$\min(\overline{v}^+) = \overline{v}_1^+ = \overline{v}_1(1-\theta\overline{\omega}^2) \geq \overline{v}_1(1-\theta) = \min(\overline{v})(1-\theta). \tag{8.19}$$

By definition and Lemma 8.2.10 we have

$$\delta(\overline{v};\overline{v}^+) = \frac{1}{2\overline{v}_1^+}\left\|\overline{v}^{-1}\left((\overline{v}^+)^2 - \overline{v}^2\right)\right\|.$$

Since $\overline{v}^+ < \overline{v}$ it holds $\overline{v}^+ + \overline{v} < 2\overline{v} \leq 2\overline{v}_n e$. Using also the definition of $\overline{v}^+$ we get

$$\left\|\overline{v}^{-1}\left((\overline{v}^+)^2 - \overline{v}^2\right)\right\| = \left\|\overline{v}^{-1}(\overline{v}^+ + \overline{v})(\overline{v}^+ - \overline{v})\right\| \leq 2\theta\overline{v}_n\left\|\overline{v}^{-1}\frac{\overline{v}^3}{\overline{v}_n^2}\right\| \leq 2\theta\overline{v}_n\sqrt{n}.$$

Using (8.19) we obtain

$$\delta(\overline{v};\overline{v}^+) \leq \frac{1}{2(1-\theta)\overline{v}_1}2\theta\overline{v}_n\sqrt{n} = \frac{1}{1-\theta}\frac{\theta\sqrt{n}}{\overline{\omega}}.$$

This proves the lemma. □

Assuming $\delta(v;\overline{v}) \leq 1/2$ and taking $\theta = \overline{\omega}/(6\sqrt{n})$, application of Theorem 8.2.6 gives $\delta(v^+;\overline{v}^+) \leq 1/2$. Since $\overline{\omega}$ increases during the course of the algorithm (Lemma 8.2.12) the default value $\theta = \overline{\omega}_0/(6\sqrt{n})$ guarantees that one Newton step per target update is sufficient. The following theorem can be proved analogous to Theorem 8.2.9.

Theorem 8.2.15 *Let $(x^{(0)}, s^{(0)})$ be a given initial point and let*

$$\overline{v}^{(0)} := \sqrt{x^{(0)}s^{(0)}} \quad \textit{and} \quad \overline{\omega}_0 := \omega(\overline{v}^{(0)}).$$

If the step size θ has its default value $\overline{\omega}_0/(6\sqrt{n})$ in every iteration then after at most

$$\mathcal{O}\left(\frac{\sqrt{n}}{\overline{\omega}_0^3}\ln\frac{(x^{(0)})^T s^{(0)}}{\epsilon}\right)$$

iterations the algorithm using Dikin steps stops with a positive primal–dual pair (x^*, s^*) satisfying $(x^*)^T s^* \leq \epsilon$.

Comparing Theorem 8.2.15 with Theorem 8.2.9 we see that this target-following algorithm has exactly the same complexity as the Dikin-path-following method analyzed before. Still, there is a major conceptual difference between the two algorithms, since one chooses its targets on one smooth path, while the other has targets on various Dikin-paths. Moreover, when starting at the same point in the v-space, a Dikin step as in the second algorithm moves the target closer to the central path than a step along the Dikin-path; this can be verified by comparing the values of $\overline{\omega}^+$ in Lemma 8.2.7(*iii*) and Lemma 8.2.12.

Improved analysis and complexity

Unfortunately, when $\overline{\omega}_0$ is smaller than $\Omega(1)$, the complexity bound of the target-following algorithms considered above is highly affected. For instance, when $\overline{\omega}_0 = \Omega(1/\sqrt{n})$ we only obtain $\mathcal{O}(n^2 \ln 1/\epsilon)$ iteration algorithms. However, the straightforward analysis given above can be improved significantly to yield a bound of

$$\mathcal{O}\left(\sqrt{n}\left(\frac{1}{\overline{\omega}_0}\ln\frac{1}{\overline{\omega}_0} + \ln\frac{(x^{(0)})^T s^{(0)}}{\epsilon}\right)\right)$$

iterations, using the fact that $\overline{\omega}$ increases in each iteration. Actually, $\overline{\omega}$ will reach a value of constant order in a limited number of steps, as is clear from Lemmas 8.2.7(*iii*) and 8.2.12. From that point on we can use this new value to bound $\overline{\omega}$ from below. The first goal is thus to bound the number of iterations to have $\overline{\omega}$ 'close to' 1. We only show this procedure and analysis for the second Dikin-type algorithm. Similar results are straightforwardly obtained for the Dikin-path-following method.

Lemma 8.2.16 *Let $\theta \le 1/3$. After at most*

$$\mathcal{O}\left(\frac{1}{\theta}\ln\frac{1}{\overline{\omega}_0}\right)$$

target updates using Dikin steps with step size θ we have $\overline{\omega}^2 \ge 1/2$.

Proof: Using (8.17) we have for $\overline{\omega}^2 \le 1/2$

$$\frac{1-\theta\overline{\omega}^2}{1-\theta} \ge \frac{1-\theta/2}{1-\theta} = 1 + \frac{\theta/2}{1-\theta}.$$

So $\overline{\omega}^2 \ge 1/2$ will certainly hold if

$$\left(1+\frac{\theta/2}{1-\theta}\right)^{2k} (\overline{\omega}_0)^2 \ge 1/2,$$

or equivalently, if

$$2k \ln\left(1+\frac{\theta/2}{1-\theta}\right) \geq \ln\left(\frac{1/2}{(\overline{\omega}_0)^2}\right).$$

Using $\ln(1+t) > t/2$ for $t < 1$, this will certainly be satisfied if

$$k\frac{\theta/2}{1-\theta} \geq \ln\left(\frac{1/2}{(\overline{\omega}_0)^2}\right).$$

Hence we find that the number of iterations required is at most

$$\frac{2(1-\theta)}{\theta}\ln\left(\frac{1}{2(\overline{\omega}_0)^2}\right),$$

which is of the order specified in the lemma. □

From the discussion succeeding Lemma 8.2.14 we know that $\theta = \overline{\omega}_0/(6\sqrt{n})$ is an acceptable choice. Thus we reach a point with $\overline{\omega}^2 \geq 1/2$ in $\mathcal{O}\left((\sqrt{n}/\overline{\omega}_0)\ln 1/\overline{\omega}_0\right)$ iterations; in that process $\overline{v}$ and hence $e^T\overline{v}^2$ decreases. From then on, we can use $\theta = 1/(6\sqrt{2n})$ and we need $\mathcal{O}(\sqrt{n}\ln((x^0)^T s^0)/\epsilon)$ more iterations to terminate. We have proved the following theorem.

Theorem 8.2.17 *The algorithm tracing targets determined by Dikin steps requires at most*

$$\mathcal{O}\left(\sqrt{n}\left(\frac{1}{\overline{\omega}_0}\ln\frac{1}{\overline{\omega}_0}+\ln\frac{(x^{(0)})^T s^{(0)}}{\epsilon}\right)\right)$$

iterations to obtain an ϵ–approximate solution.

Unfortunately, this complexity bound is not better than the one obtained for weighted path–following algorithms (see Ding and Li [45] or Section 8.2.4); still, the new algorithm has the advantage of generating, in theory and in practice, increasingly centered pairs. Let us define 'close to the central path' by requiring that the iterate is in the region of quadratic convergence of some point on the central path. We can relate 'closeness' to the value of ω as follows.

Lemma 8.2.18 *If $\omega := \omega(v) \geq n/(n+1)$, then there exists a target–point $\overline{v}$ on the central path such that $\delta := \delta(v;\overline{v}) < 1/\sqrt{2}$.*

Proof: If $\overline{v}^2 = \mu e$ for some $\mu > 0$ then δ reduces to

$$\frac{1}{2}\left\|\sqrt{\mu}v^{-1} - \frac{v}{\sqrt{\mu}}\right\|.$$

This measure is minimal for $\mu = \|v\| \,/\, \|v^{-1}\|$ with value

$$\frac{1}{\sqrt{2}}\sqrt{\|v\|\;\|v^{-1}\| - n}.$$

Hence we will have $\delta \le 1/\sqrt{2}$ if

$$\|v\|\,\|v^{-1}\| - n \le 1.$$

Using the bounds $\|v\| \le \sqrt{n}\max(v)$ and $\|v^{-1}\| \le \sqrt{n}/\min(v)$, this implies that it suffices to have $1/\omega \le n + 1/n$. □

The next lemma estimates the number of updates needed to reach a target with $\overline{\omega} \ge n/(n+1)$.

Lemma 8.2.19 *Let $\theta \le 1/3$. After at most*

$$\mathcal{O}\left(\frac{1}{\theta\overline{\omega}_0}\ln(n+1)\right)$$

iterations we have $\overline{\omega} \ge n/(n+1)$.

Proof: From equation (8.18) we need k to satisfy

$$(1-\overline{\omega}^{(k)}) \le \left(1 - \frac{\theta\overline{\omega}_0}{1-\theta}\right)^k (1-\overline{\omega}_0) \le \frac{1}{n+1}.$$

Taking logarithms and using $\ln(1-t) \le -t$ for $t < 1$ we obtain that k should satisfy

$$k \ge \frac{1-\theta}{\theta\overline{\omega}_0}\ln((n+1)(1-\overline{\omega}_0)),$$

which gives the order in the lemma. □

Other scaling factors

The analogy between the Dikin–affine scaling algorithm and a target–following algorithm using Dikin steps suggests that a family of target–following analogons for the family of algorithms derived in Section 6.3 is possible. So, we consider algorithms with ν–order scaling and target updates of the form

$$\overline{v}^{+} = \overline{v}\left(e - \theta\frac{\overline{v}^{2\nu}}{\overline{v}_n^{2\nu}}\right), \quad \nu \ge 0. \tag{8.20}$$

In this setting the Dikin step has $\nu = 1$ and weighted path–following has $\nu = 0$. Again it is easy to analyze the resulting algorithms. We assume that $\nu = \mathcal{O}(1)$, since otherwise the computations may require exponentially large or small numbers, and the step size might become exponentially small. First observe that

$$\|\overline{v}^+\| \leq \|\overline{v}\|(1 - \theta\overline{\omega}^{2\nu}).$$

It is left to the reader to verify the following lemmas, which can be proved similarly as in the case $\nu = 1$.

Lemma 8.2.20 *If $\theta \leq 1/(2\nu+1)$ then $\overline{v}^+$ has the same ranking as $\overline{v}$; moreover, $\overline{\omega}^+ \geq \overline{\omega}$ with equality only if $\overline{\omega} = 1$.*

Lemma 8.2.21 *Let $\overline{v}^+$ result from $\overline{v}$ by a target update using (8.20) with step size $\theta \leq 1/(2\nu+1)$. Then*

$$\frac{\min(\overline{v})}{\min(\overline{v}^+)} \leq \frac{1}{1-\theta} \quad \text{and} \quad \delta(\overline{v};\overline{v}^+) \leq \frac{1}{1-\theta}\frac{\theta\sqrt{n}}{\overline{\omega}}.$$

We find that the algorithm using ν–order scaling for the target update requires

$$\mathcal{O}\left(\frac{\sqrt{n}}{\overline{\omega}_0^{2\nu+1}} \ln \frac{(x^{(0)})^T s^{(0)}}{\epsilon}\right)$$

iterations to obtain an ϵ–approximate solution. In a similar way as in Lemma 8.2.16 and Theorem 8.2.17 we can improve the convergence analysis and improve the complexity bound to

$$\mathcal{O}\left(\sqrt{n}\left(\frac{1}{\overline{\omega}_0}\ln\frac{1}{\overline{\omega}_0} + \ln\frac{(x^{(0)})^T s^{(0)}}{\epsilon}\right)\right).$$

Comparing the resulting complexity bound with the ones derived for the affine versions in Section 6.3 we observe a (seemingly) striking difference. In the affine versions the complexity depends on ν and is best for $\nu = 1$. In the path–following variants the worst–case bound does not depend on the value of the scaling while the theory allows the largest possible step for $\nu = 0$. However, observe that for $\nu = 0$ the proximity in terms of $\overline{\omega}$ does not improve, hence the worst–case bound in terms of $\overline{\omega}$ is tight; for variants using $\nu > 0$ the proximity to the central path improves, which implies that in practice larger steps can be taken than predicted by the theory.

8.2.4 Other applications

We now apply the approach taken in Section 8.2.2 to various primal–dual algorithms found in the literature, and to some new primal–dual variants of pure primal or dual methods that appear in the literature. The reader should recall that the missing element to complete the convergence analysis of a target–following method is to determine the step size that can be taken. The step size is obtained from the condition that after a Newton step the iterate should be close to an updated target, in the sense that it belongs to the region of quadratic convergence around the target (cf. Theorem 8.2.6). The number of iterations required then follows from analyzing the effect of the step size on the measure of progress. Some of the applications are also given in Mizuno [174, 176].

Path–following methods

The standard path–following methods were derived and analyzed by Monteiro and Adler [180] and Kojima et al. [139], being inspired by studies on the central path by Megiddo [168] and Bayer and Lagarias [17], among others. Ding and Li [45] analyzed primal–dual weighted path–following methods [45] (see also Mizuno [174]; a primal version was studied by Den Hertog et al. [104, 208]). In the weighted path–following methods the centering phase is by–passed and the iterates keep (approximately) the distance to the path as in the initial point. Let $\overline{v}^{(0)}$ be given; define

$$\overline{v}^{(k)} = \sqrt{1-\theta}\,\overline{v}^{(k-1)}$$

for some $0 < \theta < 1$. It is evident that

$$\omega(\overline{v}^{(k)}) = \omega(\overline{v}^{(0)}), \quad \forall k.$$

Lemma 8.2.22 *Let $\overline{v}$ be given and let $\overline{\omega} = \min(\overline{v})/\max(\overline{v})$; using the target update $\overline{v}^{+} = \sqrt{1-\theta}\,\overline{v}$, we have*

$$\frac{\min(\overline{v})}{\min(\overline{v}^{+})} = \frac{1}{\sqrt{1-\theta}} \quad \textit{and} \quad \delta(\overline{v};\overline{v}^{+}) \le \frac{1}{2\sqrt{1-\theta}}\frac{\theta\sqrt{n}}{\overline{\omega}}.$$

Proof: The first statement is trivial. The second follows from

$$\begin{aligned}
\delta(\overline{v};\overline{v}^{+}) &= \frac{1}{2\sqrt{1-\theta}\min(\overline{v})}\left\|\frac{(1-\theta)\overline{v}^2-\overline{v}^2}{\overline{v}}\right\| \\
&= \frac{1}{2\sqrt{1-\theta}\min(\overline{v})}\|\theta\overline{v}\| \le \frac{1}{2\sqrt{1-\theta}}\frac{\theta\sqrt{n}}{\overline{\omega}}.
\end{aligned}$$

□

Combining Lemma 8.2.22 with Theorem 8.2.6 gives that $\delta(v^+;\overline{v}^+) < 1/2$ for $\theta = \overline{\omega}/(3\sqrt{n})$. Since $\|\overline{v}^+\|^2 = (1-\theta)\,\|\overline{v}\|^2$, we get by Lemma 8.2.5 that the number of iterations required for the algorithm is $\mathcal{O}(\sqrt{n}/\overline{\omega}\ln(x^{(0)})^T s^{(0)}/\epsilon)$. Note that for central path–following methods $\overline{\omega} = 1$, so the complexity bound is negatively influenced by non–central starting points. The bound is in accordance with [45] for weighted path–following.

Cone–affine scaling

Recently, Sturm and Zhang [224] proposed a new search–direction, which they used in a *cone–affine scaling method.* Their direction is a linear combination of the primal–dual affine scaling direction and a new centering direction (cf. Section 6.6). Here we analyze a method following a target–sequence constructed with cone–affine scaling steps. The target update is as follows. Let $\overline{v} := \overline{v}^{(k-1)}$, denote $\overline{v}^+ := \overline{v}^{(k)}$, and define

$$\overline{v}^+ := \sqrt{\theta \min(\overline{v})\overline{v}}, \tag{8.21}$$

for some $\theta < 1$. The new duality gap satisfies

$$e^T(\overline{v}^+)^2 = \theta \min(\overline{v}) e^T \overline{v} \leq \theta e^T \overline{v}^2,$$

hence the algorithm requires at most $\mathcal{O}(1/(1-\theta)\ln(e^T(\overline{v}^{(0)})^2/\epsilon))$ iterations to obtain an ϵ–accurate solution. As in the Dikin–path–following algorithms the ordering of the elements of the targets remains the same:

$$\overline{v}_1 \leq \ldots \leq \overline{v}_n \quad \Longrightarrow \quad \overline{v}_1^+ \leq \ldots \leq \overline{v}_n^+.$$

For the ratio $\omega(\overline{v}^+)$ we derive

$$\omega(\overline{v}^+) = \frac{\overline{v}_1^+}{\overline{v}_n^+} = \frac{\sqrt{\overline{v}_1}}{\sqrt{\overline{v}_n}} = \sqrt{\omega(\overline{v})} \geq \omega(\overline{v}).$$

Lemma 8.2.23 *Let $\overline{v}$ be given and let $\overline{\omega} = \min(\overline{v})/\max(\overline{v})$; using the target update (8.21) we have*

$$\frac{\min(\overline{v})}{\min(\overline{v}^+)} = \frac{1}{\sqrt{\theta}} \quad \textit{and} \quad \delta(\overline{v};\overline{v}^+) \leq \frac{1}{2\sqrt{\theta}}\left(\frac{1}{\overline{\omega}} - \theta\right)\sqrt{n}.$$

Proof: The first statement is trivial. The second follows from

$$\begin{aligned}\delta(\overline{v};\overline{v}^+) &= \frac{1}{2\sqrt{\theta}\min(\overline{v})}\left\|\frac{\theta\min(\overline{v})\overline{v}-\overline{v}^2}{\overline{v}}\right\| = \frac{1}{2\sqrt{\theta}\min(\overline{v})}\|\theta\min(\overline{v})e-\overline{v}\| \\ &\leq \frac{1}{2\sqrt{\theta}}\left(\frac{\max(\overline{v})}{\min(\overline{v})}-\theta\right)\sqrt{n} = \frac{1}{2\sqrt{\theta}}\left(\frac{1}{\overline{\omega}}-\theta\right)\sqrt{n}.\end{aligned}$$

□

Applying Theorem 8.2.6 with the bounds in Lemma 8.2.23 we can compute the minimal value of θ such that $\delta(v^+;\overline{v}^+) \leq 1/2$ will hold given $\delta(v;\overline{v}) \leq 1/2$. Unfortunately, this requires a condition on $\overline{\omega}$. If we require

$$\frac{1}{\overline{\omega}} \leq 1 + \frac{1}{5\sqrt{n}},$$

and choose $\theta = 1 - 1/(5\sqrt{n})$, then it holds $\delta(v^+;\overline{v}^+) \leq 1/2$ and the algorithm has an $\mathcal{O}(\sqrt{n}\ln 1/\epsilon)$ iteration complexity. Observe that even in a target–following framework the algorithm is required to stay in a small neighborhood of the central path.

Freund's shifted barrier method

Freund [59] analyzes a shifted barrier method for the primal LP problem, to allow the logarithmic barrier method to start with an infeasible point. We will outline his method and then analyze a primal–dual variant. Let $x^{(0)}$ be given such that $Ax^{(0)} = b$ and define $h \in \mathbb{R}^n$ and $\mu_0 \in \mathbb{R}$ such that $x^{(0)} + \mu_0 h > 0$. As Freund, we make the following assumption.

Assumption 8.2.24 *The shift h is chosen such that for all dual feasible slacks s the condition $\|hs\| \leq \sqrt{n}$ holds.*

Note that the assumption can be satisfied if the dual feasible region is bounded. Freund shows, that when an approximation $\overline{s}$ to the analytic center of the (bounded) dual feasible region is known then the algorithm can be started with this approximation, the shift $h = \overline{s}^{-1}/n$ and a suitable value for μ_0. The system to be (approximately) solved in an iteration is given by

$$\begin{aligned}Ax &= b, \quad x + \mu h \geq 0, \\ A^T y + s &= c, \quad s \geq 0, \\ (x + \mu h)s &= \mu e.\end{aligned}$$

While Freund's algorithm does not necessarily generate feasible dual iterates in each iteration, the primal–dual variant does. The main task it to estimate the effect of updating the target $\sqrt{\mu}e$ in the v–space, in which we use the distance measure

$$\delta_F(x+\mu h, s; \sqrt{\mu}e) := \frac{1}{2\sqrt{\mu}} \left\| \frac{(x+\mu h)s - \mu e}{\sqrt{(x+\mu h)s}} \right\|.$$

We define the following notion.

Definition 8.2.25 *The vectors x and s are called (μ, β)–approximate solutions if*

$$\begin{aligned} Ax &= b, \quad x+\mu h \geq 0, \\ A^T y + s &= c, \quad s \geq 0, \end{aligned}$$

and $\delta_F(x+\mu h, s; \sqrt{\mu}e) \leq \beta$ for some constant $\beta < 1$.

We have the following lemma.

Lemma 8.2.26 *Let x and s be $(\mu, 1/4)$–approximate solutions and let $\mu^+ = (1-\theta)\mu$ for $\theta = 1/(16\sqrt{n})$. Then we can compute $(\mu^+, 1/4)$–approximate solutions x^+ and s^+ with one Newton step.*

Proof: Using Lemma 8.2.2 it holds for all i

$$\frac{1}{\rho} \leq \sqrt{\frac{\mu}{(x_i+\mu h_i)s_i}} \leq \rho$$

where $\rho := 1/4 + \sqrt{1+(1/4)^2} \leq \sqrt{5/3}$. Consequently,

$$(x_i+\mu h_i)s_i \geq \frac{\mu}{\rho^2} \geq \frac{3}{5}\mu.$$

Then,

$$\begin{aligned} (x_i+\mu^+ h_i)s_i &= (x_i+\mu h_i)s_i + (\mu^+ - \mu)h_i s_i \geq \frac{3}{5}\mu - \theta\mu h_i s_i \\ &\geq \frac{3}{5}\mu - \theta\mu\sqrt{n} = \left(\frac{3}{5} - \frac{1}{16}\right)\mu = \frac{43}{80}\mu > 0, \end{aligned}$$

so we can use the pair $(x+\mu^+h, s)$ as starting point for Newton's method toward the new target $\sqrt{\mu^+ e}$. We first establish that this pair is still close to the current target $\sqrt{\mu e}$:

$$\begin{aligned}
\delta_F(x+\mu^+h, s; \sqrt{\mu e}) &= \frac{1}{2\sqrt{\mu}}\left\|\frac{(x+\mu^+h)s-\mu e}{\sqrt{(x+\mu^+h)s}}\right\| \\
&= \frac{1}{2\sqrt{\mu}}\left\|\frac{(x+\mu h)s-\mu e+(\mu^+-\mu)hs}{\sqrt{(x+\mu^+h)s}}\right\| \\
&\le \delta_F(x+\mu h, s; \sqrt{\mu e})\left\|\sqrt{\frac{(x+\mu h)s}{(x+\mu^+h)s}}\right\|_\infty \\
&\quad + \frac{\theta\mu}{2\sqrt{\mu}}\left\|\frac{hs}{\sqrt{(x+\mu^+h)s}}\right\| \\
&\le \frac{1}{4}\frac{\rho\sqrt{\mu}}{\sqrt{43\mu/80}} + \frac{\theta\mu}{2\sqrt{\mu}}\frac{\sqrt{n}}{\sqrt{43\mu/80}} \\
&\le \sqrt{\frac{80}{43}}\left(\frac{1}{4}\sqrt{\frac{5}{3}} + \frac{1}{2}\frac{1}{16}\right) < \frac{1}{2}.
\end{aligned}$$

Let $(x^+ + \mu^+h, s^+)$ result from a Newton step with respect to the new target $\sqrt{\mu^+ e}$. Then Theorem 8.2.6 implies

$$\delta_F(x^+ + \mu^+h, s^+; \sqrt{\mu^+ e}) \le \frac{\sqrt{6}}{2}\frac{\theta\sqrt{n}}{2\sqrt{1-\theta}} + \frac{1}{2\sqrt{6}}\frac{1}{\sqrt{1-\theta}}.$$

Since, $1-\theta \ge 1 - 1/(16\sqrt{2})$, we have $1/\sqrt{1-\theta} \le 44/43$, hence

$$\delta_F(x^+ + \mu^+h, s^+; \sqrt{\mu^+ e}) \le \frac{\sqrt{6}}{2}\frac{1}{16}\frac{44}{2\cdot 43} + \frac{1}{2\sqrt{6}}\frac{44}{43} < \frac{1}{4}.$$

So, the pair (x^+, s^+) is a $(\mu^+, 1/4)$–approximate solution. □

We let the algorithm run until $(x+\mu h)^T s \le \epsilon$; from the condition of approximate solutions it then follows that $n\mu \le 2\epsilon$. Hence after $\mathcal{O}(\sqrt{n}\ln(1/\epsilon))$ iterations the algorithm has generated μ^* and a pair (x^*, s^*) such that

$$(x^*)^T s^* = (x^* + \mu^* h)^T s^* - \mu^* h^T s^* \le \epsilon + \mu^*\sqrt{n}\|hs^*\| \le \epsilon + n\mu^* \le 3\epsilon,$$

and

$$x^* = x^* + \mu^* h - \mu^* h \ge -\mu^* h \ge -\frac{2\epsilon}{n}\|h\|_\infty.$$

Hence the pair (x^*, s^*) is an approximately feasible and approximately optimal solution if ϵ is chosen sufficiently small.

Efficient centering

The next application of the target–following concept is the problem of efficient centering, which is stated as follows. Given an arbitrary interior–feasible point (x, s) compute a point close to the central path. In this section we give a simple analysis of an algorithm, independently proposed by Den Hertog [100] and Mizuno [174]. The idea of the algorithm is to successively increase the smaller elements of the target–vector until they all become equal to the largest element. Let $(\overline{v}^{(0)})^2 = x^{(0)}s^{(0)}$ be given; update $\overline{v}$ to obtain $\overline{v}^+$ as follows:

$$\overline{v}_i^+ = \max(\overline{v}_i, \sqrt{1+\theta}\min(\overline{v})), \quad i = 1, \ldots, n; \tag{8.22}$$

if $\min(\overline{v}^+) > \max(\overline{v})$, then we set $\overline{v}^+ = \max(\overline{v})\, e$ which is on the central path. We denote $\overline{\omega} := \omega(\overline{v})$ and $\overline{\omega}^+ := \omega(\overline{v}^+)$. The goal of the algorithm is to obtain a vector which is a multiple of the all–one vector. Since

$$\left(\frac{\max(\overline{v}^+)}{\min(\overline{v}^+)}\right)^2 \leq \frac{1}{1+\theta}\left(\frac{\max(\overline{v})}{\min(\overline{v})}\right)^2,$$

or equivalently $(\overline{\omega}^+)^2 \geq (1+\theta)\overline{\omega}^2$, it follows that reaching this goal will require at most

$$\mathcal{O}\left(\frac{1}{\theta}\ln\frac{1}{\overline{\omega}_0}\right)$$

iterations, where $\overline{\omega}_0 = \omega(\overline{v}^{(0)})$. The appropriate value of θ is determined from the following lemma.

Lemma 8.2.27 *Let $\overline{v}$ be given; using the target update (8.22) we have*

$$\frac{\min(\overline{v})}{\min(\overline{v}^+)} \leq 1 \quad \textit{and} \quad \delta(\overline{v}; \overline{v}^+) \leq \frac{1}{2}\theta\sqrt{n}.$$

Proof: If we are not at the last iteration then from (8.22) it follows for any i

$$\overline{v}_i^+ \geq \sqrt{1+\theta}\min(\overline{v}) \geq \min(\overline{v});$$

when $\overline{v}^+ = \max(\overline{v})e$ at the last iteration we have $\overline{v}_i^+ \geq \min(\overline{v})$, hence the first bound. Let J be the set of indices for which $\overline{v}_i$ is increased. Then we have $\overline{v}_i^+ = \overline{v}_i$ for $i \notin J$ and

$$0 \leq (\overline{v}_i^+)^2 - \overline{v}_i^2 \leq \theta\min(\overline{v})^2 \quad \text{for } i \in J.$$

Consequently,

$$\delta(\overline{v}; \overline{v}^+) = \frac{1}{2\min(\overline{v}^+)}\left\|\frac{(\overline{v}^+)^2 - \overline{v}^2}{\overline{v}}\right\| \leq \frac{1}{2\min(\overline{v})}\left\|\frac{\theta\min(\overline{v})^2 e_J}{\overline{v}}\right\| \leq \frac{1}{2}\theta\sqrt{n}.$$

where e_J is the 0–1 characteristic vector of indices in J. □

Combining this result with Theorem 8.2.6 gives that we can take $\theta = 1/(3\sqrt{n})$ to have $\delta(v^+; \overline{v}^+) < 1/2$. So we obtain that the algorithm needs at most $\mathcal{O}(\sqrt{n} \ln 1/\overline{\omega}_0)$ iterations.

If we combine the above centering scheme with the standard primal–dual path–following algorithm we obtain an algorithm for the LP problem needing at most

$$\mathcal{O}\left(\sqrt{n}\left(\ln \frac{1}{\overline{\omega}_0} + \ln \frac{n \max(\overline{v}^{(0)})^2}{\epsilon}\right)\right)$$

iterations, starting from any interior–feasible point. This is done by first centering, and then working to optimality. Note that in the centering phase the duality gap in subsequent target–points increases, but is bounded by $n \max(\overline{v}^{(0)})^2$.

It is interesting to consider the seemingly equivalent scheme of moving the larger components of $\overline{v}$ downwards. One can check that the analysis does not yield as good a bound as before. Due to the asymmetry of the proximity measure, there is a factor $\overline{\omega}$ that appears in the bound on $\delta(\overline{v}; \overline{v}^+)$. It is also clear that if we combine the efficient centering scheme with a standard path–following algorithm, we can reach the target $(\min(\overline{v}^{(0)}))e$ with complexity proportional to $\sqrt{n}$ without $\overline{\omega}$ factor. So the observed asymmetry is not intrinsic to the problem.

Computing weighted centers

In this application we discuss some algorithms to find an approximate solution to the KKT–system

$$\begin{aligned} Ax &= b, \quad x \geq 0, \\ A^T y + s &= c, \quad s \geq 0, \\ xs &= w^2, \end{aligned} \tag{8.23}$$

where $w \in \mathbb{R}^n_{++}$ is a prespecified weight–vector. Approximate means that we will compute a feasible pair (x, s), such that

$$\delta(v; w) \leq 1/2,$$

where $v = \sqrt{xs}$ as usual. We make the assumption that a (specific) point on or close to the central path is available. Note that we might use the centering algorithm of the previous subsection to find such a point. This problem has interesting special cases that are considered by Atkinson and Vaidya [10], Freund

[60] and Goffin and Vial [78], namely to obtain the weighted analytic center of a polytope. If $b = 0$ and (x, y, s) is a solution to system (8.23) then y is the weighted analytic center of the dual space, if the latter is bounded; when $c = 0$ and (x, y, s) satisfies the given system then x is the weighted analytic center of the primal space, if it is bounded.

We first analyze an algorithm proposed by Mizuno [176], which is similar to the algorithm for finding a center as discussed in the previous subsection. Then we give a simplified analysis of the algorithm proposed by Atkinson and Vaidya [10] for computing weighted analytic centers. We extend their algorithm to the case of computing weighted primal and dual centers, i.e., for finding a solution to the system (8.23).

Mizuno's algorithm

Assume that we start close to the center μe, with $\mu = \max(w^2)$. The aim is to get close to the weighted center w. The first target–point is set to $\overline{v} = \max(w)e$. We then gradually decrease the elements of the vector $\overline{v}$ until they all reach the correct value. This is performed updating the target as follows:

$$\overline{v}_i^+ = \max(w_i, \sqrt{1-\theta}\,\overline{v}_i). \tag{8.24}$$

Each component $\overline{v}_i$ is decreased until it reaches its final value w_i.

Lemma 8.2.28 *Let $\overline{v}^+$ be obtained from $\overline{v}$ with an update of the target using (8.24). Then*

$$\frac{\min(\overline{v})}{\min(\overline{v}^+)} \leq \frac{1}{\sqrt{1-\theta}} \quad \textit{and} \quad \delta(\overline{v};\overline{v}^+) \leq \frac{1}{2\sqrt{1-\theta}}\theta\sqrt{n}.$$

Proof: The first bound is trivial. The components of $\overline{v}$ that are decreased by a factor $\sqrt{1-\theta}$ have not yet achieved their final value w_i. Since they all start with the same value, they have all been reduced by the same cumulated factor and thus

$$\overline{v}_i^+ = \sqrt{1-\theta}\,\overline{v}_i \quad \Longrightarrow \quad \overline{v}_i = \min(\overline{v}).$$

So for all i it holds $|(\overline{v}_i^+)^2 - \overline{v}_i^2| \leq \theta \min(\overline{v})^2$. Hence

$$\begin{aligned}\delta(\overline{v};\overline{v}^+) &= \frac{1}{2\min(\overline{v}^+)}\left\|\frac{(\overline{v}^+)^2 - \overline{v}^2}{\overline{v}}\right\| \\ &\leq \frac{1}{2\sqrt{1-\theta}\min(\overline{v})}\left\|\frac{\theta\min(\overline{v})^2 e}{\overline{v}}\right\| \leq \frac{1}{2\sqrt{1-\theta}}\theta\sqrt{n}.\end{aligned}$$

□

Using Theorem 8.2.6 gives $\delta(v^+; \overline{v}^+) < 1/2$ for $\theta = 1/(3\sqrt{n})$. The number of iterations to be performed is determined by the condition

$$(1-\theta)^k \max(w)^2 \leq \min(w)^2,$$

implying

$$k \geq \frac{2}{\theta} \ln \left(\frac{\max(w)^2}{\min(w)^2} \right).$$

Consequently the number of Newton steps to compute the weighted center is at most $\mathcal{O}(\sqrt{n} \ln 1/\omega(w))$.

Atkinson and Vaidya's algorithm (dual)

Atkinson and Vaidya [10] propose an algorithm for the special case $b = 0$. This corresponds to computing the weighted analytic center of the dual feasible region. The algorithm is completely different from the one in the previous paragraph. We give a simple analysis for the algorithm using two nested traceable target-sequences. Moreover, we extend the algorithm to the general case (i.e., solving (8.23)) and show that this algorithm has a complexity bound which is worse than the one for Mizuno's algorithm.

So, first consider the case $b = 0$. Assuming $w^2 \geq e$ and w^2 integral, Atkinson and Vaidya suggest to start with a target-vector $\overline{v}^{(0)} = e$, and to successively increase the weights using a scaling technique à la Edmonds and Karp [49]. The basic idea is to recursively solve the given problem with all weights w_i^2 replaced by the maximum of 1 and $\lfloor w_i^2/2 \rfloor$. Let $p = \lfloor \log_2 \max(w^2) \rfloor$. Then w_i^2 can be written in binary notation as

$$w_i^2 = \beta_{i_0} \beta_{i_1} \cdots \beta_{i_p},$$

where $\beta_{i_j} \in \{0, 1\}$ for all i, j. Elements of the weight-vector w^2 which do not need p digits for their binary description start by convention with a string of zeroes. Now, at iteration k the target is defined by

$$(\overline{v}_i^{(k)})^2 = \beta_{i_0} \beta_{i_1} \cdots \beta_{i_k},$$

where we set $\overline{v}_i^{(k)} = 1$ if $\beta_{i_0} \beta_{i_1} \cdots \beta_{i_k} = 0$. Note that an update of the target to get $\overline{v}^{(k)}$ from $\overline{v}^{(k-1)}$ amounts to doubling the target (i.e., adding a zero to the binary description) and possibly adding 1 (if $\beta_{i_k} = 1$) or subtracting 1 (if $\beta_{i_0} \beta_{i_1} \cdots \beta_{i_k} = 0$). This is the outer target-sequence in the algorithm. For ease of notation we denote $\overline{v} := \overline{v}^{(k-1)}$ and $\overline{v}^+ := \overline{v}^{(k)}$. Summarizing, the technique boils down to a scheme that updates $\overline{v}_i$ in the following way:

$$(\overline{v}_i^+)^2 = \begin{cases} 2\overline{v}_i^2 - 1 & \text{if } i \in I_1 = \{\, i \;:\; \beta_{i_0}\beta_{i_1}\cdots\beta_{i_k} = 0 \,\} \\ 2\overline{v}_i^2 & \text{if } i \in I_2 = \{\, i \;:\; \beta_{i_0}\beta_{i_1}\cdots\beta_{i_{k-1}} \neq 0 \text{ and } \beta_{i_k} = 0 \,\} \\ 2\overline{v}_i^2 + 1 & \text{if } i \in I_3 = \{\, i \;:\; \beta_{i_0}\beta_{i_1}\cdots\beta_{i_{k-1}} \neq 0 \text{ and } \beta_{i_k} = 1 \,\}. \end{cases} \tag{8.25}$$

Observe that

$$i \in I_1 \Longrightarrow \overline{v}_i^+ = \overline{v}_i = 1. \tag{8.26}$$

The number of updates in the outer target–sequence is determined by the condition

$$2^k \geq \max(w)^2,$$

which implies that there will be $\lfloor \log_2 \max(w^2) \rfloor + 1$ updates.

We next need to compute the complexity of one outer update. This will be done by defining an inner target–sequence that leads from $\overline{v}$ to $\overline{v}^+$. In [10] a pure dual algorithm is used which means that doubling all weights does not change the position of the dual weighted center. Hence, the only Newton steps needed are to get from $2\overline{v}^2$ to $(\overline{v}^+)^2$, which are quite close to each other. Let (x,s) and $\overline{v}$ be given such that $\delta(v;\overline{v}) < 1/2$, where $v := \sqrt{xs}$. Since $b = 0$, by setting

$$x^+ = 2x, \quad s^+ = s, \quad v^+ = \sqrt{x^+ s^+}, \tag{8.27}$$

we have a feasible pair (x^+, s^+) for which

$$\delta(v^+; \sqrt{2}\overline{v}) = \delta(v;\overline{v}) < \frac{1}{2}.$$

In the analysis we consider the target–sequence that leads from $\sqrt{2}\overline{v}$ to $\overline{v}^+$. For this purpose we use the following scheme. Let $j = 0$ and $(\tilde{v}^{(0)})^2 = 2\overline{v}^2$. Define

$$\vartheta_i = \begin{cases} \frac{\alpha}{\overline{v}_i\sqrt{n}} & \text{if } i \in I_1, \\ 0 & \text{if } i \in I_2, \\ -\frac{\alpha}{\overline{v}_i\sqrt{n}} & \text{if } i \in I_3, \end{cases} \tag{8.28}$$

where $\alpha > 0$ is a certain constant. Update $\tilde{v}^{(j)}$ for $j \geq 1$ in the following way:

$$(\tilde{v}_i^{(j)})^2 = (1 - \vartheta_i)(\tilde{v}_i^{(j-1)})^2. \tag{8.29}$$

Of course, we do not overshoot the target value. The conditions

$$\left(1 - \frac{\alpha}{\overline{v}_i\sqrt{n}}\right)^j (2\overline{v}_i^2) \leq 2\overline{v}_i^2 - 1 \qquad \text{if } i \in I_1,$$

$$\left(1 + \frac{\alpha}{\overline{v}_i\sqrt{n}}\right)^j (2\overline{v}_i^2) \geq 2\overline{v}_i^2 + 1 \qquad \text{if } i \in I_3,$$

determine the number of updates to be performed. For $i \in I_1$, it suffices to have

$$j \geq \frac{\overline{v}_i \sqrt{n}}{\alpha} \ln \left(\frac{2\overline{v}_i^2}{2\overline{v}_i^2 - 1} \right).$$

Since from (8.26) we have $\overline{v}_i = 1$ it follows that at most

$$\frac{\sqrt{n}}{\alpha} \ln 2$$

iterations are needed for $i \in I_1$. For $i \in I_3$, j satisfying

$$1 + \frac{j\alpha}{\overline{v}_i \sqrt{n}} \geq 1 + \frac{1}{2\overline{v}_i^2}$$

suffices. This leads to the condition that $j \geq \sqrt{n}/(2\alpha\overline{v}_i)$ suffices; using the fact that $\overline{v}_i \geq 1$, this proves that the number of updates to be performed is not larger than $\sqrt{n}/2\alpha$. We need to show now that the specific choice of the update guarantees that one Newton step suffices per inner update.

Lemma 8.2.29 *Define $\theta := \alpha/\sqrt{n}$. Let $\tilde{v}^{(j)}$ be obtained from $\tilde{v}^{(j-1)}$ with an update of the target using (8.28) and (8.29). Then*

$$\frac{\min(\tilde{v}^{(j-1)})}{\min(\tilde{v}^{(j)})} \leq \frac{1}{\sqrt{1-\theta}} \quad \text{and} \quad \delta(\tilde{v}^{(j-1)}; \tilde{v}^{(j)}) \leq \frac{3\alpha}{2\sqrt{1-\theta}}.$$

Proof: For ease of notation, let $\tilde{v} = \tilde{v}^{(j-1)}$ and $\tilde{v}^+ = \tilde{v}^{(j)}$. Then

$$\min(\tilde{v}^+)^2 \geq (1-\theta)\min(\tilde{v})^2,$$

and hence we have

$$\begin{aligned}
\delta(\tilde{v}; \tilde{v}^+) &= \frac{1}{2\min(\tilde{v}^+)} \left\| \frac{(\tilde{v}^+)^2 - \tilde{v}^2}{\tilde{v}} \right\| \\
&\leq \frac{1}{2\sqrt{1-\theta}\min(\tilde{v})} \left(\sum_{i \in I_1 \cup I_3} \left(\frac{\tilde{v}_i^2(1-\vartheta_i) - \tilde{v}_i^2}{\tilde{v}_i} \right)^2 \right)^{1/2} \\
&= \frac{1}{2\sqrt{1-\theta}\min(\tilde{v})} \left(\sum_{i \in I_1 \cup I_3} (\vartheta_i \tilde{v}_i)^2 \right)^{1/2} \\
&= \frac{1}{2\sqrt{1-\theta}\min(\tilde{v})} \left(\sum_{i \in I_1 \cup I_3} \left(\frac{\alpha}{\sqrt{n}} \frac{\tilde{v}_i}{\overline{v}_i} \right)^2 \right)^{1/2}
\end{aligned}$$

$$\leq \frac{\theta}{2\sqrt{1-\theta}\min(\tilde{v})}\left(\sum_{i\in I_1\cup I_3}\left(\frac{\sqrt{2\overline{v}_i^2+1}}{\overline{v}_i}\right)^2\right)^{1/2}$$
$$\leq \frac{\theta}{2\sqrt{1-\theta}\min(\tilde{v})}3\sqrt{n}=\frac{3\alpha}{2\sqrt{1-\theta}\min(\tilde{v})}.$$

Since $\min(\tilde{v}) \geq 1$ the lemma follows. □

Using Theorem 8.2.6 it follows from the lemma that for $\alpha = 1/7$ we can get close to $\tilde{v}^{(j)}$ from a point close to $\tilde{v}^{(j-1)}$ in one full Newton step. So the entire algorithm performs at most

$$\mathcal{O}(\sqrt{n}\log_2 \max(w)) \tag{8.30}$$

Newton steps, and for this pure dual algorithm we get the same complexity as in [10] using a much simpler analysis.

Atkinson and Vaidya's algorithm (primal–dual)

We now analyze the same strategy for the problem of finding the primal–dual weighted centers, i.e., the solution of system (8.23). The outer iteration is the same as before, i.e., doubling the target and subtracting or adding one if necessary, see (8.25). The number of Newton steps needed to get close to a new target is more than one now, since the update of $\overline{v}$ is big: the trick in (8.27) cannot be used anymore. Again, to compute an iterate in the quadratic convergence region of $\overline{v}^+$ another target–sequence is constructed by which we reach $\overline{v}^+$ from $\overline{v}$. The following scheme is used. Let $\tilde{v}^{(0)} = \overline{v}$ and define

$$\vartheta_i = \begin{cases} 0 & \text{if } i \in I_1, \\ -\frac{\alpha}{\overline{v}_i\sqrt{n}} & \text{if } i \in I_2 \cup I_3, \end{cases}$$

where $\alpha > 0$ is a certain constant. Update $\tilde{v}^{(j)}$ for $j \geq 1$ in the following way:

$$(\tilde{v}_i^{(j)})^2 = (1-\vartheta_i)(\tilde{v}_i^{(j-1)})^2.$$

Note that the proof of Lemma 8.2.29 is easily adapted for this sequence and that its result remains the same. Using the condition $(\tilde{v}_i^{(j)})^2 \geq (\overline{v}_i^+)^2$, or

$$\left(1+\frac{\alpha}{\overline{v}_i\sqrt{n}}\right)^j \overline{v}_i^2 \geq 2\overline{v}_i^2 \qquad \text{if } i \in I_2,$$
$$\left(1+\frac{\alpha}{\overline{v}_i\sqrt{n}}\right)^j \overline{v}_i^2 \geq 2\overline{v}_i^2+1 \qquad \text{if } i \in I_3$$

and using the fact that $\overline{v}_i \geq 1$, it follows that the number of inner updates per outer iteration must be of the order $\mathcal{O}(\max(\overline{v})\sqrt{n})$, so an upper bound expressed in the data is

$$\mathcal{O}(\max(w)\sqrt{n}).$$

Hence, the total number of Newton steps required is

$$\mathcal{O}(\max(w)\sqrt{n}\log_2 \max(w)).$$

This is a factor $\max(w)$ worse than the result in (8.30) and in [10]. This difference can be explained by noticing that doubling all weights does not have any effect in a pure primal or dual method, but has quite an effect in a primal-dual method.

8.3 LONG–STEP PRIMAL–DUAL ALGORITHMS FOR LP

8.3.1 Outline

In Section 8.2 applications are given of the target–following approach with target–sequences having the property that in each step the target is only slightly changed. Stated otherwise, short–step methods were investigated. In this section we are concerned with the analysis of medium–step and long–step methods in the spirit of Den Hertog [100], Gonzaga [85] and Jansen et al. [116], among others. From the cited literature it appears that these algorithms are much more efficient from a practical point of view, however, possess a worse theoretical complexity bound. In general, the long–step methods use a step size which is $\mathcal{O}(1)$ and converge in $\mathcal{O}(n \ln 1/\epsilon)$ iterations, whereas the medium–step methods converge in $\mathcal{O}(\sqrt{n} \ln 1/\epsilon)$ iterations with a step size $\mathcal{O}(1/\sqrt{n})$.

The main implications of using long steps are: (*i*) it is not feasible to do full Newton steps, instead damped Newton steps should be used, (*ii*) no longer one (damped) Newton step suffices to reach the region of quadratic convergence of the new target. This implies that for the analysis of long–step methods a different machinery should be used. In this section we adopt the approach by Den Hertog [100] and Jansen et al. [116]. The generic long–step algorithm is described in Figure 8.3. Observe that algorithms of this form contain two nested loops. In each *outer iteration* the target is kept constant, while a number of *inner iterations* (i.e. damped Newton steps) is performed to obtain an approximation to the target. We introduce a barrier function $f(v; \overline{v})$ that measures proximity to the target. We will establish properties of this barrier function and its relationship with the proximity measure $\delta(v;\overline{v})$ in (8.6).

Input
  $(x^{(0)}, s^{(0)})$: the initial pair of interior–feasible solutions;
Parameters
  ϵ is the accuracy parameter;
  ξ is the proximity parameter (default $\xi = 1/4$);
  θ is the step size;
begin
  $x := x^{(0)}$; $s := s^{(0)}$; $\overline{v} := \sqrt{xs}$;
  while $x^T s > \epsilon$ **do**
    $\overline{v} := (1-\theta)\overline{v}$;
    while $\delta(v;\overline{v}) > \xi$ **do**
      compute $(\Delta x, \Delta s)$ from (8.3);
      find α s.t. $f(x+\alpha\Delta x, s+\alpha\Delta s;\overline{v}) - f(x, s;\overline{v})$ is sufficiently negative;
      $x := x + \alpha\Delta x$;
      $s := s + \alpha\Delta s$;
    end
  end
end.

Figure 8.3 Long–step algorithm.

We give some applications of the long–step approach. We first analyze the family of path–following algorithms of which both the weighted logarithmic barrier method and the Dikin–path–following algorithm are special cases. Then we use the present methodology to analyze long–step algorithms for the problem of computing a point on the central path and for the computation of a weighted center.

8.3.2 Barrier function and its properties

We use the weighted logarithmic barrier function defined by

$$f(x, s;\overline{v}) = \frac{x^T s}{\max(\overline{v})^2} - \sum_{i=1}^{n} \frac{\overline{v}_i^2}{\max(\overline{v})^2} \ln x_i s_i - \frac{\|\overline{v}\|^2}{\max(\overline{v})^2} + \sum_{i=1}^{n} \frac{\overline{v}_i^2}{\max(\overline{v})^2} \ln \overline{v}_i^2 .$$

We can rewrite this function in terms of $v = \sqrt{xs}$,

$$\begin{aligned}
\phi(v;\overline{v}) &= \frac{1}{\max(\overline{v})^2}\left(\|v\|^2 - \sum_{i=1}^{n}\overline{v}_i^2 \ln v_i^2 - \|\overline{v}\|^2 + \sum_{i=1}^{n}\overline{v}_i^2 \ln \overline{v}_i^2\right) \\
&= \frac{1}{\max(\overline{v})^2}\left(e^T(v^2 - \overline{v}^2) + \sum_{i=1}^{n}\overline{v}_i^2 \ln\left(\frac{\overline{v}_i^2}{v_i^2}\right)\right) \qquad (8.31) \\
&= \sum_{i=1}^{n}\frac{\overline{v}_i^2}{\max(\overline{v})^2}\left(\frac{v_i^2}{\overline{v}_i^2} - \ln\left(\frac{v_i^2}{\overline{v}_i^2}\right) - 1\right). \qquad (8.32)
\end{aligned}$$

A slightly different weighted barrier function was used by Ding and Li [45]. Observe that $\phi(v;\overline{v}) = f(x,s;\overline{v})$ and that ϕ has n parameters, namely the weights $\overline{v}_1,\ldots,\overline{v}_n$. Note also that it is homogeneous in the sense that $\phi(\lambda v;\lambda\overline{v}) = \phi(v;\overline{v})$ for $0 < \lambda \in \mathbb{R}$. Moreover, each term in the summation in (8.32) is minimal for $v_i = \overline{v}_i$, hence

$$\phi(v;\overline{v}) \geq \phi(\overline{v};\overline{v}) = 0 \quad \forall\, v \in \mathbb{R}^n_{++}.$$

Observe that in case the target is on the central path, say $\overline{v} = \sqrt{\mu}e$, the function $f(\cdot)$ reduces to

$$f(x,s;\sqrt{\mu}e) = \frac{x^T s}{\mu} - \sum_{i=1}^{n}\ln x_i s_i - n + n\ln\mu,$$

which is (up to a constant) the primal–dual logarithmic barrier function used in [116]. In this case the number of parameters reduces to one. We will show that a damped Newton step gives a sufficient decrease in the value of the barrier function. Recall from (8.6) that

$$\delta(v;\overline{v}) = \frac{1}{2\min(\overline{v})}\|p_v\|, \qquad (8.33)$$

where p_v is defined in (8.4), and from (8.9) $\rho(\delta) = \delta + \sqrt{1+\delta^2}$. We define

$$r := \sqrt{\left\|\frac{p_x}{v}\right\|^2 + \left\|\frac{p_s}{v}\right\|^2} \qquad (8.34)$$

and observe

$$r \leq \frac{1}{\min(v)}\sqrt{\|p_x\|^2 + \|p_s\|^2} = \frac{\|p_v\|}{\min(v)}. \qquad (8.35)$$

Lemma 8.3.1 *Let $\delta := \delta(v;\overline{v})$ and $\overline{\omega} := \omega(\overline{v})$. Let the step size be given by*

$$\alpha = \frac{1}{r} - \frac{\max(\overline{v})^2}{\|p_v\|^2 + r\max(\overline{v})^2}, \tag{8.36}$$

where r is defined by (8.34). Then

$$\Delta f(\alpha) := f(x + \alpha\Delta x, s + \alpha\Delta s; \overline{v}) - f(x, s; \overline{v}) \leq -\frac{2\delta^2\overline{\omega}^4}{\rho(\delta)^2 + 2\delta\rho(\delta)\overline{\omega}^2}.$$

Proof: First observe that $0 < \alpha r < 1$, so it holds $x + \alpha\Delta x > 0$ and $s + \alpha\Delta s > 0$. It holds

$$\begin{aligned}
\Delta f(\alpha) &= f(x + \alpha\Delta x, s + \alpha\Delta s; \overline{v}) - f(x, s; \overline{v}) \\
&= \frac{1}{\max(\overline{v})^2}\left(\alpha(\Delta s^T x + s^T\Delta x) - \sum_{i=1}^n \overline{v}_i^2 \ln\left(1 + \alpha\frac{\Delta x_i}{x_i}\right)\left(1 + \alpha\frac{\Delta s_i}{s_i}\right)\right) \\
&= \frac{1}{\max(\overline{v})^2}\left(\alpha e^T(\overline{v}^2 - v^2) - \sum_{i=1}^n \overline{v}_i^2 \ln\left(1 + \alpha\frac{(p_x)_i}{v_i}\right)\left(1 + \alpha\frac{(p_s)_i}{v_i}\right)\right).
\end{aligned}$$

Applying Lemma A.1 to the concatenation of the vectors $\alpha p_x/v$ and $\alpha p_s/v$ we obtain

$$\begin{aligned}
\Delta f(\alpha) &\leq \frac{1}{\max(\overline{v})^2}\left(\alpha e^T(\overline{v}^2 - v^2) - \alpha\sum_{i=1}^n \frac{\overline{v}_i^2}{v_i^2}(v_i(p_x)_i + v_i(p_s)_i)\right. \\
&\quad \left. -\alpha r\max(\overline{v})^2 - \max(\overline{v})^2\ln(1 - \alpha r)\right) \\
&= \frac{\alpha}{\max(\overline{v})^2}\left(e^T(\overline{v}^2 - v^2) - e^T\left(\frac{\overline{v}^2}{v^2}(\overline{v}^2 - v^2)\right)\right) - \alpha r - \ln(1 - \alpha r) \\
&= -\alpha\frac{\|p_v\|^2}{\max(\overline{v})^2} - \alpha r - \ln(1 - \alpha r),
\end{aligned} \tag{8.37}$$

where the last equality follows from the observation

$$e^T(\overline{v}^2 - v^2) - e^T\left(\frac{\overline{v}^2}{v^2}(\overline{v}^2 - v^2)\right) = -e^T\left(v^{-2}\left(\overline{v}^2 - v^2\right)^2\right) = -e^T p_v^2.$$

The right-hand side of (8.37) is minimal for the value of α defined in (8.36). By substitution of this value we obtain

$$\Delta f(\alpha) \leq -\frac{\|p_v\|^2}{r\max(\overline{v})^2} + \ln\left(1 + \frac{\|p_v\|^2}{r\max(\overline{v})^2}\right).$$

Since this bound is monotonically decreasing in $\|p_v\|^2/(r\max(\overline{v})^2)$, we may replace this quotient by a smaller value. Using (8.33) and (8.35) it follows

$$\frac{\|p_v\|^2}{r\max(\overline{v})^2} \geq \frac{\min(v)\|p_v\|}{\max(\overline{v})^2} = 2\delta\overline{\omega}^2 \frac{\min(v)}{\min(\overline{v})} \geq \frac{2\delta\overline{\omega}^2}{\rho(\delta)},$$

where the last inequality uses the fact that for any i we have $v_i/\min(\overline{v}) \geq v_i/\overline{v}_i \geq 1/\rho(\delta)$ by Lemma 8.2.2. So we find

$$\Delta f(\alpha) \leq -\frac{2\delta\overline{\omega}^2}{\rho(\delta)} + \ln\left(1 + \frac{2\delta\overline{\omega}^2}{\rho(\delta)}\right).$$

Using the inequality

$$-x + \ln(1+x) \leq -\frac{x^2}{2(1+x)}, \qquad x > 0,$$

the bound for $\Delta f(\alpha)$ follows. □

As a result we have the following corollary.

Corollary 8.3.2 *If* $\delta := \delta(v;\overline{v}) \geq 1/4$ *then* $\Delta f(\alpha) \leq -\overline{\omega}^4/(14 + 6\overline{\omega}^2)$.

We relate the proximity measure $\delta(v;\overline{v})$ to the value of the barrier function for points close to a target.

Lemma 8.3.3 *If* $\delta := \delta(v;\overline{v}) \leq 1/4$ *then*

$$\phi(v;\overline{v}) \leq -\sigma - \ln(1-\sigma),$$

where $\sigma := 2\delta\rho(\delta) < 13/20$.

Proof: Observe that for $t > 0$ it holds $t - 1 - \ln t > 0$. Hence,

$$\begin{aligned}\phi(v;\overline{v}) &= \sum_{i=1}^n \frac{\overline{v}_i^2}{\max(\overline{v})^2}\left(\frac{v_i^2}{\overline{v}_i^2} - 1 - \ln\frac{v_i^2}{\overline{v}_i^2}\right) \leq e^T\left(\frac{v^2}{\overline{v}^2} - e\right) - \sum_{i=1}^n \ln\frac{v_i^2}{\overline{v}_i^2} \\ &= e^T h - \sum_{i=1}^n \ln(1+h_i),\end{aligned}$$

where $h := \overline{v}^{-2}v^2 - e$. Using Lemma 8.2.2 and definition (8.6) of δ, we get

$$\begin{aligned} \|h\| &= \left\| \frac{v}{\overline{v}^2}\frac{e}{v}(v^2 - \overline{v}^2) \right\| \leq \left\| \frac{v}{\overline{v}^2} \right\|_\infty \|p_v\| \\ &\leq \rho(\delta)\|\overline{v}^{-1}\|_\infty 2\min(\overline{v})\delta = 2\delta\rho(\delta) < \frac{13}{20} < 1. \end{aligned} \tag{8.38}$$

So upon application of Lemma A.1 we find

$$\phi(v;\overline{v}) \leq -\|h\| - \ln(1 - \|h\|) \leq -2\delta\rho(\delta) - \ln(1 - 2\delta\rho(\delta)).$$

This completes the proof. □

Corollary 8.3.4 *If $\delta := \delta(v;\overline{v}) \leq 1/4$ then $\phi(v;\overline{v}) \leq 2/5$.*

The next lemma relates the duality gap in an arbitrary positive primal–dual pair to the duality gap in (arbitrary) targets.

Lemma 8.3.5 *Let $\delta := \delta(v;\overline{v})$. It holds $\|v\|^2 < \rho(\delta)^2\|\overline{v}\|^2$. If moreover $\delta \leq 1/4$ then*

$$\|v\|^2 \leq \|\overline{v}\|^2 + \|\overline{v}^2\| \leq \|\overline{v}\|^2 + \max(\overline{v})^2\sqrt{n}.$$

Proof: By Lemma 8.2.2 it holds $v \leq \rho(\delta)\overline{v}$. This makes the first statement trivial. We have

$$\begin{aligned} \left| \|v\|^2 - \|\overline{v}\|^2 \right| &= |e^Tv^2 - e^T\overline{v}^2| = \left| (\overline{v}^2)^T \left(\frac{v^2}{\overline{v}^2} - e \right) \right| \\ &\leq \|\overline{v}^2\| \left\| \frac{v^2}{\overline{v}^2} - e \right\| < \|\overline{v}^2\|, \end{aligned}$$

where the last inequality follows from (8.38). So

$$\|v\|^2 \leq \|\overline{v}\|^2 + \|\overline{v}^2\| < \|\overline{v}\|^2 + \max(\overline{v})^2\sqrt{n}.$$

This proves the second bound. □

Assume now that an iterate v and target $\overline{v}$ are given such that $\delta(v;\overline{v}) \leq 1/4$. The target is updated to $\overline{v}^+$. We proceed by deriving a bound for the number of inner iterations needed to compute a primal–dual approximation to $\overline{v}^+$. Note that during this outer iteration the target is fixed and $\delta(\cdot;\overline{v}^+)$ is greater than 1/4.

Theorem 8.3.6 *Let v and $\overline{v}$ be such that $\delta := \delta(v;\overline{v}) \leq 1/4$. For arbitrary $\overline{v}^+ \in \mathbb{R}^n_{++}$ the number of inner iterations in the outer iteration with respect to $\overline{v}^+$ is bounded by*

$$\frac{14+6(\overline{\omega}^+)^2}{(\overline{\omega}^+)^4}\left(\phi(\overline{v};\overline{v}^+) + \frac{2}{5}\frac{\max(\overline{v})^2}{\max(\overline{v}^+)^2} + 4\delta\rho(\delta)\sqrt{n}\max_{1\leq i\leq n}\left(\left|\frac{\overline{v}_i^2}{(\overline{v}_i^+)^2} - 1\right|\right)\right).$$

Proof: Assume that k iterations have been performed without any further updating of the target; denote the corresponding iterates in the v–space by $v^{(j)}$, where $v^{(0)} = v$. Then we have $\delta(v^{(j)};\overline{v}^+) > 1/4$, for all $0 \leq j < k$. Since the hypothesis of Corollary 8.3.2 holds at each inner iteration we can bound the total decrease of the barrier function by

$$\phi(v^{(0)};\overline{v}^+) - \phi(v^{(k)};\overline{v}^+) \geq \frac{k(\overline{\omega}^+)^4}{14+6(\overline{\omega}^+)^2}.$$

Since $\phi(v^{(k)};\overline{v}^+) \geq 0$ we thus obtain

$$k \leq \frac{14+6(\overline{\omega}^+)^2}{(\overline{\omega}^+)^4}\phi(v^{(0)};\overline{v}^+).$$

So, it remains to bound $\phi(v;\overline{v}^+)$. Using definition (8.31)

$$\begin{aligned}
\phi(v;\overline{v}^+) &= \sum_{i=1}^n \frac{(\overline{v}_i^+)^2}{\max(\overline{v}^+)^2}\left(\frac{\overline{v}_i^2}{(\overline{v}_i^+)^2}\ln\frac{\overline{v}_i^2}{(\overline{v}_i^+)^2} - 1 + \frac{v_i^2 - \overline{v}_i^2}{(\overline{v}_i^+)^2} - \ln\frac{v_i^2}{\overline{v}_i^2}\right) \\
&= \phi(\overline{v};\overline{v}^+) + \sum_{i-1}^n \frac{(\overline{v}_i^+)^2}{\max(\overline{v}^+)^2}\left(\frac{\overline{v}_i^2}{(\overline{v}_i^+)^2}\left(\frac{v_i^2}{\overline{v}_i^2} - \ln\frac{v_i^2}{\overline{v}_i^2} - 1\right)\right. \\
&\quad \left. + \left(\frac{\overline{v}_i^2}{(\overline{v}_i^+)^2} - 1\right)\ln\frac{v_i^2}{\overline{v}_i^2}\right) \\
&\leq \phi(\overline{v};\overline{v}^+) + \frac{\max(\overline{v})^2}{\max(\overline{v}^+)^2}\phi(v;\overline{v}) + \sum_{i=1}^n \frac{(\overline{v}_i^+)^2}{\max(\overline{v}^+)^2}\left(\frac{\overline{v}_i^2}{(\overline{v}_i^+)^2} - 1\right)\ln\frac{v_i^2}{\overline{v}_i^2}.
\end{aligned}$$

From Corollary 8.3.4 it follows $\phi(v;\overline{v}) \leq 2/5$. Note that for any i the term in the remaining sum is positive only if the quantities

$$\frac{\overline{v}_i^2}{(\overline{v}_i^+)^2} - 1 \quad \text{and} \quad \ln\frac{v_i^2}{\overline{v}_i^2} \tag{8.39}$$

have the same sign. Let I be the set of indices for which both are positive. Then

$$\sum_{i\in I}\frac{(\overline{v}_i^+)^2}{\max(\overline{v}^+)^2}\left(\frac{\overline{v}_i^2}{(\overline{v}_i^+)^2} - 1\right)\ln\frac{v_i^2}{\overline{v}_i^2} \leq \max_{i\in I}\left(\left|\frac{\overline{v}_i^2}{(\overline{v}_i^+)^2} - 1\right|\right)\sum_{i\in I}\ln\frac{v_i^2}{\overline{v}_i^2}.$$

Notice that,

$$\sum_{i\in I} \ln\left(1+\frac{v_i^2}{\overline{v}_i^2}-1\right) \leq \sum_{i\in I}\left(\frac{v_i^2}{\overline{v}_i^2}-1\right) \leq \sqrt{I}\,\|\overline{v}^{-2}v^2 - e\| \leq \sqrt{n}\,2\delta\rho(\delta),$$

where the second inequality follows from Cauchy–Schwarz and the last from (8.38). The same bound holds for the indices for which both quantities in (8.39) are negative. After combining, we obtain the required bound. □

Observe that this theorem serves a similar role as Theorem 8.2.6 in short-step algorithms. To compute the number of damped Newton steps we need to investigate $\overline{\omega}^+$ and the value $\phi(\overline{v};\overline{v}^+)$. The number of outer iterations depends on the number of target updates needed to finish the algorithm.

8.3.3 Applications

Path-following

We apply the obtained results to a method using Dikin steps with ν-order scaling as in (8.20), assuming $\nu = \mathcal{O}(1)$. The main task is to establish the effect of a target update

$$\overline{v}^+ = \overline{v}\left(e - \theta\frac{\overline{v}^{2\nu}}{\max(\overline{v})^{2\nu}}\right) \tag{8.40}$$

with step size $0 < \theta < 1/(2\nu+1)$. Recall that $\nu = 1$ corresponds to the usual Dikin step and $\nu = 0$ to weighted path-following. From Lemma 8.2.20 it is assured that $\omega(\overline{v}^+) \geq \omega(\overline{v})$ and that we may assume $\overline{v}$ to be ordered as in Lemma 8.2.10, so $\overline{v}_1 \leq \cdots \leq \overline{v}_n$ and $\overline{v}_1^+ \leq \cdots \leq \overline{v}_n^+$. In the next lemma we estimate the quantities in Theorem 8.3.6.

Lemma 8.3.7 *Let $\overline{v}^+$ result from (8.40) with $0 < \theta \leq 1/(2\nu+1)$ and assume that $\delta := \delta(v;\overline{v}) \leq 1/4$. Then*

$$\phi(\overline{v};\overline{v}^+) \leq \frac{3n\theta^2}{(1-\theta)^2}, \quad \frac{\max(\overline{v})^2}{\max(\overline{v}^+)^2} \leq \frac{1}{(1-\theta)^2},$$

$$\max_{1\leq i\leq n}\left(\left|\frac{\overline{v}_i^2}{(\overline{v}_i^+)^2}-1\right|\right) \leq \frac{\theta(2-\theta)}{(1-\theta)^2}.$$

Proof: Note that

$$\frac{\overline{v}_i^2}{(\overline{v}_i^+)^2} = \frac{1}{(1-\theta\overline{v}_i^{2\nu}/\overline{v}_n^{2\nu})^2} \leq \frac{1}{(1-\theta)^2},$$

wherefrom the second and third bound easily follow. It remains to bound $\phi(\overline{v};\overline{v}^+)$. Straightforward calculus shows

$$\begin{aligned}
&\phi(\overline{v};\overline{v}^+)\\
&=\sum_{i=1}^n \frac{\overline{v}_i^2(1-\theta\overline{v}_i^{2\nu}/\overline{v}_n^{2\nu})^2}{\overline{v}_n^2(1-\theta)^2}\left(\frac{1}{(1-\theta\overline{v}_i^{2\nu}/\overline{v}_n^{2\nu})^2} - \ln\frac{1}{(1-\theta\overline{v}_i^{2\nu}/\overline{v}_n^{2\nu})^2} - 1\right)\\
&=\frac{1}{(1-\theta)^2}\sum_{i=1}^n \frac{\overline{v}_i^2}{\overline{v}_n^2}\left(1+2\left(1-\theta\frac{\overline{v}_i^{2\nu}}{\overline{v}_n^{2\nu}}\right)^2 \ln\left(1-\theta\frac{\overline{v}_i^{2\nu}}{\overline{v}_n^{2\nu}}\right) - \left(1-\theta\frac{\overline{v}_i^{2\nu}}{\overline{v}_n^{2\nu}}\right)^2\right)\\
&\leq\frac{1}{(1-\theta)^2}\sum_{i=1}^n \frac{\overline{v}_i^2}{\overline{v}_n^2}\left(2\left(1-\theta\frac{\overline{v}_i^{2\nu}}{\overline{v}_n^{2\nu}}\right)^2 \frac{-\theta\overline{v}_i^{2\nu}}{\overline{v}_n^{2\nu}} + 2\theta\frac{\overline{v}_i^{2\nu}}{\overline{v}_n^{2\nu}} - \theta^2\frac{\overline{v}_i^{4\nu}}{\overline{v}_n^{4\nu}}\right)\\
&=\frac{1}{(1-\theta)^2}\sum_{i=1}^n \frac{\overline{v}_i^2}{\overline{v}_n^2}\left(3\theta^2\frac{\overline{v}_i^{4\nu}}{\overline{v}_n^{4\nu}} - 2\theta^3\frac{\overline{v}_i^{6\nu}}{\overline{v}_n^{6\nu}}\right) \leq \frac{3n\theta^2}{(1-\theta)^2},
\end{aligned}$$

completing the proof. □

Combining Lemma 8.3.7 with Theorem 8.3.6 we deduce the following corollary.

Corollary 8.3.8 *Let $\overline{v}^+$ result from (8.40) with $0 < \theta \leq 1/(2\nu+1)$ and assume that $\delta(v;\overline{v}) \leq 1/4$. Then at most*

$$\mathcal{O}\left(\frac{1}{(\overline{\omega}^+)^4}\left(n\theta^2 + \sqrt{n}\theta\right)\right)$$

damped Newton steps are required to obtain a primal–dual feasible pair (x^+, s^+) such that $\delta(v^+;\overline{v}^+) \leq 1/4$, where $v^+ = \sqrt{x^+s^+}$.

Typically, long–step methods apply a large update $\theta = \mathcal{O}(1)$. In this case Corollary 8.3.8 reveals that $\mathcal{O}(n/\overline{\omega}^4)$ (damped) Newton steps will be needed within an outer iteration. For medium–step algorithms using $\theta = \mathcal{O}(1/\sqrt{n})$ the number of steps is $\mathcal{O}(1)$ per outer iteration. Next we have to compute the number of outer iterations, i.e., the number of target updates.

Theorem 8.3.9 *Let $(x^{(0)}, s^{(0)})$ be a given initial point and let*

$$\overline{v}^{(0)} := \sqrt{x^{(0)}s^{(0)}} \quad \textit{and} \quad \overline{\omega}_0 := \omega(\overline{v}^{(0)}).$$

After at most

$$\mathcal{O}\left(\frac{1}{\theta\overline{\omega}_0^{2\nu}} \ln \frac{2(x^{(0)})^T s^{(0)}}{\epsilon}\right)$$

target updates the algorithm stops with a primal–dual pair (x^*, s^*) satisfying $(x^*)^T s^* \leq \epsilon$.

Proof: At the start of the algorithm the duality gap is given by

$$(x^{(0)})^T s^{(0)} = \|\overline{v}^{(0)}\|^2.$$

Since the duality gap in an approximate center is closely related to the duality gap in the exact center (Lemma 8.3.5), and at the end of each outer iteration we have a point whose proximity to the target is less than 1/4, we may consider the reduction in the gap in exact centers. If, as before, the target–point at the beginning of some iteration is denoted as $\overline{v}$ and at the end of the same iteration as $\overline{v}^+$, then we have by (8.16)

$$\|\overline{v}^+\| \leq \|\overline{v}\| \left(1 - \theta\overline{\omega}^{2\nu}\right),$$

where $\overline{\omega} = \omega(\overline{v})$. Hence, in the iteration under consideration the duality gap in the exact center is reduced by at least the factor

$$\left(1 - \theta\overline{\omega}^{2\nu}\right)^2.$$

Since $\overline{\omega} \geq \overline{\omega}_0$, this factor is smaller than $\left(1 - \theta(\overline{\omega}_0)^{2\nu}\right)^2$, and we deduce that after the given number of iterations it holds $\|\overline{v}\|^2 \leq \epsilon/2$. Using Lemma 8.3.5 and $\delta \leq 1/4$ it follows that

$$\|v\|^2 \leq \rho(\delta)^2 \|\overline{v}\|^2 \leq 2\|\overline{v}\|^2 \leq \epsilon,$$

completing the proof. □

Combining Theorem 8.3.9 and Corollary 8.3.8 we obtain the total iteration bound for the general long–step and medium–step algorithm using update (8.40).

Corollary 8.3.10 *For step size $\theta \leq 1/(2\nu+1)$ the path–following algorithm with ν–order scaling needs at most*

$$\mathcal{O}\left(\frac{n}{\overline{\omega}_0^{2\nu+4}} \ln \frac{2(x^{(0)})^T s^{(0)}}{\epsilon}\right)$$

inner iterations (damped Newton steps) in total. If, moreover, $\theta = \mathcal{O}(1/\sqrt{n})$ then only

$$\mathcal{O}\left(\frac{\sqrt{n}}{\overline{\omega}_0^{2\nu+4}} \ln \frac{2(x^{(0)})^T s^{(0)}}{\epsilon}\right)$$

are needed.

The iteration bound in the corollary is rather bad in terms of $\overline{\omega}_0$. We can improve it by computing the number of iterations needed to have $\overline{\omega}^2 \geq 1/2$, cf. Lemma 8.2.16. This gives the following corollary.

Corollary 8.3.11 *For step size $\theta \leq 1/(2\nu+1)$ the path–following algorithm with ν–order scaling needs at most*

$$\mathcal{O}\left(n\left(\frac{1}{\overline{\omega}_0^4} \ln \frac{1}{\overline{\omega}_0} + \ln \frac{2(x^{(0)})^T s^{(0)}}{\epsilon}\right)\right)$$

inner iterations in total.

Computing a central point

We will now consider the problem of computing a point on the central path , given an arbitrary positive primal–dual feasible pair (cf. Section 8.2.4). Here we propose a long–step algorithm. Observe, that we only need to quantify the potential value in the current point with respect to a point on the central path. Let (x, s) be an arbitrary positive primal–dual feasible pair, and let $v^2 = xs$. The target on the central path will be defined as the one with the same duality gap as (x, s).

Lemma 8.3.12 *Define $\mu := e^T v^2/n$ and the target on the central path $\overline{v} := \sqrt{\mu}\, e$. Then,*

$$\phi(v; \overline{v}) \leq 2n \ln \frac{1}{\omega(v)}.$$

Proof: Using (8.31) and the definition of $\overline{v}$ it holds

$$\begin{aligned}\phi(v;\overline{v}) &= \frac{e^T v^2 - e^T \overline{v}^2}{\max(\overline{v})^2} + \sum_{i=1}^{n} \frac{\overline{v}_i^2}{\max(\overline{v})^2} \ln \frac{\overline{v}_i^2}{v_i^2} = \sum_{i=1}^{n} \ln \frac{e^T v^2}{n v_i^2} \\ &\leq n \ln \frac{1}{\omega(v)^2} = 2n \ln \frac{1}{\omega(v)},\end{aligned}$$

which proves the result. □

Combining Lemma 8.3.12 with Theorem 8.3.6 it follows that the centering can be done by minimizing the barrier function in at most $\mathcal{O}(n \ln 1/\omega)$ iterations. Observe that in the long–step algorithm we may choose the point $\overline{v} := \sqrt{e^T v^2/n}\, e$ on the central path, whereas in the small–step algorithm in Section 8.2.4 we had to choose $\overline{v} := \max(v)\, e$. The following lemma analyzes this and still another variant. The different strategies are depicted in Figure 8.4.

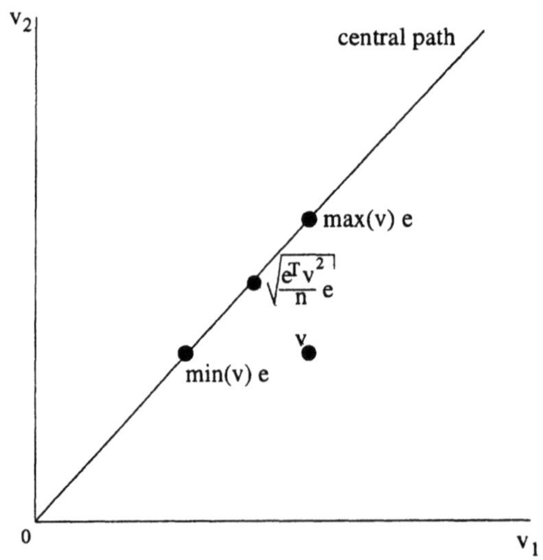

Figure 8.4 Different centering strategies in long–step algorithms.

Lemma 8.3.13 *(i) If* $\overline{v} := \max(v)\, e$ *then* $\phi(v;\overline{v}) \leq 2n \ln 1/\omega(v)$.
(ii) If $\overline{v} := \min(v)\, e$ *then* $\phi(v;\overline{v}) \leq n(1-\omega(v)^2)/\omega(v)^2$.

Proof: Left to the reader. □

Computing a weighted center

We consider the problem of computing the weighted center corresponding to $\overline{v}$, given a positive primal–dual pair (x, s). Let $v^2 := xs$ and define $\mu := e^T v^2/n$ and $\overline{\mu} := e^T \overline{v}^2/n$. We propose the following three–step procedure: the first target is $\sqrt{\mu}e$ on the central path; the second target is $\sqrt{\overline{\mu}}\,e$ on the central path; the final target is $\overline{v}$.

Lemma 8.3.14 *With v and $\overline{v}$ as given above, $\omega := \omega(v)$ and $\overline{\omega} := \omega(\overline{v})$, the three–step algorithm requires at most*

$$\mathcal{O}\left(n\left(\ln\frac{1}{\omega}+\frac{\mu}{\overline{\mu}}-1+\left|\ln\frac{\overline{\mu}}{\mu}\right|+\frac{1}{\overline{\omega}^4}\ln\frac{1}{\overline{\omega}}\right)\right.$$
$$\left.+\sqrt{n}\left(\max\left(1-\frac{\mu}{\overline{\mu}},\frac{\mu}{\overline{\mu}}-1\right)+\frac{1}{\overline{\omega}^4}\left(\frac{1}{\overline{\omega}^2}-1\right)\right)\right)$$

damped Newton steps.

Proof: By Lemma 8.3.12 we have $\phi(v; \sqrt{\mu}e) \leq 2n \ln 1/\omega$. For the second stage, it holds

$$\phi(\sqrt{\mu}e; \sqrt{\overline{\mu}}e) = \frac{n\mu - n\overline{\mu}}{\overline{\mu}} - \sum_{i=1}^{n} \ln\frac{\mu}{\overline{\mu}} \leq n\left(\frac{\mu}{\overline{\mu}}-1\right) + n\left|\ln\frac{\mu}{\overline{\mu}}\right|.$$

For the final stage, we have

$$\begin{aligned}\phi(\sqrt{\overline{\mu}}e; \overline{v}) &= \frac{n\overline{\mu} - e^T\overline{v}^2}{\max(\overline{v}^2)} + \sum_{i=1}^{n} \frac{\overline{v}_i^2}{\max(\overline{v}^2)} \ln\frac{\overline{v}_i^2}{\overline{\mu}} = \sum_{i=1}^{n} \frac{\overline{v}_i^2}{\max(\overline{v}^2)} \ln\frac{n\overline{v}_i^2}{e^T\overline{v}^2} \\ &\leq \sum_{i=1}^{n} \frac{\overline{v}_i^2}{\max(\overline{v}^2)} \ln\frac{1}{\overline{\omega}^2} \leq n\ln\frac{1}{\overline{\omega}^2} = 2n\ln\frac{1}{\overline{\omega}}.\end{aligned}$$

The other quantities in Theorem 8.3.6 are easy to compute. Combining leads to the complexity bound. □

Of course, other applications of the long–step target–following approach can be proposed and analyzed in the general framework approached in this section.

8.4 CONCLUDING REMARKS

The target–following framework in this chapter offers the opportunity of simplifying the complexity proofs for many old as well as new interior point methods.

The key–idea is to analyze the complementarity of nonoptimal solutions and to check how a method forces the complementarity to its final value. This chapter summarizes many pages of literature on interior point methods into a few, and gives tools for some newly proposed methods to be analyzed. Some methods we considered are path–following variants of the Dikin–affine scaling method derived in Chapter 5; these methods have the property of simultaneously driving the iterates to optimality as well as to the central path when started at an arbitrary interior–feasible solution.

Both small–step as well as long–step primal–dual methods for LP have been studied. For the long–step interior point methods it was shown that they yield a significantly worse complexity bound with the use of targets not close to the central path as compared to central path–following methods.

The target–following framework is also interesting from a computational point of view, since it may provide insight how problems like 'recentering' and 'warm–starting' could be tackled. The latter aspect has been picked up by Gondzio [82], who improved his interior point code HOPDM using target–following ideas.

9

TARGET–FOLLOWING FOR NONLINEAR PROGRAMMING

9.1 INTRODUCTION

For convex nonlinear programming (NLP) Nesterov and Nemirovskii [193] gave an analysis of the standard (primal) short–step logarithmic barrier method. They introduced the self–concordance condition (Definition 6.5.3), which generalizes earlier smoothness conditions by Zhu [258] and Jarre [118], and is generally accepted as the one suitable for the analysis of short–step methods. In this chapter we show that the target–following framework can be applied to convex programming problems as well, thereby generalizing the analysis of Nesterov and Nemirovskii [193] to interior point methods not necessarily following the central path, see Section 9.2. Here, we use the fact that the primal–dual system (8.1) is also the KKT–system for minimizing the primal (or dual) barrier function over the primal (dual) feasible region. The main difference of our analysis compared to the one in [193] is that in our case the self–concordance parameters are not constant but change from one iteration to another, depending on the change in the targets. As far as we know, many of the methods we analyze in this section have been analyzed and applied to LP, but not analyzed for NLP problems.

Nesterov and Nemirovskii [193] also consider *variational inequalities*, which is a class of problems including as special cases many equilibrium problems in economics, transportation planning and game–theory, as well as convex programming itself (for a survey see Harker and Pang [98]). The variational inequality is not an optimization problem, but a feasibility problem. Still, in [193] it is shown how an interior point method can be adopted to (approximately) solve the problem. In Section 9.3 we generalize these results to the target–following setting. The observation that for the analysis of variational inequalities no objective function is involved, implies that we do not need to consider a barrier function explicitly. Instead, we show that it suffices to apply mappings that have the same self–concordance type properties as gradients of a barrier function; we will call such mappings *self–concordant barrier–operators.* The advantage of this approach is that it provides a way of dealing with non–central path–following methods for cones other than the nonnegative orthant, as the cone of positive semidefinite matrices and the second order cone.

In NLP complexity of algorithms means the number of arithmetic operations required to compute a feasible solution with objective value within ϵ of the optimal value (Judin and Nemirovskii [124]). Since all methods in this section require $\mathcal{O}(n^3)$ operations per (Newton) step, we are mainly interested in the number of Newton steps to be performed.

9.2 CONVEX PROGRAMMING

In this section we analyze target–following methods for convex programming. We first introduce required backgroundon self–concordant barriers, then analyze the Newton process, and finally show how to apply the obtained results.

9.2.1 Self–concordant barriers

The problem we consider is

$$\text{(CP)} \quad \min_y \{ \, f_0(y) \; : \; -f_i(y) \geq 0, \; i = 1, \ldots, n, \; y \in \mathbb{R}^m \, \}.$$

We assume that the functions $f_i(y)$, $i = 0, \ldots, n$, are convex and three times continuously differentiable. Without loss of generality we may assume $f_0(y)$ to be linear. The feasible set is denoted by

$$\mathcal{F} := \{ \, y \in \mathbb{R}^m \; : \; -f_i(y) \geq 0, \; i = 1, \ldots, n \, \},$$

and the set of positive solutions in $\mathcal{F}$ as

$$\mathcal{F}^0 := \{ \, y \in \mathbb{R}^m \; : \; -f_i(y) > 0, \; i = 1, \ldots, n \, \}.$$

We make the following assumption.

Assumption 9.2.1 *$\mathcal{F}^0$ is not empty and the level–sets of (CP) are bounded.*

The Wolfe–dual of (CP) is given by

$$\text{(CD)} \ \min_{x,y}\{\ f_0(y) - \sum_{i=1}^{n} x_i f_i(y)\ :\ \sum_{i=1}^{n} x_i \nabla f_i(y) = \nabla f_0(y),\ x_i \geq 0\ i = 1,\ldots,n\ \}.$$

Let us consider the following system of nonlinear equations:

$$\begin{array}{rcl} f_i(y) & \leq & 0, \quad x_i \geq 0, \quad i = 1,\ldots,n, \\ \sum_{i=1}^{n} x_i \nabla f_i(y) & = & \nabla f_0(y) \\ -x_i f_i(y) & = & w_i, \end{array} \tag{9.1}$$

for some $w \in \mathbb{R}^n_{++}$. The following theorem is analogous to Theorem 2.1.5 and follows from Monteiro and Zhou [185].

Theorem 9.2.2 *If Assumption 9.2.1 holds then system (9.1) has a unique solution for any* $w \in \mathbb{R}^n_{++}$.

The proof of the theorem uses the fact that (9.1) is the KKT–system for minimizing the weighted logarithmic barrier

$$f(y; w) = f_0(y) - \sum_{i=1}^{n} w_i \ln(-f_i(y)) \tag{9.2}$$

over $\mathcal{F}$. This permits us to interpret target–following in the v–space as using different weights w in a weighted logarithmic barrier method. Henceforth, speaking of a target w will have the interpretation of a weight w in a barrier function. Since (9.1) contains nonlinear and nonconvex dual constraints it is hard to analyze primal–dual (feasible) algorithms. Instead only primal or dual target–following methods will be considered. Nesterov and Todd [194] are able to analyze primal–dual potential reduction methods for convex problems by transforming the problem in a conic formulation. However, the practical merits of such reformulations are unclear.

In the analysis we use self–concordant barriers (Definition 6.5.3). We use the self–concordance properties of the barrier function to measure proximity to a target (cf. Den Hertog [100] and Section 8.3). The following lemma is a fundamental result for constructing self–concordant functions for intersections of convex spaces.

Lemma 9.2.3 *Let* G_i, $i = 1,\ldots,n$, *be closed convex domains in* $\mathbb{R}^m$, *such that* $G = \cap_{i=1}^{n} G_i$ *has a nonempty interior, and let* $F_i(y)$ *be a* $(1,\vartheta_i)$*–self–concordant* *barrier for G_i. Let $w \in \mathbb{R}^n_+$. Then*

$$F(y;w) := \sum_{i=1}^{n} w_i F_i(y)$$

is a $(\min(w), \sum_{i=1}^{n} w_i \vartheta_i)$-self-concordant barrier for G.

Proof: Elementary; cf. Propositions 2.1.1 and 2.3.1 in [193]. □

To analyze target–following methods for (CP) we assume that the functions $-\ln(-f_i)$ are $(1, \vartheta_i)$–self–concordant barriers for the spaces $\{ y : f_i(y) \leq 0 \}$. As in [100] we restrict ourselves to the use of logarithmic barriers. Although this may seem to be a real restriction, up to now (as far as we know) all convex programming problems that have been handled by the notion of self–concordance use some type of logarithmic barrier. Moreover, the general case will be handled in the context of variational inequalities in Section 9.3. Furthermore, the existence of weighted centers has only been extensively studied in the logarithmic case (Monteiro and Zhou [185]). From Remark 2.3.1 and Corollary 2.3.3 in [193] it follows that we may assume $\vartheta_i \geq 1$. As mentioned, we assume $f_0(y)$ to be linear (following Nesterov and Nemirovskii [193] and Den Hertog et al. [101] we could use the notion of *compatible* objective functions, which would unnecessarily complicate the discussion). The self–concordance property has been shown for many classes of convex programming problems as: linear and convex quadratic programming with convex quadratic constraints, primal geometric programming, ℓ_p-approximation, matrix–norm minimization and finding a maximal inscribed ellipsoid (Nesterov and Nemirovskii [193]) and dual geometric programming, extended entropy programming and ℓ_p–programming (Den Hertog et al. [101]).

Motivated by (9.2) and Lemma 9.2.3 we use the following weighted barrier function

$$\phi(y;w) := \frac{f_0(y)}{\min(w)} - \sum_{i=1}^{n} \frac{w_i}{\min(w)} \ln(-f_i(y)),$$

where $w \in \mathbb{R}^n_{++}$. Upon application of Lemma 9.2.3 it follows that $\phi(y;w)$ is an $(a(w), \vartheta(w))$–self–concordant barrier, with

$$a(w) = 1, \quad \vartheta(w) = \sum_{i=1}^{n} \frac{w_i}{\min(w)} \vartheta_i. \tag{9.3}$$

Since $a(w)$ is independent of w, we henceforth ignore it and just talk about $\phi(\cdot;w)$ as a $\vartheta(w)$–self–concordant barrier. We denote by $g(y;w)$ and $H(y;w)$

the gradient respectively Hessian of $\phi(y;w)$. We use the notation

$$\psi(y;w) = -\sum_{i=1}^{n} \frac{w_i}{\min(w)} \ln(-f_i(y)) \tag{9.4}$$

for the barrier term only. Because of the linearity of $f_0(y)$, $\psi(y;w)$ is also a $\vartheta(w)$–self–concordant barrier. Observe that the Hessians of $\phi(y;w)$ and $\psi(y;w)$ are equal, so

$$H(y;w) := \nabla^2\phi(y,w) = \nabla^2\psi(y;w).$$

We denote $H_i := \nabla^2(-\ln(-f_i))$.

9.2.2 Analysis of the Newton process

As in Chapter 8 the major task is to analyze the Newton process. Since the function $\phi(y;w)$ is a self–concordant barrier it can be minimized efficiently by a (damped) Newton–type method. Some basic results are obtained from Nesterov and Nemirovskii [193, Chapter 2] and Den Hertog [100, Chapter 2] and are quoted below. Then we study the effect of the Newton step on the objective value and proximity measure. In the analysis we denote the current iterate and weight by y and w respectively, while the next iterate and new weight are denoted by y^+ and w^+. We denote by $p(y;w)$ the Newton step at y with respect to the barrier function $\phi(y;w)$, so

$$p(y;w) = -H(y;w)^{-1}g(y;w).$$

The proximity measure used is the *Hessian norm of the Newton step*

$$\delta(y;w) := \|p(y;w)\|_{H(y,w)} = \sqrt{p(y;w)^T H(y;w)p(y;w)}.$$

The first result is important for self–concordant barriers and is called the *semi-boundedness property*.

Lemma 9.2.4 *Let $x \in \mathcal{F}^0$, $y \in \mathcal{F}$. Then $\nabla\psi(x;w)^T(y-x) \leq \vartheta(w)$.*

Proof: See Proposition 2.3.2 in [193]. □

The following lemma concerns feasibility and quadratic convergence of the Newton step.

Lemma 9.2.5 *Let $p := p(y; w)$ be the Newton step at $y \in \mathcal{F}^0$ and assume that $\delta(y; w) < 1$. Then*
(i) $y + p \in \mathcal{F}^0$;
(ii) The proximity after the Newton step satisfies

$$\delta(y + p; w) \leq \frac{\delta(y; w)^2}{(1 - \delta(y; w))^2}.$$

Proof: See Lemmas 2.20 and 2.21 in [100]. □

The next lemma gives a bound for the value of the barrier function in an approximate minimizer of $\phi(y; w)$.

Lemma 9.2.6 *Let $y(w)$ denote the minimizer of $\phi(y; w)$, and assume that $\delta := \delta(y; w) < 1/3$. Denote $\gamma(\delta) := 1 - (1 - 3\delta)^{1/3}$. Then*

$$\phi(y; w) - \phi(y(w); w) \leq \frac{1}{2}\gamma(\delta)^2 \frac{1 + \gamma(\delta)}{1 - \gamma(\delta)}.$$

Proof: See Theorem 2.2.2 in [193]. □

The next result enables us to estimate the duality gap in points close to the minimizer of $\phi(y; w)$. Its contents and proof are similar to p.68(A) and Proposition 3.2.4 in [193].

Lemma 9.2.7 *Let y^* be an optimal solution of (CP), $y(w)$ the minimizer of $\phi(y; w)$ and y such that $\delta(y; w) < 1/3$. Let $\gamma(\delta)$ be as defined in Lemma 9.2.6. Then*
(i) $f_0(y(w)) - f_0(y^*) \leq \min(w)\vartheta(w)$;
(ii) $f_0(y) - f_0(y(w)) \leq \min(w)\left(\vartheta(w) + 1/2\,\gamma(\delta)^2(1 + \gamma(\delta))/(1 - \gamma(\delta))\right)$;
(iii) $f_0(y) \leq f_0(y^*) + \min(w)\left(2\vartheta(w) + 1/2\,\gamma(\delta)^2(1 + \gamma(\delta))/(1 - \gamma(\delta))\right)$.

Proof: *(i)* Since $f_0(y)$ is linear and $y(w)$ minimizes $\phi(y; w)$ we have

$$\begin{aligned} f_0(y(w)) - f_0(y^*) &= \nabla f_0(y(w))^T (y(w) - y^*) \\ &= -\min(w)(\nabla\psi(y(w); w))^T (y(w) - y^*) \leq \min(w)\vartheta(w), \end{aligned}$$

where the inequality follows from Lemma 9.2.4.
(ii) We have

$$f_0(y) = \min(w)(\phi(y; w) - \psi(y; w)) + f_0(y(w))$$

$$
\begin{aligned}
& -\min(w)(\phi(y(w);w) - \psi(y(w);w)) \\
= \;& f_0(y(w)) + \min(w)(\phi(y;w) - \phi(y(w);w)) \\
& +\min(w)(\psi(y(w);w) - \psi(y;w)) \\
\le \;& f_0(y(w)) + \min(w)\frac{1}{2}\gamma(\delta)^2\frac{1+\gamma(\delta)}{1-\gamma(\delta)} \\
& +\min(w)(\nabla\psi(y(w);w))^T(y(w)-y),
\end{aligned}
$$

where the inequality follows from the convexity of $\psi(y;w)$ and Lemma 9.2.6. To complete the proof of (i), we use that [193, Th. 2.1.1] shows that for $z := y(w) + (y(w) - y)$ it holds $z \in \mathcal{F}^0$. Then Lemma 9.2.4 gives

$$
\nabla\psi(y(w);w))^T(y(w)-y) = \nabla\psi(y(w);w))^T(z - y(w)) \le \vartheta(w).
$$

(iii) Follows from (i) and (ii). □

The following lemma will be important in the proof of the main result.

Lemma 9.2.8 *Let $\psi(y;w)$ be as defined in (9.4) and denote $H := H(y;w)$. Let $\alpha \in \mathbb{R}^n_{++}$ and consider*

$$
G(y;w,\alpha) := -\sum_{i=1}^{n} \frac{\alpha_i w_i}{\min(\alpha)\min(w)} \ln(-f_i(y)).
$$

Let $y \in \mathcal{F}^0$ and $w \in \mathbb{R}^n_{++}$ be arbitrary. Define $\chi_G := \nabla G(y;w,\alpha)$ and $H_G := \nabla^2 G(y;w,\alpha)$. Then

$$
\chi_G^T H^{-1} \chi_G \le \left(\frac{\max(\alpha)}{\min(\alpha)}\right)^2 \vartheta(w).
$$

Proof: The self-concordance parameters of $G(y;w,\alpha)$ satisfy

$$
a_G(w,\alpha) \ge 1, \quad \vartheta_G(w,\alpha) \le \sum_{i=1}^{n} \frac{\alpha_i w_i}{\min(\alpha)\min(w)}\vartheta_i,
$$

as follows from Lemma 9.2.3. Observe that

$$
H = \sum_{i=1}^{n} \frac{w_i}{\min(w)} H_i = \sum_{i=1}^{n} \frac{\alpha_i w_i}{\min(\alpha)\min(w)} H_i \frac{\min(\alpha)}{\alpha_i} \succeq \frac{\min(\alpha)}{\max(\alpha)} H_G,
$$

since H_i is positive definite for any i. Then we have

$$\begin{aligned}\chi_G^T H^{-1}\chi_G &\le \frac{\max(\alpha)}{\min(\alpha)}\chi_G^T H_G^{-1}\chi_G \le \frac{\max(\alpha)}{\min(\alpha)}\vartheta_G(w,\alpha)\\ &\le \frac{\max(\alpha)}{\min(\alpha)}\sum_{i=1}^{n}\frac{\alpha_i w_i}{\min(\alpha)\min(w)}\vartheta_i \le \left(\frac{\max(\alpha)}{\min(\alpha)}\right)^2\vartheta(w),\end{aligned}$$

where the second inequality can be shown as follows[1]. Let $p_G := -H_G^{-1}\chi_G$. Then

$$\chi_G^T H_G^{-1}\chi_G = -p_G^T\chi_G \le \sqrt{\vartheta_G(\alpha,w)}\sqrt{p_G^T H_G p_G} = \sqrt{\vartheta_G(\alpha,w)}\sqrt{\chi_G^T H_G^{-1}\chi_G},$$

where the inequality follows from Definition 6.5.3. Hence

$$\sqrt{\chi_G^T H_G^{-1}\chi_G} \le \sqrt{\vartheta_G(\alpha,w)}.$$

This completes the proof. □

We estimate the effect of a target update on the proximity measure. For that purpose we impose the following condition on the new target w^+.

Condition 9.2.9 *There exist a constant $\tau \in (0,1)$ such that w^+ satisfies*

$$\tau \le \frac{w_i^+}{w_i} \le \frac{1}{\tau}, \quad i = 1,\ldots,n.$$

Observe that the condition implies $w^+ > 0$; moreover, for any i it holds

$$\tau^2\frac{w_i}{\min(w)} \le \frac{w_i^+}{\min(w^+)} \le \frac{1}{\tau^2}\frac{w_i}{\min(w)}. \tag{9.5}$$

Also, we have

$$\min(w^+) = w_j^+ \ge \tau w_j \ge \tau\min(w). \tag{9.6}$$

Whenever the target is updated, the self–concordance parameter of the barrier function changes. This is the major difference with the analysis in [193]. The new parameter can be bounded as follows.

[1] Den Hertog [100, Lemma 2.25] uses an argument based on eigenvalues in a similar proof for a different result.

Lemma 9.2.10 *Let w^+ satisfy Condition 9.2.9. For the self-concordance parameters of $\phi(y;w)$ and $\phi(y;w^+)$ it holds*

$$\tau^2 \vartheta(w) \leq \vartheta(w^+) \leq \frac{1}{\tau^2}\vartheta(w).$$

Proof: By Lemma 9.2.3 we have $\vartheta(w^+) = \sum_{i=1}^n (w_i^+/\min(w^+))\vartheta_i$. The desired inequalities immediately follow from (9.5). □

To bound the proximity after a Newton step followed by a target update, we first relate the new Hessian matrix to the old one (for simplicity we write f_i for $f_i(y)$):

$$\begin{aligned} H(y;w^+) &= \sum_{i=1}^n \frac{w_i^+}{\min(w^+)} \nabla^2(-\ln(-f_i)) \\ &\succeq \min\left(\frac{\min(w)}{\min(w^+)}\frac{w^+}{w}\right)\left(\sum_{i=1}^n \frac{w_i}{\min(w)}\nabla^2(-\ln(-f_i))\right) \\ &\succeq \tau\frac{\min(w)}{\min(w^+)} H(y;w), \end{aligned} \tag{9.7}$$

where $y \in \mathcal{F}^0$ is arbitrary. We now prove the main theorem.

Theorem 9.2.11 *Let $g := g(y;w)$, $H := H(y;w)$ and $p := -H^{-1}g$, such that $\delta(y;w) = \|p\|_H \leq 1/3$. Let w^+ be a target satisfying Condition 9.2.9. Define $g^{++} := g(y^+;w^+)$, $H^{++} := H(y^+;w^+)$ and let $p^{++} := p(y^+;w^+) := -(H^{++})^{-1}g^{++}$ be the Newton step in y^+ with respect to $\phi(\cdot;w^+)$. Then, it holds*

$$\delta(y^+;w^+) \leq \frac{1}{\tau}\left(\frac{\delta(y;w)^2}{(1-\delta(y;w))^2} + \left(\frac{1}{\tau} - \tau\right)\sqrt{\vartheta(w)}\right). \tag{9.8}$$

Proof: Let us further define $g^+ := g(y^+;w)$, $H^+ := H(y^+;w)$ and $p^+ := p(y^+;w) = -(H^+)^{-1}g^+$. From (9.7) we derive

$$(H^{++})^{-1} \preceq \frac{1}{\tau}\frac{\min(w^+)}{\min(w)}(H^+)^{-1}.$$

Writing f_i and ∇f_i for $f_i(y^+)$ and $\nabla f_i(y^+)$ respectively, we obtain

$$\|p^{++}\|_{H^{++}} = \sqrt{(g^{++})^T(H^{++})^{-1}g^{++}}$$

$$\le \frac{1}{\sqrt{\tau}}\sqrt{\frac{\min(w^+)}{\min(w)}}\sqrt{(g^{++})^T(H^+)^{-1}g^{++}}$$
$$= \frac{1}{\sqrt{\tau}}\sqrt{\frac{\min(w^+)}{\min(w)}}\left\|\frac{\nabla f_0}{\min(w^+)}+\sum_{i=1}^n \frac{w_i^+}{\min(w^+)}\frac{\nabla f_i}{-f_i}\right\|_{(H^+)^{-1}}$$
$$= \frac{1}{\sqrt{\tau}}\sqrt{\frac{\min(w)}{\min(w^+)}}\left\|\frac{\nabla f_0}{\min(w)}+\sum_{i=1}^n \frac{w_i}{\min(w)}\frac{\nabla f_i}{-f_i}+\sum_{i=1}^n \frac{w_i^+-w_i}{\min(w)}\frac{\nabla f_i}{-f_i}\right\|_{(H^+)^{-1}}$$
$$\le \frac{1}{\tau}\left(\|g^+\|_{(H^+)^{-1}}+\left\|\sum_{i=1}^n \frac{w_i^+-w_i}{\min(w)}\frac{\nabla f_i}{-f_i}\right\|_{(H^+)^{-1}}\right). \tag{9.9}$$

In the first inequality we use (9.7), in the second we apply (9.6) and the triangle–inequality. Observe that the first part in the last term is bounded by the quadratic convergence result of Lemma 9.2.5. We apply Lemma 9.2.8 to estimate the norm in the second part. Let us define $I_1 := \{i : w_i^+ < w_i\}$, $I_2 := \{i : w_i^+ = w_i\}$ $I_3 := \{i : w_i^+ > w_i\}$, and split $H^+ = \sum_{i=1}^n H_i(y^+;w) =: H^{1+} + H^{2+} + H^{3+}$ according to these sets. Then it holds

$$\begin{aligned}\left\|\sum_{i=1}^n \frac{w_i^+-w_i}{\min(w)}\frac{\nabla f_i}{-f_i}\right\|_{(H^+)^{-1}} &\le \left\|\sum_{i\in I_1} \frac{w_i^+-w_i}{\min(w)}\frac{\nabla f_i}{-f_i}\right\|_{(H^+)^{-1}}\\ &\quad+\left\|\sum_{i\in I_2} \frac{w_i^+-w_i}{\min(w)}\frac{\nabla f_i}{-f_i}\right\|_{(H^+)^{-1}}+\left\|\sum_{i\in I_3} \frac{w_i^+-w_i}{\min(w)}\frac{\nabla f_i}{-f_i}\right\|_{(H^+)^{-1}}\\ &\le \left\|\sum_{i\in I_1} \frac{w_i^+-w_i}{\min(w)}\frac{\nabla f_i}{-f_i}\right\|_{(H^{1+})^{-1}}+\left\|\sum_{i\in I_3} \frac{w_i^+-w_i}{\min(w)}\frac{\nabla f_i}{-f_i}\right\|_{(H^{3+})^{-1}}\end{aligned}$$

For the indices in I_1 we let $\alpha_i := 1 - w_i^+/w_i > 0$, and using Lemma 9.2.8 it holds

$$\begin{aligned}\left\|\sum_{i\in I_1} \frac{w_i^+-w_i}{\min(w)}\frac{\nabla f_i}{-f_i}\right\|_{(H^{1+})^{-1}} &= \min(\alpha)\left\|\sum_{i\in I_1} \frac{\alpha_i w_i}{\min(\alpha)\min(w)}\frac{\nabla f_i}{-f_i}\right\|_{(H^{1+})^{-1}}\\ &\le \min(\alpha)\frac{\max(\alpha)}{\min(\alpha)}\sqrt{\vartheta(w)} \le (1-\tau)\sqrt{\vartheta(w)}.\end{aligned}$$

For the indices in I_3 we define $\alpha_i := w_i^+/w_i - 1$, and the inequality

$$\left\|\sum_{i\in I_3} \frac{w_i^+-w_i}{\min(w)}\frac{\nabla f_i}{-f_i}\right\|_{(H^{3+})^{-1}} \le \left(\frac{1}{\tau}-1\right)\sqrt{\vartheta(w)}.$$

follows similarly. Combining these bounds, we have proved (9.8). □

The following corollary can be derived.

Corollary 9.2.12 *If* $\delta(y;w) \leq 1/4$ *and* $\tau \geq 1 - 1/(18\sqrt{\vartheta(w)})$ *then* $\delta(y^+;w^+) \leq 1/4$.

Proof: If $n \geq 2$ then $\vartheta(w) \geq 2$. The result now follows from Theorem 9.2.11. □

Observe, that we have an easy to check condition on the new target (Condition 9.2.9) to analyze the Newton process for a specific updating scheme of the weights w. The maximal step that can be taken with the scheme can be determined with the help of the parameter τ as in Corollary 9.2.12. We mention that the condition seems more easily than the one in Theorems 8.2.6 and 8.3.6 for the linear case. However, Lemma 8.2.2 and the discussion preceding it show that they are almost equivalent. The analysis developed here is related to the use of self–concordant families (Definition 3.1.1 in Nesterov and Nemirovskii [193]). Their analysis is more general in the sense that the barriers need not be logarithmic barriers. On the other hand, their families are parametrized by only one parameter.

9.2.3 Applications

Many interior point algorithms for LP can, when suitably adjusted, be applied to NLP as well. In this section we analyze just two: the path–following and the Dikin–path–following algorithm.

Central and weighted path–following

The weighted path–following algorithm for (CP) was already given in the book of Fiacco and McCormick [53]. Here we give a first proof of polynomiality. The target update is given by

$$w^+ = (1-\theta)w, \tag{9.10}$$

where θ is a suitable step size; let $w^{(0)}$ denote the initial target, and $\omega_0^2 := \omega(w^{(0)}) = \min(w)/\max(w)$. Using (9.3) it holds

$$\vartheta(w^+) = \vartheta(w) = \vartheta(w^{(0)}) \leq \frac{1}{\omega_0^2}\sum_{i=1}^{n}\vartheta_i. \tag{9.11}$$

We impose the condition that the algorithm computes approximations y to the centers $y(w)$ such that $\delta(y;w) \le 1/4$ for all iterates. Observe that the update satisfies Condition 9.2.9 with $\tau = 1-\theta$, hence Corollary 9.2.12 guarantees that we can take $\theta \le 1/(18\sqrt{\vartheta(w)})$. Observe that θ can be bounded from below by

$$\theta := \frac{\omega_0}{18\sqrt{\vartheta(e)}}, \tag{9.12}$$

whose validity follows from (9.11). We compute the number of iterations in terms of θ needed to solve the problem.

Lemma 9.2.13 *The weighted path-following algorithm for (CP) with update (9.10) stops after at most*

$$\frac{1}{\theta}\ln\left(\frac{3\min(w^{(0)})\vartheta(w^{(0)})}{\epsilon}\right)$$

iterations with $y \in \mathcal{F}$ for which $f_0(y) - f_0(y^*) \le \epsilon$.

Proof: Denote the final iterate by $y^{(k)}$. Since $\delta := \delta(y^{(k)}; w^{(k)}) \le 1/4$, applying Lemma 9.2.6 with $y^{(k)}$ gives

$$\gamma(\delta) < \frac{3}{8}, \quad \frac{1}{2}\gamma(\delta)^2\frac{1+\gamma(\delta)}{1-\gamma(\delta)} < \frac{1}{5}.$$

From Lemma 9.2.7 we obtain

$$\begin{aligned} f_0(y^{(k)}) - f_0(y^*) &\le \min(w^{(k)})\left(2\vartheta(w^{(k)}) + \frac{1}{5}\right) \le 3\min(w^{(k)})\vartheta(w^{(k)}) \\ &= 3\min(w^{(k)})\vartheta(w^{(0)}). \end{aligned}$$

Hence it suffices to require

$$\min(w^{(k)}) \le \frac{\epsilon}{3\vartheta(w^{(0)})}.$$

Using $\min(w^{(k)}) = (1-\theta)^k \min(w^{(0)})$ it follows that the number of iterations needed is certainly not larger than

$$\frac{1}{\theta}\ln\left(\frac{3\min(w^{(0)})\vartheta(w^{(0)})}{\epsilon}\right).$$

□

Using (9.12) the complexity of the algorithm can be derived and is given as a corollary.

Corollary 9.2.14 *The weighted path-following algorithm for (CP) requires at most*

$$\mathcal{O}\left(\frac{\sqrt{\vartheta(e)}}{\omega_0}\ln\left(\frac{3\min(w^{(0)})\vartheta(w^{(0)})}{\epsilon}\right)\right)$$

iterations when started from an interior-feasible point $y^{(0)}$ and an initial weight vector $w^{(0)}$ for which $\delta(y^{(0)}; w^{(0)}) \leq 1/4$.

Dikin-path-following

We analyze the Dikin-path-following algorithm introduced in Section 8.2.3. Recall from (8.14) that the path starting at some given $w^{(0)}$ is given by

$$\Psi(t; \sqrt{w^{(0)}})^2 = \frac{w^{(0)}}{w^{(0)}t + e}, \qquad t \geq 0,$$

and satisfies (Lemma 8.2.7)

$$\Psi(t_1 + t_2; \sqrt{w^{(0)}}) = \Psi(t_2; \Psi(t_1; \sqrt{w^{(0)}})).$$

The Dikin-path converges to the central path, in the sense that $\omega(w)$ increases along the path (Lemma 8.2.7). We denote $\omega_0^2 := \min(w^{(0)})/\max(w^{(0)})$. With (9.3), it holds for any w on the Dikin-path

$$\vartheta(w) = \sum_{i=1}^n \frac{w_i}{\min(w)}\vartheta_i \leq \sum_{i=1}^n \frac{w_i^{(0)}}{\min(w^{(0)})}\vartheta_i = \vartheta(w^{(0)}) \leq \frac{1}{\omega_0^2}\sum_{i=1}^n \vartheta_i,$$

since the ratio $w_i/\min(w)$ is decreasing along the Dikin-path. We define a step from w to w^+ along the path, with step size θ, as

$$w^+ := \frac{w}{\theta\, w + e}.$$

It is easily seen that

$$\frac{1}{\theta\, w_i^{(0)} + 1} \leq \frac{w_i^+}{w_i} = \frac{1}{\theta\, w_i + 1} \leq 1,$$

hence Condition 9.2.9 is satisfied with $\tau = 1/(\theta\max(w)+1)$. Applying Corollary 9.2.12 it follows that θ should satisfy

$$\theta \leq \frac{1}{\max(w)}\frac{1}{18\sqrt{\vartheta(w)}},$$

which is guaranteed by taking

$$\theta = \frac{\omega_0}{\max(w)} \frac{1}{18\sqrt{\vartheta(e)}}. \tag{9.13}$$

Using Lemma 9.2.7 we obtain the sufficient condition for termination

$$f_0(y) - f_0(y^*) \leq 3\min(w^{(k)})\vartheta(w^{(k)}) \leq 3\min(w^{(k)})\vartheta(w^{(0)}) \leq \epsilon.$$

It holds

$$\begin{aligned}
\min(w^{(k)}) &= \frac{\min(w^{(k-1)})}{\theta\min(w^{(k-1)}) + 1} \\
&\leq \frac{\min(w^{(k-1)})}{\omega_0\min(w^{(k-1)})/(18\sqrt{\vartheta(e)}\max(w^{(k-1)})) + 1} \\
&\leq \left(\frac{1}{\omega_0^3/(18\sqrt{\vartheta(e)}) + 1}\right)^k \min(w^{(0)}),
\end{aligned}$$

plugging in the relevant choice for θ, and using Lemma 8.2.7(iii) in the last inequality. Hence, it suffices to do at most

$$\mathcal{O}\left(\frac{\sqrt{\vartheta(e)}}{\omega_0^3} \ln \frac{3\min(w^{(0)})\vartheta(w^{(0)})}{\epsilon}\right)$$

iterations. Note that the iteration bound is exactly the same as in Theorem 8.2.9. Using a similar strategy as in Lemma 8.2.16 the denominator can be improved from ω_0^3 to ω_0. Then we have a complexity bound that is exactly the same as in the weighted path–following method. Although we know that the deviation of the central path (measured via $\omega(w)$) decreases along the Dikin–path, in the analysis above it is not possible to use this extra information for a better complexity result.

Discussion

Other applications of the target–following concept in NLP are omitted as they follow from the results given above. Notice that a natural choice for the weights could be $w_i = 1/\vartheta_i$, since this leads to $\vartheta(w) = n\max_i(\vartheta_i)$, independent of w. It is not clear how the target–following approach can be generalized to the theory of self–concordant barriers for general convex cones as in [193]. For instance, it is not evident how to include weights in the barrier $-\ln\det X$ for the cone of positive semidefinite matrices. However, it is straightforward to include weights in its derivative $-X^{-1}$, for instance using

$$-W^{1/2}X^{-1}W^{1/2},$$

for a diagonal matrix $W \succ 0$. This idea will be exploited in the next section, where variational inequalities are analyzed.

9.3 VARIATIONAL INEQUALITIES WITH MONOTONE OPERATORS

In this section we extend the use of weighted interior point methods (the target–following approach) to monotone variational inequalities, hence extending the analysis in Chapter 7 of Nesterov and Nemirovskii [193]. Variational inequalities have many applications, e.g., in equilibrium problems, saddle–point problems and game theory. For a survey we refer to Harker and Pang [98]. A nice aspect of the approach we take in this section is that we do not have to deal with (weighted) barrier functions themselves, but only with operators (mappings) having similar properties as gradients and Hessians of self–concordant barriers. This circumvents the problem noted at the end of Section 9.2.3.

9.3.1 Problem statement and definitions

The problem we consider is stated as follows. Let E be a real vector space[2] with conjugate E^* and inner product $\langle \cdot, \cdot \rangle$. Let G be a closed convex subset of E with nonempty interior $Q := \text{int}(G)$, and let $S : G \to E^*$ be a C^2 smooth single–valued monotone operator, i.e., it holds

$$\langle x - y, S(x) - S(y) \rangle \geq 0, \quad \forall x, y \in G.$$

The variational inequality associated with G and S is as follows

$$\text{(VI)} \quad \text{Find } x \in G \text{ such that } \langle S(x), y - x \rangle \geq 0 \ \ \forall y \in G.$$

Using monotonicity, a solution of (VI) also solves

$$(\overline{\text{VI}}) \quad \text{Find } x \in G \text{ such that } \langle S(y), y - x \rangle \geq 0 \ \ \forall y \in G. \tag{9.14}$$

[2] We allow problems other than defined on $\mathbf{R}^n$.

It is easy to see that the reverse is also true, i.e., a solution to ($\overline{\mathrm{VI}}$) is a solution to (VI). For a discussion on this type of problems, its solvability and the relation with other fields in optimization we refer to Chapter 7 in [193]. One result needed here is that (VI) has a solution if G is bounded which we assume to be the case in this section.

Note that (VI) is not an optimization problem, but a feasibility problem. Still, interior point techniques can be adopted to (approximately) solve the problem. To give an illustration we show that the monotone nonlinear complementarity problem

(MNCP) Given $f : \mathbb{R}^n_+ \to \mathbb{R}^n$, find $x \in \mathbb{R}^n_+$ such that $f(x) \geq 0$, and $\langle f(x), x \rangle = 0$,

where f is a continuous monotone mapping, is a specific instance of (9.14).

Lemma 9.3.1 *(MNCP) is equivalent to ($\overline{VI}$) where $G = \mathbb{R}^n_+$ and $S = f$.*

Proof: We first show that a solution $\overline{x} \geq 0$ to (MNCP) also solves ($\overline{\mathrm{VI}}$). Let $y \in \mathbb{R}^n_+$ be arbitrary. Substituting $\overline{x}^T f(\overline{x}) = 0$ in the monotonicity condition and using $y \geq 0$ and $f(\overline{x}) \geq 0$ it follows

$$f(y)^T(y - \overline{x}) \geq y^T f(\overline{x}) \geq 0.$$

The reverse implication is shown by contradiction. Suppose $\overline{x} \geq 0$ solves ($\overline{\mathrm{VI}}$), but $f_i(\overline{x}) < 0$ for some i. Define

$$\overline{y}_j = \begin{cases} \overline{x}_j & j \neq i \\ \overline{x}_j + \epsilon & j = i. \end{cases}$$

By the continuity of f there exists a positive value for ϵ for which $\overline{y} \geq 0$ and $f_i(\overline{y}) < 0$. Then

$$\langle f(\overline{y}), \overline{y} - \overline{x} \rangle = f_i(\overline{y})\epsilon < 0,$$

which gives a contradiction. Hence $f(\overline{x}) \geq 0$. Suppose now that $\overline{x}^T f(\overline{x}) > 0$, then there is an i for which $\overline{x}_i > 0$ and $f_i(\overline{x}) > 0$. Define

$$\overline{y}_j = \begin{cases} \overline{x}_j & j \neq i \\ \overline{x}_j - \epsilon & j = i. \end{cases}$$

By the continuity of f there exists a positive value for ϵ for which $\overline{y} \geq 0$ and $f_i(\overline{y}) > 0$. Then

$$\langle f(\overline{y}), \overline{y} - \overline{x} \rangle = f_i(\overline{y})(-\epsilon) < 0,$$

which gives a contradiction. This completes the proof. □

For single–valued monotone operators, the first derivative $\nabla S(x)$ defines a linear mapping; from the monotonicity it follows that

$$\langle \nabla S(x)h, h\rangle \geq 0, \quad \forall h \in E.$$

There is a natural conjugate operator $(\nabla S)^*(x) : E \to E^*$, which can be used to define a symmetric and positive semidefinite operator

$$\hat{S}(x) = \frac{1}{2}\left(\nabla S(x) + (\nabla S)^*(x)\right).$$

We now define the following Euclidean seminorm

$$\|h\|_{S,x} = \langle \hat{S}(x)h, h\rangle^{1/2}.$$

In case ∇S is nondegenerate, the conjugate norm is

$$\|\eta\|^*_{S,x} = \langle \eta, \hat{S}^{-1}(x)\eta\rangle^{1/2}, \quad \eta \in E^*.$$

We introduce the following definition (cf. Definition 6.5.3).

Definition 9.3.2 *Let G be a closed convex domain in a finite–dimensional real vector space E, with nonempty interior $Q :=$ int(G), and let $a, \vartheta \geq 0$. A single–valued monotone operator $F : Q \to E^*$ is called a–self–concordant if it is C^2–smooth and if the following relation holds for all $x \in Q$ and $h^{(i)} \in E$, $i = 1, 2, 3$:*

$$|\nabla^2 F(x)[h^{(1)}, h^{(2)}, h^{(3)}]| \leq 2\,a^{-1/2} \prod_{i=1}^{3} \|h^{(i)}\|_{F,x};$$

an a–self–concordant operator F is called strongly self–concordant if the sequence of operators $\hat{F}(x^{(i)})$ is unbounded whenever $x^{(i)} \in Q$ form a sequence converging to a boundary point of Q. A single–valued monotone operator $F : Q \to E^*$ is called an (a, ϑ)–self–concordant barrier–operator if it is strongly a–self–concordant and for all $h \in E$:

$$|\langle F(x), h\rangle| \leq \sqrt{\vartheta}\|h\|_{F,x}.$$

The first statement is as [193, Definition 7.2.1][3]. We introduce self–concordant barrier–operators since we prefer not to deal with self–concordant barriers

[3] For self–concordant functions the given inequality follows from the first inequality in Definition 6.5.3 [193, Appendix 1]. Requiring the inequality here explicitly is necessary, since $\nabla^2 F$ need not be symmetric.

themselves, but only with mappings having 'self–concordance properties'. From the definitions it is immediate that if $f(x)$ is an (a, ϑ)–self–concordant barrier then $F(x) \equiv \nabla f(x)$ is an (a, ϑ)–self–concordant barrier–operator. To solve the variational inequality (9.14) we state the notion of compatibility (cf. [193, Definition 7.3.1]).

Definition 9.3.3 *Let F be an (a, ϑ)–self–concordant barrier–operator for G and let $\beta \geq 0$. A C^2–smooth monotone operator $S : Q \to E^*$ is called β–compatible with F if the inequality*

$$|\nabla^2 S(x)[h^{(1)}, h^{(2)}, h^{(3)}]| \leq \beta \prod_{i=1}^{3} \left(3\langle \hat{S}(x)h^{(i)}, h^{(i)}\rangle\right)^{1/3} \left(3a^{-1}\langle \hat{F}(x)h^{(i)}, h^{(i)}\rangle\right)^{1/6}$$

holds for every $x \in Q$ and all $h^{(1)}, h^{(2)}, h^{(3)} \in E$.

Section 7.2 in [193] gives a comprehensive discussion and proof of the solvability of the equation $F(x) = 0$ for a strongly self–concordant operator F. Solving equations of this type lies at the heart of interior point methods for $(\overline{\text{VI}})$ to be discussed. While Section 7.3 in [193] analyzes a central path–following method, the purpose of this section is to introduce weights $w \in E_+$ and analyze target–following methods for solving $(\overline{\text{VI}})$. Hence, we assume the existence of a mapping $F : E \times E_+ \to E^*$ and parameter functions $a(w)$ and $\vartheta(w)$ such that $F(x; w)$ is an $(a(w), \vartheta(w))$–self–concordant barrier–operator for G, and such that $S(x)$ is β–compatible with $F(x; e)$ where e is the identity in E_+. For such a mapping we consider the following family of operators:

$$\Sigma(x; w) = \frac{(1+\beta)^2}{a(w)} \left(S(x) + F(x; w)\right) \; : \; Q \to E^*. \tag{9.15}$$

9.3.2 Analysis of the Newton process

The analysis provided here is closely related to the analysis in [193] and follows a similar line of reasoning. The outlined procedure fits in the target–following framework. First, we show that $\Sigma(x; w)$ defined in (9.15) is a strongly self–concordant operator. Then we show that for a fixed value of the weights $w \in E_+$ a solution to the equation $\Sigma(x; w) = 0$ provides certain information about the original problem. In fact, the solution of this equation can be considered as the weighted center corresponding to w. By adjusting w in the 'right direction' a series of solutions is obtained that converges to a solution of ($\overline{\text{VI}}$). Successive solutions are computed by the (damped) Newton–type step

$$x^+ = x + \alpha(\hat{\Sigma}(x;w))^{-1}\Sigma(x;w) \tag{9.16}$$

for some $\alpha \in (0,1)$. We need to show that approximate solutions to successive problems defined by weights w can be efficiently computed. We will give conditions on the weights (cf. Condition 9.2.9), which guarantee this to be possible. First, we show that $\Sigma(x;w)$ is a self–concordant operator.

Lemma 9.3.4 *Let $S(x)$ be compatible with an $(a(w),\vartheta(w))$–self–concordant barrier–operator $F(x;w)$ and let $\Sigma(x;w)$ be as defined in (9.15). Then $\Sigma(x;w)$ is a strongly 1–self–concordant operator.*

Proof: Let $x \in Q$ and $h^{(i)} \in E$, $i = 1,2,3$. Then from Definitions 9.3.2 and 9.3.3 we have

$$\begin{aligned}
|\nabla^2\Sigma(x;w)[h^{(1)},h^{(2)},h^{(3)}]| \;\le\; & \frac{(1+\beta)^2}{a(w)}\left(\beta\prod_{i=1}^{3}\left(3\langle\hat{S}(x)h^{(i)},h^{(i)}\rangle\right)^{1/3}\right.\\
& \left.\left(3a(w)^{-1}\langle\hat{F}(x;w)h^{(i)},h^{(i)}\rangle\right)^{1/6} + 2\,a(w)^{-1/2}\prod_{i=1}^{3}\langle\hat{F}(x;w)h^{(i)},h^{(i)}\rangle^{1/2}\right)\\
= \;& (1+\beta)^2a(w)^{-3/2}\left(\beta\prod_{i=1}^{3}\left(3\langle\hat{S}(x)h^{(i)},h^{(i)}\rangle\right)^{1/3}\right.\\
& \left.\left(3\langle\hat{F}(x;w)h^{(i)},h^{(i)}\rangle\right)^{1/6} + 2\prod_{i=1}^{3}\langle\hat{F}(x;w)h^{(i)},h^{(i)}\rangle^{1/2}\right).
\end{aligned}$$

It needs to be shown that the right–hand side is not larger than

$$\begin{aligned}
2\prod_{i=1}^{3} & \left(\langle\hat{\Sigma}(x;w)h^{(i)},h^{(i)}\rangle\right)^{1/2}\\
& = 2a(w)^{-3/2}(1+\beta)^3\prod_{i=1}^{3}\left(\langle\hat{S}(x)h^{(i)},h^{(i)}\rangle + \langle\hat{F}(x;w)h^{(i)},h^{(i)}\rangle\right)^{1/2}.
\end{aligned}$$

The proof of this fact follows along the same line as in [193, Prop. 7.3.2] and is therefore omitted. □

The next lemma shows that we can solve ($\overline{\text{VI}}$) by computing (approximate) solutions of the equality $\Sigma(x;w) = 0$ for a sequence of weights w for which $a(w)\vartheta(w)$ converges to zero.

Lemma 9.3.5 *Let $S(x)$ be β–compatible with an $(a(w), \vartheta(w))$–self–concordant barrier–operator $F(x; w)$. Let $x(w)$ denote the solution of the equation $\Sigma(x; w) = 0$. Then*

$$\forall y \in G: \quad \langle S(x(w)), x(w) - y \rangle \leq a(w)\vartheta(w).$$

Proof: Since $\Sigma(x(w); w) = 0$, it follows $S(x(w)) = -F(x(w); w)$. Since F is a self–concordant barrier–operator it holds (cf. Lemma 9.2.4) for all $x \in Q$ and $y \in G$ that $\langle F(x; w), y - x \rangle \leq a(w)\vartheta(w)$; hence the lemma follows. □

Note that for $x, y \in Q$ the monotonicity of S implies

$$\langle S(y), x - y \rangle \leq \langle S(x), x - y \rangle.$$

The variational inequality is (approximately) solved if the left term is sufficiently small. The criterion in Lemma 9.3.5 guarantees that this will happen if we choose a series of weights w such that $a(w)$ and $\vartheta(w)$ converge to zero and exact centers are computed. In the algorithm only approximations to exact centers will be computed. We refer to Nesterov and Nemirovskii [193, Section 7.3.5], where it is shown that under mild conditions (e.g., Lipschitz continuity of the operator S) using approximations is valid.

We impose the following condition on the target updates.

Condition 9.3.6 *Let $w \in E_+$ be a given weight. We say that $w^+ \in E_+$ is a valid updated weight, if there exists $\overline{w} \in E_+$ such that*

$$F(x; w) - F(x; w^+) = F(x; \overline{w}), \quad \forall x \in G, \tag{9.17}$$

and constants $\tau_0, \tau_1, \tau_2 \in (0, 1)$ such that
(i) $\tau_0 a(w) \leq a(w^+) \leq a(w)$;
(ii) $\vartheta(\overline{w}) \leq \tau_1 \vartheta(w)$;
(iii) $\hat{F}(x; \overline{w}) \preceq \tau_2 \hat{F}(x; w)$.

In the analysis we use the proximity measure

$$\delta(x; w) := \|\Sigma(x; w)\|^*_{\Sigma, x}.$$

If $\delta(x; w)$ is sufficiently small then the Newton–type iterate (9.16) with respect to equation $\Sigma(x; w) = 0$ is feasible and the proximity decreases [193, Theorem 7.2.1]. The next lemma shows that the proximity remains small after target updates satisfying Condition 9.3.6.

Lemma 9.3.7 *Let $S(x)$ be β-compatible with an $(a(w), \vartheta(w))$-self-concordant barrier-operator $F(x;w)$. Let $\Sigma(x;w)$ be as defined in (9.15). Assume that we have $x \in Q$ and $w \in E_+$ satisfying*

$$\delta(x;w) = \|\Sigma(x;w)\|^*_{\Sigma,x} \leq \lambda,$$

and that w^+ is a valid updated weight in the sense of Condition 9.3.6. Define $\Sigma^+ := \Sigma(x;w^+)$. Then it holds

$$\delta(x;w^+) = \|\Sigma(x;w^+)\|^*_{\Sigma^+,x} \leq \frac{1}{\tau_0\sqrt{1-\tau_2}}\left(\lambda + (1+\beta)\sqrt{\tau_1\tau_2}\sqrt{\frac{\vartheta(w)}{a(w)}}\right).$$

Proof: Using Condition 9.3.6(i), (iii) and (9.17) we have

$$\begin{aligned}
\hat{\Sigma}(x;w^+) &= \frac{(1+\beta)^2\left(\hat{S}(x) + \hat{F}(x;w^+)\right)}{a(w^+)} \\
&\succeq \frac{(1+\beta)^2\left(\hat{S}(x) + \hat{F}(x;w) - \hat{F}(x;\overline{w})\right)}{a(w)} \\
&\succeq \frac{(1+\beta)^2\left(\hat{S}(x) + \hat{F}(x;w) - \tau_2\hat{F}(x;w)\right)}{a(w)} \succeq (1-\tau_2)\hat{\Sigma}(x;w).
\end{aligned}$$

Hence for all $h \in E$ it holds

$$\|h\|_{\Sigma^+,x} \geq \sqrt{1-\tau_2}\|h\|_{\Sigma,x},$$

which implies for all $\eta \in E^*$

$$\|\eta\|^*_{\Sigma^+,x} \leq \frac{1}{\sqrt{1-\tau_2}}\|\eta\|^*_{\Sigma,x}.$$

Applying this inequality with $\eta := \Sigma(x;w^+)$ we obtain

$$\begin{aligned}
\|\Sigma(x;w^+)\|^*_{\Sigma^+,x} &\leq \frac{1}{\sqrt{1-\tau_2}}\|\Sigma(x;w^+)\|^*_{\Sigma,x} \\
&= \frac{(1+\beta)^2}{\sqrt{1-\tau_2}a(w^+)}\|S(x) + F(x;w^+)\|^*_{\Sigma,x} \\
&\leq \frac{(1+\beta)^2}{\sqrt{1-\tau_2}\tau_0 a(w)}\left(\|S(x) + F(x;w)\|^*_{\Sigma,x} + \|F(x;\overline{w})\|^*_{\Sigma,x}\right) \\
&\leq \frac{(1+\beta)^2}{\sqrt{1-\tau_2}\tau_0}\left(\frac{\lambda}{(1+\beta)^2} + \frac{\|F(x;\overline{w})\|^*_{\Sigma,x}}{a(w)}\right),
\end{aligned}$$

where the last inequality follows from the assumption on δ in the lemma. To estimate the norm, observe that the monotonicity of both S and F gives

$$\hat{\Sigma}(x;w) = \frac{(1+\beta)^2}{a(w)}(\hat{S}(x)+\hat{F}(x;w)) \succeq \frac{(1+\beta)^2}{a(w)}\hat{F}(x;w) \succeq \frac{(1+\beta)^2}{\tau_2 a(w)}\hat{F}(x;\overline{w}).$$

Consequently,

$$\begin{aligned} \|F(x;\overline{w})\|^*_{\Sigma,x} &\le \frac{\sqrt{a(w)\tau_2}}{1+\beta}\|F(x;\overline{w})\|^*_{F(\cdot;\overline{w}),x} \\ &\le \frac{\sqrt{a(w)\tau_2}}{1+\beta}\sqrt{\vartheta(\overline{w})} \\ &\le \frac{\sqrt{a(w)\tau_1\tau_2}}{1+\beta}\sqrt{\vartheta(w)}, \end{aligned}$$

where the second inequality follows from the definition of self–concordant barrier–operators. Combining the bounds gives the desired result. □

As in the linear and convex case 'recentering' with Newton's method is a quadratically convergent process. As shown by Nesterov and Nemirovskii [193, pp.296,Th.7.2.1] to stay in the region of quadratic convergence we have to require

$$\|\Sigma(x;w^+)\|^*_{\Sigma^+,x} \le 5\lambda.$$

We still need to establish the effect of the target update on the self–concordance parameters $a(w)$ and $\vartheta(w)$. This dependence is captured within Condition 9.3.6.

Lemma 9.3.8 *Let $S(x)$ be β–compatible with an $(a(w),\vartheta(w))$–self-concordant barrier-operator $F(x;w)$. Let w^+ be a valid updated weight satisfying Condition 9.3.6. Then $F(x;w^+)$ is an $(a(w^+),\vartheta(w^+))$–self-concordant barrier-operator, with*

$$a(w^+) \le a(w) \quad \textit{and} \quad (1-\tau_1)\vartheta(w) \le \vartheta(w^+) \le \vartheta(w).$$

Moreover, $S(x)$ is β–compatible with $F(x;w^+)$.

Proof: The statement on $a(w^+)$ follows immediately from the condition on w^+. Using (9.17) we have $F(x;w) = F(x;w^+) + F(x;\overline{w})$; applying Lemma 9.2.3 it holds

$$\vartheta(w) = \vartheta(w^+) + \vartheta(\overline{w}),$$

from which the bounds on $\vartheta(w^+)$ follow. Since β–compatibility does not depend on ϑ and $a(w^+) \le a(w)$ the last statement in the lemma follows. □

9.3.3 Applications

We give three important spaces for which weighted self-concordant barrier-operators can be derived: the non-negative orthant, the second order cone and the cone of positive semidefinite matrices. These are the same cones as considered by Nesterov and Todd [194]; they are *self-scaled cones.*

Non-negative orthant

Let $G \equiv \mathbb{R}^n_+$. The logarithmic barrier function $f(x) = -\sum_{i=1}^n \ln(x_i)$ is a strongly $(1, n)$-self-concordant barrier for G. Its gradient is $\nabla f(x) = -X^{-1}e$. It is easy to see that ∇f is a $(1, n)$-self-concordant barrier-operator. Let $w \in \mathbb{R}^n_{++}$ and consider

$$F(x; w) = -WX^{-1}e,$$

where $W = \text{Diag}\,(w)$. Then F is a $(\min(w), \sum_{i=1}^n w_i)$-self-concordant barrier-operator, which follows as in the proof of Lemma 9.2.3. Observe, that in this case $F(x; w)$ is the gradient of the weighted barrier function

$$f(x; w) = -\sum_{i=1}^n w_i \ln(x_i).$$

Consider the weighted path-following variant with $w^+ = (1-\theta)w$. We check the statements in Condition 9.3.6. Note that

$$F(x; w) - F(x; w^+) = -WX^{-1}e + W^+X^{-1}e = -\theta WX^{-1}e = F(x; \theta W).$$

Consequently, it follows that $\tau_0 = 1 - \theta$ and $\tau_1 = \tau_2 = \theta$. Using these bounds in Lemma 9.3.7 reveals that we may choose $\theta = \mathcal{O}(a(w)/\vartheta(w))$. The 'usual' complexity bound of $\mathcal{O}(\sqrt{n}/\sqrt{\omega(w)} \ln 1/\epsilon)$ iterations is obtained using $a(w) = \min(w)$, $\vartheta(w) = \sum_{i=1}^n w_i \vartheta_i$ and Lemma 9.3.5.

Second order cone

Let $G \equiv \{\ (t, x) \in \mathbb{R}^{n+1}_+ : t \geq ||x||\ \}$. The function $f(t, x) = -\ln(t^2 - ||x||^2)$ is a 2-self-concordant barrier for G [193, Prop.5.4.3]. The gradient of f is given by

$$\nabla f(t, x) = \frac{2}{t^2 - ||x||^2} \begin{pmatrix} -t \\ x \end{pmatrix},$$

which is a $(1, 2)$-self-concordant barrier-operator. Let $w^{(1)} \in \mathbb{R}$, $w^{(2)} \in \mathbb{R}^n_+$ and $w = (w^{(1)}, w^{(2)})$; let $W = \text{Diag}\,(w)$. Consider

$$F(t, x; w) = W\nabla f(t, x).$$

Lemma 9.3.9 *F is a self-concordant barrier-operator with parameters*

$$a(w) = \frac{\min(w)^3}{\max(w)^2}, \quad \textit{and} \quad \vartheta(w) = 2\frac{\max(w)^2}{\min(w)}.$$

Proof: We let $y := (t, x)$ and define the notation

$$\nabla_i f := \frac{\partial}{\partial y_i} f(y), \quad \nabla^2_{ij} f := \frac{\partial^2}{\partial y_i \partial y_j} f(y), \quad \text{etc.}$$

Then

$$\begin{aligned}
\left|\nabla^2 F(t;w)[h,h,h]\right| &\equiv \sum_{i,j,k=1}^{n+1} \nabla^2_{jk} F_i(y) h_i h_j h_k = \sum_{i,j,k=1}^{n+1} w_i \nabla^3_{ijk} f(y) h_i h_j h_k \\
&\leq \max(w)\, 2 \left(\sum_{i,j=1}^{n+1} \nabla^2_{ij} f(y) h_i h_j \right)^{3/2} \\
&\leq 2\frac{\max(w)}{\min(w)^{3/2}} \left(\sum_{i,j=1}^{n+1} w_i \nabla^2_{ij} f(y) h_i h_j \right)^{3/2} \\
&= 2\frac{\max(w)}{\min(w)^{3/2}} \left(\sum_{i,j=1}^{n+1} \nabla F(y) h_i h_j \right)^{3/2},
\end{aligned}$$

from which the expression for $a(w)$ follows. The one for $\vartheta(w)$ can be shown similarly. □

Observe, that for instance for the weighted target update $w^+ = (1-\theta)w$ the complexity becomes worse than 'usual', namely with a factor $1/\omega(w)$. This is due to the fact that the barrier–operator is not the derivative of a separable barrier function. By letting $\max(w)$ converge to zero, it is ascertained that $a(w)\vartheta(w) \to 0$ as required by Lemma 9.3.5.

Cone of positive semidefinite matrices

Consider $G := \mathcal{S}^n$, the cone of symmetric positive semidefinite $n \times n$ matrices. The standard logarithmic barrier for G is $f(X) = -\ln\det X$, with gradient $\nabla f(X) = -X^{-1}$; it is an n–self–concordant barrier [193]. Let us now consider

$$F(X;W) = -W^{1/2} X^{-1} W^{1/2},$$

where $W = \mathrm{Diag}(w)$ for some vector $w \in \mathbb{R}^n_{++}$. Then $F(X; W)$ is a self-concordant barrier-operator, whose parameters can be derived as in Lemma 9.2.3 and 9.3.9. Algorithms and their analysis based on this barrier-operator are obtained as in the other examples given in this section.

9.4 CONCLUDING REMARKS

The target-following framework developed in Chapter 8 for LP was extended to NLP. It was shown that the theory of interior point methods for structured NLP problems satisfying the self-concordance condition, as developed primarily by Nesterov and Nemirovskii [193], can be enriched with the analysis of methods other than central path-following. Here we analyzed some methods that are well-known in LP and/or NLP, but not theoretically analyzed for convex programming before. In our study on variational inequalities we gave a new approach of treating non-central path-following methods, by introducing mappings having properties of gradients of self-concordant barriers. To provide an alternative for this approach, it can be investigated whether it is possible to include weights in the term of the barrier function corresponding to the original objective function.

10

SEMIDEFINITE PROGRAMMING

10.1 INTRODUCTION

One of the major implications of the work by Nesterov and Nemirovskii [193] is the applicability of efficient interior point methods to problems other than those being defined on the space of real numbers. Later, Nesterov and Todd [194] developed efficient primal–dual techniques for problems defined on certain convex cones. An important type of such a cone is the space of symmetric positive semidefinite matrices

$$\mathcal{S}^n := \{ X \in \mathbb{R}^{n\times n} \;:\; X = X^T,\ X \succeq 0 \}.$$

Many problems in mathematical programming can be modeled as *semidefinite programming* (SDP) problems, that is, problems defined over $\mathcal{S}^n$. Some of these applications have been known for a long time, see e.g., Lovász [153], Lovász and Schrijver [154], Cullum [36] and Fletcher [54]; however, the lack of an efficient solution technique has prevented the theoretical opportunities to become worthwhile in practice. This has dramatically changed due to [193] and the work of Alizadeh [5], who brought SDP to the foreground as a major research topic. The general form of a semidefinite program is

$$\text{(SP)}\quad \max_y \{ b^T y \;:\; F(y) \preceq F_0 \},$$

where $y, b \in \mathbb{R}^m$, $F(y) = \sum_{i=1}^m y_i F_i$ and $F_i = F_i^T \in \mathbb{R}^{n\times n}$ for $i = 0, \dots, m$. It is easy to see that (SP) is a convex programming problem. A duality theory for this type of problems was developed by Alizadeh [5], see also [193, 205]. The natural inner product for $\mathcal{S}^n$ is the 'trace of the matrix–product', so for $X, Y \in \mathcal{S}^n$ we have $\langle X, Y\rangle = \mathrm{tr}(XY)$. The dual of (SP) is given by

$$\text{(SD)}\quad \min_X \{ \mathrm{tr}(F_0 X) \;:\; \mathrm{tr}(F_i X) = b_i\ \ i = 1, \dots, m,\ X \succeq 0 \}.$$

Nesterov and Nemirovskii [193] showed that the function $f(X) := -\ln\det(X)$ is a self-concordant barrier function for $\mathcal{S}^n$. Some interior point methods using this barrier were analyzed in Alizadeh [5], primal–dual methods were developed by Boyd and Vandenberghe [236], Helmberg et al. [99] and Freund [61], among others.

Many new applications of SDP are found in system and control theory (see e.g., Boyd and Vandenberghe [236] and Boyd et al. [236]), structural optimization (e.g., Ben–Tal and Bendsœ[18]), geometrical problems involving quadratic forms or ellipsoids (e.g., Boyd et al. [236]), statistics (e.g., minimum trace factor analysis, Watson [243], and optimal experiment design, Pukelsheim [204]), linear algebra (e.g., problems involving eigenvalues, Alizadeh [5], Ramana [205]). In recent years applications in combinatorial optimization have gained extensive attention. Here the semidefinite methodology is used to approximate NP–hard problems (see e.g., Alizadeh [5], Goemans and Williamson [72], Poljak et al. [200], Laurent and Poljak [147], Karger et al. [125], Helmberg et al. [99] and the references therein). Lovász and Schrijver [154] pioneered this field and gave some theoretical opportunities with the approach. These developments have lead to new approximation heuristics (often with an improved worst case bound) for problems as MAXCUT, MAXSAT, graph–coloring and graph–partitioning.

In this chapter we give two applications of SDP. We first consider the problem of optimizing a nonconvex quadratic function over ellipsoids. The motivation for studying this problem is that ellipsoids can be used to approximate hypercubes, which themselves form relaxations of $0-1$ constraints. The general problem where the ellipsoidal constraints are replaced by (nonconvex) quadratic functions has been studied by Shor [217], who derived a Lagrangian dual for it which appears to be a semidefinite problem; see Boyd and Vandenberghe [236] for a discussion. Here we derive a (nonlinear) convex semidefinite problem being equivalent to the two relaxations mentioned above. Second, we analyze an interior point method for computing the smallest eigenvalue of a symmetric matrix, which is an application of SDP in linear algebra. We show that the method is polynomial and possesses an asymptotic quadratic convergence rate, without assumption on the multiplicity of the smallest eigenvalue.

10.2 NONCONVEX QUADRATIC OPTIMIZATION OVER ELLIPSOIDS

Consider the problem of optimizing a (general) quadratic form over the intersection of $m \geq 1$ ellipsoids:

$$\text{(EP)} \quad \min_x \left\{ \frac{1}{2} x^T Q x + c^T x \ : \ x^T A^{(i)} x \leq 1, \quad i = 1, \ldots, m \right\},$$

where $A^{(i)} \in \mathcal{S}^n$ for $i = 1, \ldots, m$ and $Q = Q^T \in \mathbb{R}^{n \times n}$. We assume that the optimal value of (EP) is negative, since otherwise the trivial solution $x = 0$ would be optimal. In case Q is positive semidefinite, the problem is convex and easily solvable with interior point methods in polynomial time (see Nesterov and Nemirovskii [193], Vial [240]). Here, we assume Q to be not positive semidefinite, hence (EP) is a nonconvex programming problem.

The special case $m = 1$ has been extensively studied in the literature, mainly in the context of trust region methods, see e.g., Moré [186], Sörensen [219]. Recently, Ye [250] gave an efficient polynomial time algorithm for this problem. It has gained increasing attention in the last years, see e.g., Flippo and Jansen [55], Stern and Wolkowicz [220] and Ben–Tal and Teboulle [20], since in this case the problem (EP) is equivalent to a convex programming problem. Specifically, the Lagrangian dual is a concave programming problem, being strongly dual to the primal problem. The one–dimensional case reveals that the dual problem can be written as an SDP problem. Hence, the dual of the dual is also (convex) semidefinite.

Consider the following nonlinear SDP problem

$$(\overline{\text{EP}}) \quad \min_U \left\{ \frac{1}{2} \text{tr}(QU) - \sqrt{c^T U c} \ : \ \text{tr}(A^{(i)} U) \leq 1, \ i = 1, \ldots, m, \quad U \in \mathcal{S}^n \right\}.$$

We show that $(\overline{\text{EP}})$ is a relaxation of (EP) and derive some other properties.

Lemma 10.2.1 *The following statements concerning problem $(\overline{EP})$ hold:*
(i) $(\overline{EP})$ is a convex programming problem;
(ii) $(\overline{EP})$ is a relaxation of (EP);
(iii) The feasible set of $(\overline{EP})$ contains an interior (Slater) point;
(iv) If both U^* and $\overline{U}^*$ are optimal in (EP) then $\text{tr}(QU^*) = \text{tr}(Q\overline{U}^*)$ *and* $c^T U^* c = c^T \overline{U}^* c$.

Proof: (i) First, $\mathcal{S}^n$ is convex, while the trace–operator is linear on this space; so the feasible set is convex. The square root function being concave, the

objective is convex as well.
(*ii*) Let $x \in \mathbb{R}^n$ and let $U = xx^T \in \mathcal{S}^n$. Observe that for any matrix M

$$\mathrm{tr}(MU) = \mathrm{tr}(xx^T M) = \mathrm{tr}(x^T M x) = x^T M x.$$

Hence, if x is feasible in (EP) then U is feasible in ($\overline{\mathrm{EP}}$). Since

$$\sqrt{c^T U c} = \sqrt{c^T x x^T c} = \sqrt{(c^T x)^2} = |c^T x|,$$

the objective value attained by U equals

$$\frac{1}{2} x^T Q x - |c^T x|.$$

Observing that in any optimal solution of (EP) it holds $c^T x \leq 0$, completes the proof.
(*iii*) Take $U = \epsilon I$, where $I \in \mathcal{S}^n$ is the identity matrix and $\epsilon > 0$. Obviously $U \in \mathcal{S}^n$. Also, since

$$\mathrm{tr}(A^{(i)}U) = \mathrm{tr}(\epsilon A^{(i)}) = \epsilon \sum_{j=1}^{n} A_{jj}^{(i)},$$

with $\epsilon < \min_i(1/\sum_{j=1}^n A_{jj}^{(i)})$ the constraints are satisfied with strict inequality.
(*iv*) Due to convexity every linear combination of U^* and $\overline{U}^*$ is also optimal. Using the concavity of the square root function it easily follows that $c^T U^* c = c^T \overline{U}^* c$. Since the objective values of U^* and $\overline{U}^*$ are equal, it then also holds $\mathrm{tr}(QU^*) = \mathrm{tr}(Q\overline{U}^*)$. □

Note from the proof of (*ii*) that the relaxation will be tight if there is an optimal solution U to ($\overline{\mathrm{EP}}$) that is a rank–one matrix. We will use this later, but first derive the necessary and sufficient KKT–conditions for ($\overline{\mathrm{EP}}$).

Lemma 10.2.2 *A matrix $U \in \mathcal{S}^n$ is optimal in ($\overline{EP}$) if and only if there exist multipliers $\mu \in \mathbb{R}^n_+$ and $V \in \mathcal{S}^n$, such that*

$$\begin{array}{rcl} \mu_i(\mathrm{tr}(A^{(i)}U) - 1) & = & 0, \quad i = 1, \ldots, m, \\ \mathrm{tr}(A^{(i)}U) & \leq & 1, \quad i = 1, \ldots, m, \\ \mathrm{tr}(UV) & = & 0 \\ \textit{if } c^T U c > 0 & \textit{then} & Q + \sum_{i=1}^m \mu_i A^{(i)} - V - cc^T/\sqrt{c^T U c} = 0, \\ \textit{if } c^T U c = 0 & \textit{then} & Q + \sum_{i=1}^m \mu_i A^{(i)} - V = 0. \end{array} \tag{10.1}$$

Proof: First consider the case $c^T Uc > 0$. The Lagrange–function of $(\overline{\text{EP}})$ is given by

$$L_{\overline{EP}}(U, \mu, V) = \frac{1}{2}\text{tr}(QU) - \sqrt{c^T Uc} + \sum_{i=1}^{m} \frac{\mu_i}{2}(\text{tr}(A^{(i)}U) - 1) - \frac{1}{2}\text{tr}(UV).$$

The proof follows from the general KKT–theory. If $c^T Uc = 0$ then U also solves the problem

$$\min_U \left\{ \frac{1}{2}\text{tr}(QU) \ : \ \text{tr}(A^{(i)}U) \leq 1, \ i = 1, \ldots, m, \quad U \in \mathcal{S}^n \right\}.$$

Writing down the KKT–conditions for this problem the lemma follows. □

We may derive the following interesting implication of the KKT–conditions.

Lemma 10.2.3 *Let U, μ, V satisfy the KKT-conditions (10.1) for $(\overline{EP})$. If $Q + \sum_i \mu_i A^{(i)}$ has rank $k \leq n$, then* rank $(U) \leq n - k + 1$. *In particular, if the matrix $Q + \sum_i \mu_i A^{(i)}$ is positive definite, then if $c^T Uc > 0$ it holds that U is a rank-one matrix, while if $c^T Uc = 0$ then $U = 0$.*

Proof: If $c^T Uc > 0$, define $\alpha := 1/\sqrt{c^T Uc}$. Observe that

$$Q + \sum_{i=1}^{m} \mu_i A^{(i)} = V + \alpha cc^T,$$

hence $V + \alpha cc^T$ has rank k by assumption. This implies rank $(V) \geq k - 1$. Applying Corollary A.5 to U, V and condition $\text{tr}(UV) = 0$ from the KKT–system, gives rank $(U) \leq n - k + 1$. In case $k = n$, it holds rank $(U) = 1$, since $U \neq 0$ by assumption. The case $c^T Uc = 0$ follows similarly. □

The case $k = n$ considered in Lemma 10.2.3 can be referred to as the 'easy case'. For $m = 1$ the opposite 'hard case' (i.e., $Q + \mu_1 A^{(1)}$ is not of full rank) is a well–known obstacle (e.g. [20, 55, 220]) in both the analysis of the problem as well as in certain methods for its solution (see e.g. [186, 187]). However, Wolkowicz[1] reports that his interior point implementation typically does not suffer from the hard case. The next result establishes a situation where the relaxation is tight.

[1] Personal communication 1994.

Theorem 10.2.4 *If Q and $A^{(i)}$, $i = 1, \ldots, m$ are simultaneously diagonalizable, then $(\overline{EP})$ is equivalent to (EP). In particular, let U, μ, V satisfy the KKT-conditions for $(\overline{EP})$, then there is an optimal solution $\widetilde{U}$ of $(\overline{EP})$ such that $\widetilde{U}$ is a rank-one matrix.*

Proof: Without loss of generality, we assume that Q and $A^{(i)}$ are all diagonal. Let μ be an optimal multiplier. Define

$$\lambda_j = Q_{jj} + \sum_{i=1}^{m} \mu_i A_{jj}^{(i)}.$$

If $\lambda_j > 0$, $\forall j$, then the result follows from Lemma 10.2.3. Otherwise, we may assume that $k < n$ is such that

$$\lambda_1 = \cdots = \lambda_k = 0, \quad \lambda_j > 0, \quad j = k+1, \ldots, n.$$

First we consider the case $c^T U c > 0$. Define $\alpha := 1/\sqrt{c^T U c}$. Observe that for $j \leq k$ the KKT-conditions (10.1) imply

$$0 = \lambda_j = V_{jj} + \alpha c_j^2,$$

wherefrom $V_{jj} = 0$ and $c_j = 0$. Consequently, cc^T can be written as a matrix of the form

$$\begin{pmatrix} 0 & 0 \\ 0 & \bar{c}\bar{c}^T \end{pmatrix},$$

where $\bar{c} = (c_{k+1} \cdots c_n)^T$. Consequently, V is of the form

$$V = \begin{pmatrix} 0 & 0 \\ 0 & \overline{V} \end{pmatrix},$$

with $\overline{V} \in \mathcal{S}^{(n-k)\times(n-k)}$. Partitioning U in a similar way as

$$\begin{pmatrix} U_1 & U_2 \\ U_3 & \overline{U} \end{pmatrix}$$

and using the optimality condition $\operatorname{tr}(UV) = 0$ it follows $\operatorname{tr}(\overline{U}\,\overline{V}) = 0$. Hence we have the conditions

$$\operatorname{tr}(\overline{U}\,\overline{V}) = 0, \quad \operatorname{Diag}(\lambda_{k+1}, \ldots, \lambda_n) = \overline{V} + \frac{1}{\alpha}\bar{c}\bar{c}^T.$$

Consequently, $\operatorname{rank}(\overline{V}) \geq n - k + 1$ and from Corollary A.5 $\operatorname{rank}(\overline{U}) \leq 1$. Since $c^T U c = c^T \overline{U} c \neq 0$ we have $\operatorname{rank}(\overline{U}) = 1$. Now, it is easy to see that the rank-one matrix

$$\widetilde{U} = \begin{pmatrix} 0 & 0 \\ 0 & \overline{U} \end{pmatrix}$$

is feasible in $(\overline{\mathrm{EP}})$ and has the same objective value as U, so is an alternative optimal solution. The case $c^T Uc = 0$ follows along the same line of reasoning. This proves the theorem. □

To gain more insight into problem $(\overline{\mathrm{EP}})$ we compute a Lagrangian dual of (EP). The Lagrange–function is given by

$$\begin{aligned} L_{EP}(x,\mu) &= \frac{1}{2}x^T Qx + c^T x + \sum_{i=1}^m \frac{\mu_i}{2}(x^T A^{(i)}x - 1) \\ &= \frac{1}{2}x^T(Q + \sum_{i=1}^m \mu_i A^{(i)})x + c^T x - \frac{e^T\mu}{2}, \end{aligned}$$

where $\mu \in \mathbb{R}^n_+$. It is easy to see that this function is unbounded from below if the matrix $Q + \sum_{i=1}^m \mu_i A^{(i)}$ has a negative eigenvalue. Hence, we require $Q + \sum_{i=1}^m \mu_i A^{(i)} \succeq 0$. Differentiating the Lagrangian and using the generalized inverse we derive the following dual problem

$$\text{(ED)} \sup_{\mu} \left\{ -\frac{1}{2}c^T(Q + \sum_{i=1}^m \mu_i A^{(i)})^+ c - \frac{1}{2}e^T\mu \; : \; Q + \sum_{i=1}^m \mu_i A^{(i)} \succeq 0, \; \mu \geq 0 \right\}.$$

We make the following technical assumption (cf. Ben–Tal and Teboulle [20]) guaranteeing the optimum to be attained.

Assumption 10.2.5 *There exists $\mu \in \mathbb{R}^m_+$ such that $Q + \sum_{i=1}^m \mu_i A^{(i)} \succ 0$.*

Instead of (ED) we will consider the following problem

$$(\overline{\mathrm{ED}}) \qquad \sup_{\mu} \left\{ -\frac{1}{2}c^T \left(Q + \sum_{i=1}^m \mu_i A^{(i)}\right)^{-1} c - \frac{1}{2}e^T\mu \; : \; Q + \sum_{i=1}^m \mu_i A^{(i)} \succ 0, \; \mu \geq 0 \right\}.$$

Due to Assumption 10.2.5 the feasible set contains a Slater point; furthermore, the problem is a convex programming problem. We will show that $(\overline{\mathrm{EP}})$ is a Lagrangian dual of $(\overline{\mathrm{ED}})$. First, we rewrite $(\overline{\mathrm{ED}})$ as

$$\sup_{\mu,Y} \left\{ -\frac{1}{2}c^T Y^{-1} c - \frac{1}{2}e^T\mu \; : \; Y = Q + \sum_{i=1}^m \mu_i A^{(i)}, \; Y \succ 0, \; \mu \geq 0 \right\}.$$

The Lagrangian is

$$\begin{aligned} L_{\overline{ED}}(\mu, Y, U, V) &= -\frac{1}{2}c^T Y^{-1} c - \frac{1}{2}e^T \mu + \frac{1}{2}\mathrm{tr}(YV) \\ &\quad + \frac{1}{2}\mathrm{tr}\left(\left(Q + \sum_{i=1}^{m} \mu_i A^{(i)} - Y\right) U\right), \end{aligned}$$

where $U, V \in \mathcal{S}^n$. The KKT–conditions give

$$\begin{aligned} Y^{-1}cc^T Y^{-1} + V - U &= 0, \\ \mathrm{tr}(A^{(i)}U) &\leq 1, \quad i = 1, \ldots, m, \\ \mu_i(\mathrm{tr}(A^{(i)}U) - 1) &= 0, \quad i = 1, \ldots, m, \\ \mathrm{tr}(YV) &= 0, \\ Y - Q - \sum_{i=1}^{m} \mu_i A^{(i)} &= 0, \quad Y \succ 0, \ \mu \geq 0. \end{aligned}$$

From the first equation we obtain $(c^T Y^{-1} c)^2 = c^T(U - V)c \geq 0$, and $c^T Y^{-1} c = \mathrm{tr}(UY) + \mathrm{tr}(UV)$. This gives the following dual problem

$$\inf_{U,V} \left\{ \frac{1}{2}\mathrm{tr}(QU) - \sqrt{c^T(U - V)c} \ : \ \mathrm{tr}(A^{(i)}U) \leq 1, \ U \in \mathcal{S}^n \right\}.$$

It is easy to see that in any optimal solution we have $V = 0$, while Lemma 10.2.1(*iii*) shows that the optimum is attained. Hence we arrive at problem $(\overline{\mathrm{EP}})$.

Shor [217] considered a relaxation method for minimizing a quadratic function subject to general quadratic constraints. Simplifying to the setting of (EP) this leads to the following problem:

$$\text{(RP)} \quad \max_{t,\mu} \left\{ t \ : \ \begin{pmatrix} Q/2 & c/2 \\ c^T/2 & -t \end{pmatrix} + \sum_{i=1}^{m} \mu_i \begin{pmatrix} A^{(i)} & 0 \\ 0 & -1 \end{pmatrix} \succeq 0, \ \mu \geq 0 \right\}.$$

If x is feasible in (EP) and (t, μ) in (RP) then

$$\begin{aligned} 0 &\leq \begin{pmatrix} x^T & 1 \end{pmatrix} \left(\begin{pmatrix} Q/2 & c/2 \\ c^T/2 & -t \end{pmatrix} + \sum_{i=1}^{m} \mu_i \begin{pmatrix} A^{(i)} & 0 \\ 0 & -1 \end{pmatrix} \right) \begin{pmatrix} x \\ 1 \end{pmatrix} \\ &= \frac{1}{2}x^T Q x + c^T x - t + \sum_{i=1}^{m} \mu_i (x^T A^{(i)} x - 1) \leq \frac{1}{2}x^T Q x + c^T x - t, \end{aligned}$$

which shows that Shor's semidefinite problem (RP) is a relaxation. The semidefinite dual of (RP) is (after some elementary simplifications of the general form) given by

$$\text{(RD)} \qquad \min_{X,y} \left\{ \frac{1}{2}\text{tr}(QX) + c^T y \ : \ \text{tr}(A^{(i)}X) \leq 1,\ i = 1, \ldots, m, \right.$$
$$\left. \begin{pmatrix} X & y \\ y^T & 1 \end{pmatrix} \succeq 0 \right\}.$$

The relationship between (RP), (RD) and (EP) in the general form is investigated by Boyd and Vandenberghe [236], see also Fujie and Kojima [63]. Here we consider the connection between (RD) and ($\overline{\text{EP}}$).

Lemma 10.2.6 *The optimal values of (RD) and ($\overline{EP}$) are equal.*

Proof: We first show that the optimal value of ($\overline{\text{EP}}$) is not larger than the one of (RD). If (X, y) is feasible in (RD) then X is feasible in ($\overline{\text{EP}}$). From Lemma A.6 it follows that $c^T Xc - c^T yy^T c \geq 0$, so

$$\sqrt{c^T Xc} \geq |c^T y|.$$

and X in ($\overline{\text{EP}}$) yields an objective value not larger than the one for (X, y) in (RD). Conversely, if X is feasible in ($\overline{\text{EP}}$) then we can find $y \in \mathbb{R}^n$ such that $X \succeq yy^T$ and $(X - yy^T)c = 0$ as follows: if $Xc = 0$ then we take $y = 0$, otherwise $y = Xc/\sqrt{c^T Xc}$, since

$$X - \frac{Xcc^T X}{c^T Xc} = X^{1/2} \left(I - \frac{X^{1/2}cc^T X^{1/2}}{c^T Xc} \right) X^{1/2} \succeq 0.$$

Hence (X, y) is feasible in (RD) and has the same objective value as X in ($\overline{\text{EP}}$). To complete the proof, we show that if (X, y) is an arbitrary optimal solution of (RD) then $c^T Xc - (c^T y)^2 = 0$. By contradiction, suppose this equality does not hold. Consider again $(X, Xc/\sqrt{c^T Xc})$, which is feasible in (RD), moreover, by assumption

$$c^T \left(\frac{Xc}{\sqrt{c^T Xc}} \right) = \sqrt{c^T Xc} > |c^T y|,$$

which implies that (X, y) cannot be optimal. This completes the proof. □

To summarize, we have shown that the relaxation ($\overline{\text{EP}}$) is equivalent to other relaxations (RP) and (RD) proposed for (EP). The difference lies in the fact that ($\overline{\text{EP}}$) is a nonlinear problem, while the other two are linear but higher dimensional. At the heart of the equivalence in Lemma 10.2.6 is the Schur–complement.

10.3 COMPUTING THE SMALLEST EIGENVALUE OF A SYMMETRIC MATRIX

A classical problem in numerical linear algebra is the computation of the smallest eigenvalue of a symmetric matrix. This problem can be formulated in an obvious way as a SDP problem, see also Ramana [205] and Nemirovskii and Nesterov[193, p.238]. In this section we develop a specific primal–dual interior point method for this problem. We show that it has polynomial convergence and is quadratically convergent in the limit, independently of the multiplicity of the smallest eigenvalue.

Let A be a symmetric $n \times n$ matrix, whose eigenvalues are $\lambda_1 \leq \lambda_2 \leq \cdots \leq \lambda_n$. We denote by k the multiplicity of λ_1. The problem is modeled as follows:

$$(\Lambda\mathrm{P}) \quad \lambda_1 = \max_{\lambda} \{ \lambda \; : \; \lambda I \preceq A \},$$

where I is the identity matrix. The dual problem is given by

$$(\Lambda\mathrm{D}) \quad \min_{X} \{ \operatorname{tr}(AX) \; : \; \operatorname{tr}(X) = 1, \; X \succeq 0 \}.$$

Note that both the primal and dual feasible set admit strictly feasible solutions. The following lemma justifies duality between (ΛP) and (ΛD).

Lemma 10.3.1 *(i) If λ is feasible for (ΛP) and X for (ΛD) then $\lambda \leq \operatorname{tr}(AX)$. (ii) The pair (λ, X) is optimal if and only if $\lambda = \lambda_1$ and the column space of X is a subspace of the eigenspace of A with respect to λ_1.*

Proof: (i) Using feasibility of λ and X we write

$$\operatorname{tr}(AX) - \lambda = \operatorname{tr}(AX) - \lambda \operatorname{tr}(X) = \operatorname{tr}((A - \lambda I)\, X) \geq 0,$$

using Lemma A.4.
For (ii), observe that the pair (λ, X) is optimal if and only if it is feasible and

$$\operatorname{tr}((A - \lambda I)\, X) = 0.$$

Since $X \neq 0$ the latter equation is equivalent to $\lambda = \lambda_1$ and $(A - \lambda_1 I)\, X = 0$ (Lemma A.4), which proves the statement. □

For any $\lambda < \lambda_1$ we define

$$S(\lambda) := A - \lambda I,$$

which is a symmetric positive definite matrix; defining

$$\mu = \frac{1}{\operatorname{tr}(S(\lambda)^{-1})}, \qquad X(\lambda) := \frac{S(\lambda)^{-1}}{\operatorname{tr}(S(\lambda)^{-1})}, \tag{10.2}$$

we see that $X(\lambda)$ is feasible in (ΛD) and moreover

$$X(\lambda)S(\lambda) = S(\lambda)X(\lambda) = \mu I. \tag{10.3}$$

Note that if λ runs from $-\infty$ to λ_1 then μ runs from ∞ to 0. Since μ depends continuously on λ, for each positive μ there exists a unique $\lambda < \lambda_1$ (denoted by $\lambda(\mu)$) such that (10.3) holds; the dependence between μ and $\lambda(\mu)$ is given by

$$\frac{1}{\mu} = \sum_{i=1}^{n} \frac{1}{\lambda_i - \lambda(\mu)}. \tag{10.4}$$

In Jansen et al. [115] it is shown that $\lambda(\mu)$ is concave and monotonically decreasing in μ. The sets

$$\{ (\lambda(\mu), S(\lambda(\mu))) \ : \ \mu > 0 \} \quad \text{and} \quad \{ X(\lambda(\mu)) \ : \ \mu > 0 \}$$

are the central paths of (ΛP) and (ΛD) respectively. Since $\lim_{\mu \to 0} \lambda(\mu) = \lambda_1$, the limit of $S(\lambda(\mu))$ exists:

$$\lim_{\mu \to 0} S(\lambda(\mu)) = S(\lambda_1) = A - \lambda_1 I.$$

We now develop a primal–dual interior point algorithm, which starts with an arbitrary strictly feasible pair $(\lambda, S(\lambda), X(\lambda))$ on the central path and uses the central path as a guideline to optimality. The search–directions in the primal and the dual spaces are denoted by $\Delta\lambda$ and ΔX, respectively. Denoting $X := X(\lambda)$, our goal is to find these directions in such a way that the pair $(\lambda + \Delta\lambda, X + \Delta X)$ is optimal. This amounts to solving the system

$$S(\lambda + \Delta\lambda) \succeq 0, \quad X + \Delta X \succeq 0, \quad \operatorname{tr}(X + \Delta X) = 1, \quad S(\lambda + \Delta\lambda)(X + \Delta X) = 0.$$

Using $S(\lambda + \Delta\lambda) = S(\lambda) - (\Delta\lambda)I$ and $S(\lambda)X = \mu I$ the last equation can be rewritten as

$$\mu I + S(\lambda)\Delta X - (\Delta\lambda)X - (\Delta\lambda)\Delta X = 0.$$

Since this equation is nonlinear and hard to solve we linearize it by omitting the second order term $(\Delta\lambda)\Delta X$. Neglecting the inequality constraints for the moment we are left with the system

$$S(\lambda)\Delta X - (\Delta\lambda)X = -\mu I, \quad \operatorname{tr}(\Delta X) = 0,$$

which has the unique solution

$$\Delta\lambda = \frac{1}{\operatorname{tr}(S(\lambda)^{-1}X)} = \frac{\mu}{\operatorname{tr}(X^2)}, \tag{10.5}$$

and

$$\Delta X = -X + (\Delta\lambda)S(\lambda)^{-1}X = -X + \frac{X^2}{\operatorname{tr}(X^2)}.9 \tag{10.6}$$

The search–directions $(\Delta\lambda, \Delta X)$ obtained in this way are the primal–dual affine scaling directions at the pair (λ, X). Note that the calculation of the search–directions requires the inversion of the matrix $S(\lambda)$ and the multiplication of the matrix X by itself. Since

$$\operatorname{tr}(S(\lambda)^{-2}) = \sum_{i=1}^{n}\left(\frac{1}{\lambda_i - \lambda}\right)^2 \le \frac{1}{\lambda_1 - \lambda}\sum_{i=1}^{n}\frac{1}{\lambda_i - \lambda} = \frac{1}{\lambda_1 - \lambda}\operatorname{tr}(S(\lambda)^{-1}),$$

it holds

$$\lambda + \Delta\lambda = \lambda + \frac{\operatorname{tr}(S(\lambda)^{-1})}{\operatorname{tr}(S(\lambda)^{-2})} \ge \lambda + \lambda_1 - \lambda = \lambda_1;$$

consequently the full Newton step will in general be infeasible. Hence, we use a damped Newton step with step size α, such that

$$\lambda_1 - \lambda_\alpha := \lambda_1 - (\lambda + \alpha\Delta\lambda) \le (1-\alpha)(\lambda_1 - \lambda). \tag{10.7}$$

In our algorithm α is chosen to be $\alpha := \|X\|$. Observe, that $\Delta\lambda \in \mathbb{R}$, so in fact the *search–direction* for λ is not important, instead, it is the step size that gives the difference between algorithms. Moreover, in the algorithm we will not use ΔX, since given λ we can compute a point on the central path from (10.2)[2]. However, we will use ΔX later to justify the quadratic convergence of a variant of the algorithm.

Note that it is easy to find an initial point using Gershgorin's lemma (e.g., [223]):

$$\lambda^{(0)} = \min_{1 \le i \le n}\left(A_{ii} - \sum_{j \ne i}|A_{ij}|\right) < \lambda_1. \tag{10.8}$$

Observe that

$$\lambda^{(0)} \ge -n\|A\|_\infty. \tag{10.9}$$

Our algorithm is summarized in Figure 10.1.

[2] This is the centering step in predictor–corrector methods for LP; in our case the centering can be done exactly.

Input

$\lambda^{(0)} := \min_{1\leq i\leq n}\{A_{ii} - \sum_{j\neq i}|A_{ij}|\}$: the initial solution;

Parameters

ϵ is the accuracy parameter;

begin

$\lambda := \lambda^{(0)}$;

while $\mathrm{tr}(A-\lambda I)^{-1} < n/\epsilon$ **do**

$S(\lambda) := A - \lambda I$;

$X := S(\lambda)^{-1}/\mathrm{tr}(S(\lambda)^{-1})$;

$\lambda := \lambda + \|X\|/\mathrm{tr}(S(\lambda)^{-1}X)$;

end

end.

Figure 10.1 Smallest eigenvalue algorithm.

In Jansen et al. [115] the algorithm is analyzed using 'traditional' interior point proof–techniques. Later, Bosch and Torenbeek [27] gave a simple and elementary proof of its polynomiality. Here we combine the approaches. First, observe that

$$\frac{\|X\|}{\mathrm{tr}(S(\lambda)^{-1}X)} = \frac{\|X\|}{\mathrm{tr}(X^2)\mathrm{tr}(S(\lambda)^{-1})} = \frac{1}{\sqrt{\mathrm{tr}(X^2)\mathrm{tr}(S(\lambda)^{-1})}} = \frac{1}{\sqrt{\mathrm{tr}(S(\lambda)^{-2})}}. \quad (10.10)$$

The following lemma is partly due to [27].

Lemma 10.3.2 *If $\lambda < \lambda_1 = \ldots = \lambda_k < \lambda_{k+1} \leq \cdots \leq \lambda_n$ and $q \geq 1$ is a positive integer then the following inequalities hold:*

$$k(\lambda_1-\lambda)^{-q} \leq \mathrm{tr}((A-\lambda I)^{-q}) \leq n(\lambda_1-\lambda)^{-q}, \quad (10.11)$$

$$\lambda + \frac{1}{\sqrt{\mathrm{tr}((A-\lambda I)^{-2})}} < \lambda_1, \quad (10.12)$$

$$\lambda_1 - \left(\lambda + \frac{1}{\sqrt{\mathrm{tr}((A-\lambda I)^{-2})}}\right) < \left(1-\frac{1}{\sqrt{n}}\right)(\lambda_1-\lambda). \quad (10.13)$$

Moreover, if $\lambda(\mu)$ is defined by (10.4) then

$$k\mu \leq \lambda_1 - \lambda(\mu) \leq n\mu. \quad (10.14)$$

Proof: To prove (10.11) note that

$$\frac{k}{(\lambda_1-\lambda)^q} < \frac{1}{(\lambda_1-\lambda)^q} + \cdots + \frac{1}{(\lambda_n-\lambda)^q} < \frac{n}{(\lambda_1-\lambda)^q}.$$

The result follows from the observation that the summation between inequality signs is exactly $\mathrm{tr}((A-\lambda I)^{-q})$. Now (10.12) follows from $k \geq 1$ and (10.11) with $q=2$. Taking again $q=2$ in (10.11) gives (10.13). Finally, (10.14) follows from (10.11) with $q=1$ and $\lambda=\lambda(\mu)$. □

The next theorem gives the number of iterations required by the method, as well as an estimate for the accuracy of the final iterate.

Theorem 10.3.3 *Suppose that $\lambda^{(0)}$ is given by (10.8) and that the step size at each iteration equals $\alpha = \|X\|$. Then after at most*

$$\sqrt{n}\ln\frac{n(n+1)\|A\|_\infty}{k\,\epsilon}$$

iterations the algorithm has generated a feasible primal–dual pair $(\overline{\lambda}, \overline{X})$ such that

$$0 < \lambda_1 - \overline{\lambda} \leq \epsilon, \quad \textit{and} \quad \|A\overline{X} - \overline{\lambda}\,\overline{X}\| \leq \frac{\epsilon}{\sqrt{n}}.$$

Moreover, the first k eigenvalues η_i of $\overline{X}$ satisfy

$$\frac{1}{k} - \frac{(n-k)\epsilon}{nk\,(\lambda_{k+1}-\lambda_1)} \leq \eta_1 = \eta_2 = \ldots = \eta_k \leq \frac{1}{k},$$

whereas the remaining eigenvalues satisfy

$$0 \leq \eta_i \leq \frac{\epsilon}{n\,(\lambda_{k+1}-\lambda_1)}, \qquad i > k.$$

Proof: Note that each iteration of the algorithm constructs a strictly feasible λ by (10.10) and (10.12). From (10.13) it holds in the Nth iteration

$$\lambda_1 - \lambda^{(N)} < \left(1 - \frac{1}{\sqrt{n}}\right)^N (\lambda_1 - \lambda^{(0)}).$$

The algorithm continues as long as $n\mu \geq \epsilon$, where μ is defined by (10.4); so in the Nth iteration it holds $\lambda_1 - \lambda^{(N)} \geq k\epsilon/n$ by (10.14). Hence

$$N \leq \sqrt{n}\ln\frac{n(\lambda_1 - \lambda^{(0)})}{k\,\epsilon}.$$

Since I/n is feasible in $(\Lambda\mathrm{D})$

$$\lambda_1 \leq \mathrm{tr}\left(A\left(\frac{1}{n}I\right)\right) = \frac{1}{n}\mathrm{tr}(A) \leq \|A\|_\infty;$$

using inequalities (10.8) and (10.9) the statement on the iteration number follows. Let us denote

$$\overline{\mu} := \frac{1}{\mathrm{tr}(S(\overline{\lambda})^{-1})}.$$

From the stopping criterion we derive $\overline{\mu} \leq \epsilon/n$, so

$$\left\|(A - \overline{\lambda}I)\,\overline{X}\right\| = \|\overline{\mu}I\| = \overline{\mu}\sqrt{n} \leq \frac{\epsilon}{\sqrt{n}}.$$

Finally, we establish that $k\overline{X}$ is a good approximation for the matrix of the projection onto the eigenspace of A with respect to λ_1. To this end we show that k of these eigenvalues are approximately one, and the other $n-k$ approximately zero. We denote the eigenvalues of $\overline{X}$ by

$$\eta_i := \frac{\overline{\mu}}{\lambda_i - \overline{\lambda}}, \quad 1 \leq i \leq n.$$

Note that $\eta_1 = \eta_2 = \ldots = \eta_k$. Since $\overline{\lambda} < \lambda_1$ and $\overline{\mu} \leq \epsilon/n$ we have for $i > k$

$$\frac{\overline{\mu}}{\lambda_i - \overline{\lambda}} \leq \frac{\overline{\mu}}{\lambda_i - \lambda_1} \leq \frac{\epsilon}{n\,(\lambda_i - \lambda_1)} \leq \frac{\epsilon}{n\,(\lambda_{k+1} - \lambda_1)}.$$

So, the sum of the $n-k$ last eigenvalues of $\overline{X}$ is at most

$$\frac{(n-k)\epsilon}{n\,(\lambda_{k+1} - \lambda_1)}.$$

The trace of $\overline{X}$ being one, the first k eigenvalues (which are mutually equal) can be bounded from below by

$$\frac{1}{k}\left(1 - \frac{(n-k)\epsilon}{n\,(\lambda_{k+1} - \lambda_1)}\right).$$

This completes the proof. □

The algorithm can be made quadratically convergent by carefully analyzing the maximal step size. Here we do use the definitions (10.5) and (10.6) of $\Delta\lambda$ and ΔX and interior point methodology. Define

$$\overline{\alpha} := \arg\sup_{\alpha}\{\, S - \alpha(\Delta\lambda)I \succeq 0,\ X + \alpha\Delta X \succeq 0 \,\}.$$

Lemma 10.3.4 *Let $\sigma := \|\Delta X\| / \mathrm{tr}(X^2)$. The step size*

$$\alpha = \frac{2}{1 + \sqrt{1 + 4\sigma}} \tag{10.15}$$

is feasible, i.e., $\alpha \leq \overline{\alpha}$.

Proof: Consider the matrix

$$M(\alpha) := (S - \alpha(\Delta\lambda)I)(X + \alpha\Delta X) = (1 - \alpha)\mu I - \alpha^2(\Delta\lambda)\Delta X,$$

where μ is defined by (10.2). Since the matrices $S - \alpha(\Delta\lambda)I$ and $X + \alpha\Delta X$ commute, they are simultaneously diagonalizable; since the eigenvalues of $M(\alpha)$ depend continuously on α, the step size α will be feasible as long as

$$\frac{M(\alpha)}{\mu} = (1 - \alpha)I - \frac{\alpha^2 \Delta X}{\mathrm{tr}(X^2)} \succeq 0,$$

which certainly holds if

$$1 - \alpha - \frac{\alpha^2 \|\Delta X\|}{\mathrm{tr}(X^2)} \geq 0. \tag{10.16}$$

Using the definition of σ the result follows. □

The step size in Lemma 10.3.4 has the nice feature that it converges to 1 when reaching optimality, since ΔX converges to zero. Hence it guarantees superlinear convergence of λ to λ_1. We can even prove that the asymptotical convergence rate is quadratic.

Theorem 10.3.5 *Let $\tau := \lambda_{k+1} - \lambda_1$ and assume that $\mu \leq \tau/n$. Using step size (10.15) the iterates $\lambda^{(k)}$ converge quadratically to λ_1.*

Proof: First observe that (10.15) implies that (10.16) holds with equality, whence $\alpha \leq 1$ and

$$1 - \alpha = \alpha^2\sigma \leq \sigma. \tag{10.17}$$

We proceed by estimating σ. Let η_i be the eigenvalues of X, with

$$\eta_1 = \eta_2 \ldots = \eta_k > \eta_{k+1} \geq \cdots \geq \eta_n.$$

In view of Theorem 10.3.3 we call the first k eigenvalues the large eigenvalues of X and the remaining eigenvalues the small eigenvalues. It follows from the proof of the cited theorem that

$$0 \leq \eta_i \leq \frac{\mu}{\tau}, \quad i > k,$$

and

$$\frac{1}{k} - \frac{(n-k)\mu}{k\tau} \leq \eta_1 = \eta_2 = \ldots = \eta_k \leq \frac{1}{k}.$$

Using the assumption on μ the small eigenvalues are smaller than the large eigenvalues. The eigenvalues of ΔX being given by $\eta_i^2/\mathrm{tr}(X^2) - \eta_i$, $\forall i$, we may write

$$\|\Delta X\|^2 = \sum_{i=1}^{n} \left(\frac{\eta_i^2}{\mathrm{tr}(X^2)} - \eta_i \right)^2 = \frac{1}{(\mathrm{tr}(X^2))^2} \sum_{i=1}^{n} \left(\eta_i^2 - \eta_i \sum_{j=1}^{n} \eta_j^2 \right)^2.$$

Since $\mathrm{tr}(X) = 1$, it holds

$$\begin{aligned} \sum_{i=1}^{n} \left(\eta_i^2 - \eta_i \sum_{j=1}^{n} \eta_j^2 \right)^2 &= \sum_{i=1}^{n} \left(\eta_i^2 \sum_{j=1}^{n} \eta_j - \eta_i \sum_{j=1}^{n} \eta_j^2 \right)^2 \\ &= \sum_{i=1}^{n} \left(\sum_{j=1}^{n} \eta_i \eta_j \left(\eta_j - \eta_i \right) \right)^2. \end{aligned}$$

Note that if η_i and η_j are both large then they are equal, so $\eta_i \eta_j (\eta_i - \eta_j) = 0$. We consider the sum between the last pair of brackets for the cases that η_i is large and small separately. If η_i is large we have

$$\begin{aligned} \left| \sum_{j=1}^{n} \eta_i \eta_j \left(\eta_j - \eta_i \right) \right| &= \sum_{j=k+1}^{n} \eta_i \eta_j \left| \eta_j - \eta_i \right| \leq \frac{1}{k^2} \sum_{j=k+1}^{n} \eta_j \\ &\leq \frac{1}{k^2} \frac{(n-k)\mu}{\tau} = \frac{(n-k)\mu}{k^2 \tau}. \end{aligned}$$

Now let η_i be small. Then

$$\left| \sum_{j=1}^{n} \eta_i \eta_j \left(\eta_j - \eta_i \right) \right| \leq \sum_{j=1}^{k} \eta_i \eta_j \left(\eta_j - \eta_i \right) + \left| \sum_{j=k+1}^{n} \eta_i \eta_j \left(\eta_j - \eta_i \right) \right|. \tag{10.18}$$

Since the small eigenvalues are less than the large eigenvalues the terms in the first sum in the right–hand side are positive. In the first sum η_j is large, so we have

$$\sum_{j=1}^{k} \eta_i \eta_j \left(\eta_j - \eta_i \right) \leq \eta_i \sum_{j=1}^{k} \frac{1}{k^2} \leq \frac{\mu}{\tau} \frac{k}{k^2} = \frac{\mu}{k\tau}.$$

In the second sum of (10.18) both η_i and η_j are small, hence each term is in absolute value less than μ^3/τ^3. Therefore, this sum will be order of magnitudes smaller than the other terms and can be neglected in this asymptotic analysis. Combining the estimates we find

$$\sum_{i=1}^{n}\left(\eta_i^2-\eta_i\sum_{j=1}^{n}\eta_j^2\right)^2 \le k\left(\frac{(n-k)\mu}{k^2\tau}\right)^2+(n-k)\left(\frac{\mu}{k\tau}\right)^2=\frac{n(n-k)}{k^3}\left(\frac{\mu}{\tau}\right)^2.$$

We get an upper bound for σ^2 by dividing the last expression by $(\mathrm{tr}(X^2))^4$. Asymptotically $\mathrm{tr}(X^2)\approx 1/k$, which follows from the last two formulas in the proof of Theorem 10.3.3 if we notice that $n\mu$ converges to zero. Thus we may bound σ^2 by

$$\sigma^2 \le nk(n-k)\left(\frac{\mu}{\tau}\right)^2,$$

whence, using (10.17),

$$1-\alpha\le\sigma\le\frac{\sqrt{nk(n-k)}}{\tau}\mu\le\frac{\sqrt{nk(n-k)}}{\tau}\frac{\lambda_1-\lambda}{k},$$

where the last inequality follows from (10.14). Using (10.7) it follows that the gap between λ_1 and λ converges quadratically to zero. □

Note that condition on μ implies that the quadratic convergence may expose itself earlier if τ is larger. The new step size given by (10.15) may not always be larger than the old step size $||X||$. One easily verifies that this is true if

$$||X||+||\Delta X||\le 1.$$

Therefore, taking always the largest of the two step sizes the algorithm will be both polynomial–time and asymptotically quadratically convergent.

To conclude this section we motivate the proposed problem formulation and algorithm. It is well–known that the smallest eigenvalue can be obtained from the following optimization problem using the Raleigh–quotient:

$$(\mathrm{PR})\quad \min_x\ \left\{\ x^TAx\ :\ x^Tx=1\ \right\};$$

this is a nonconvex problem, since the feasible set is not convex and A may be indefinite. Note that we may rewrite (PR) as

$$(\overline{\mathrm{PR}})\quad \min_x\ \left\{\ \mathrm{tr}(Axx^T)\ :\ \mathrm{tr}(xx^T)=1\ \right\}.$$

This makes clear that $(\Lambda\mathrm{D})$ considers a larger feasible set than $(\overline{\mathrm{PR}})$. However, the problem $(\Lambda\mathrm{D})$ is a convex optimization problem for which an efficient (interior point) method is available. Moreover, in this special case the relaxed problem yields the same optimal value as the original one, but does not have the same optimal solution set. In fact, the 'relaxed' problem gives even more information than (PR). This suggests that (PR) is not a good formulation for the problem under consideration. A drawback of the proposed algorithm is that it requires solving a system of equations of the size of the matrix A in any iteration of the algorithm. However, if A is tridiagonal (note that any symmetric matrix can be brought to tridiagonal form, Golub and Van Loan [81]) then the algorithm requires only $\mathcal{O}(n)$ arithmetic operations per iteration.

10.4 CONCLUDING REMARKS

Without doubt the most popular area in mathematical programming at the moment is semidefinite programming. We proposed a new relaxation for the problem of optimizing a (nonconvex) quadratic form over ellipsoids. Unfortunately, we also showed that the relaxation is equivalent to other known relaxations. These relaxations play an important role in the theoretical and computational analysis of nonlinear approaches to combinatorial optimization problems. The paper by Goemans and Williamson [72] shows the effectiveness *and* efficiency of these nonlinear relaxations for certain graph problems, and has triggered a lot of research in this area.

As an application in linear algebra we analyzed a Newton–Raphson type method for computing the smallest eigenvalue of a symmetric matrix. The importance of interior point methodology here is that it enables us to prove polynomiality and derive a step size guaranteeing quadratic convergence, independent of the multiplicity of the smallest eigenvalue. Some computational results with this type of method is given in Bosch and Toorenbeek [27].

Both applications show that optimization can play an important role in different fields of mathematics. Hence a good operations researcher should always keep putting out his or her feelers for possible applications of optimization methodology.

11

INTERIOR POINT METHODS IN DECOMPOSITION

11.1 METHODOLOGY

Interior point techniques have not only shown their applicability in barrier methods for linear and nonlinear optimization, but also in cutting plane methods. Pioneers in this area are Goffin and Vial and co–workers (e.g., [74, 75, 77]). We briefly outline their approach. Consider a problem of the form

$$\text{(NP)}\quad \min_x \left\{\, c^T x \;:\; f_i(x) \le 0,\; i \in I \,\right\},$$

where the functions $f_i(x)$ are convex. While the number of constraints can be huge (even exponential) it is only required to have an efficient procedure (called *oracle*) returning either that x is feasible or, if x is not feasible, a hyperplane separating x from the feasible set. Now it is possible to use relaxations of the form

$$\text{(RNP)}\quad \min_x \left\{\, c^T x \;:\; a_j^T x \le b_j,\; j \in J \,\right\}.$$

Instead of solving the relaxations to optimality (as in e.g., Kelley's cutting plane method [129]) Goffin and Vial propose to compute the *analytic center* (see [218] or Definition 2.2.11) of a level set of (RNP) as follows

$$\arg\min_x \left\{ -\ln(\bar{z} - c^T x) - \sum_{j \in J} \ln(b_j - a_j^T x) \;:\; c^T x \le \bar{z},\; a_j^T x \le b_j,\; j \in J \right\},$$

and feed this to the oracle; here $\bar{z}$ is an upper bound for the optimal value of (NP). In practice this gives a stable algorithm, since the analytic centers passed to the oracle typically are close to each other from iteration to iteration, and the new analytic center can be computed in a few Newton steps, even when multiple cuts have been added to the formulation. A similar idea was developed in the 1970s by Elzinga and Moore [50], who use the *ball*-center of the level set; to compute the ball-center an auxiliary LP problem has to be solved. The analytic center approach has been applied to various NLP problems (as the geometric programming problem in Bahn et al. [12]) and is called the *analytic center cutting plane method* (ACCPM). An analysis of the theoretical efficiency of a variant of the method has been given by Nesterov [192] and Goffin et al. [76]. A phenomenon observed in practicing the method is that a good localization of the feasible set is obtained in only a few iterations.

It is important to observe that various decomposition techniques fall in the outlined methodology. Examples include Lagrange relaxation and Lagrange decomposition (e.g., Guignard and Kim [93]), Dantzig–Wolfe decomposition [41] and Benders decomposition [22]. In these cases the oracle is an LP problem. Goffin, Vial and co-workers have developed and implemented an extremely efficient decomposition method for very large but structured problems, based on analytic centers. Applications that have been considered include stochastic programming problems [13], network design problems [151], and multi-commodity flow problems [73].

For long, it was generally believed that Benders and Dantzig–Wolfe decomposition behaved rather poorly in practice, showing a large number of iterations and numerical instability. In the past some attempts been made to improve on this, for instance, the development of cross decomposition by Roy [211] or Kornai–Liptak decomposition [141]. Another strategy is the use of *Pareto-optimal cuts*, introduced by Magnanti and Wong [160] in the context of Benders decomposition. They showed that Pareto–optimal cuts can be obtained at the cost of solving an extra LP problem per iteration. Later, Magnanti et al. [158] used these cuts in computational experiments on difficult network design problems, see also Magnanti and Wong [161]. It appeared that a method using Pareto–optimal cuts tremendously outperformed standard Benders decomposition, while in many test–problems the results were also better compared to the use of other 'strong Benders–cuts'.

In this chapter we show that ACCPMs in a Benders decomposition framework render Pareto–optimal cuts for free. This might attribute to understanding of the computational efficiency of ACCPMs. Moreover, outside the framework of ACCPMs we give a sufficient condition for the subproblem to generate a Pareto–optimal cut for free; we show that this condition is weakest when a strictly complementary solution of the subproblem is computed.

11.2 BENDERS DECOMPOSITION

We will shortly explain the ACCPM for Benders decomposition. A similar analysis can be done for Dantzig–Wolfe decomposition. We deal with the problem

$$\text{(BP)} \quad \min_{x,y} \left\{ c^T x + d^T y \; : \; Ax + Dy = b, \; x \geq 0, \; y \in Y \right\},$$

where Y is assumed to contain a Slater point. Let z_{BP} denote the optimal value of (BP), and define

$$\mathcal{D} := \{ u \; : A^T u \leq c \}.$$

Using *projection* on the variables y and *outer linearization* (Geoffrion [70]) we may write

$$z_{BP} = \min_{y \in Y, \xi} \left\{ \xi \; : \; d^T y + (b - Dy)^T u \leq \xi \quad \forall u \in \mathcal{D} \right\}.$$

We use the following *relaxation* (called master problem) of the latter problem:

$$\begin{aligned} \text{(MP)} \qquad \min_{y \in Y, \xi} \; \{ \; \xi \; : \; & d^T y + (b - Dy)^T u \leq \xi \quad \forall u \in \mathcal{SD}, \\ & (b - Dy)^T v \geq 0 \quad \forall v \in \mathcal{RD} \; \}, \end{aligned}$$

where $\mathcal{SD}$ denotes a subset of $\mathcal{D}$, and $\mathcal{RD}$ is a subset of the extreme rays of $\mathcal{D}$. To generate a new cut (i.e., extending $\mathcal{SD}$ with a new u from $\mathcal{D}$ or $\mathcal{RD}$ with a new ray) the subproblem

$$\text{(S}(y)) \quad \max_{u} \left\{ d^T y + (b - Dy)^T u \; : \; A^T u \leq c \right\}$$

is solved for a given $y \in Y$; this is the oracle. Benders decomposition algorithm iterates between (MP) and (S(y)), where each solution of (MP) feeds a new $y \in Y$ to (S(y)) and solving (S(y)) renders a new cut in (MP); if the subproblem is unbounded, then a ray is found which can be added in the master problem. Observe that any $y \in Y$ can be used to feed the subproblem. As outlined in Section 11.1 ACCPMs differ from the simplex based Benders decomposition method in the sense that they do not solve the problem (MP) to optimality, which would give an extreme point of Y. Instead the analytic center (i.e., a special *interior point* of Y) of the level set

$$\begin{aligned} \{ \, (y, \xi) \; : \; & d^T y + (b - Dy)^T u \leq \xi \;\; \forall u \in \mathcal{SD}, \; (b - Dy)^T v \geq 0 \;\; \forall v \in \mathcal{RD}, \\ & \xi \leq \bar{z}, \; y \in Y \, \}, \end{aligned}$$

is computed, where $\bar{z}$ is an upper bound on the optimal value of (BP).

11.3 PARETO–OPTIMAL CUTS

Pareto–optimal cuts were introduced by Magnanti and Wong [160]. Before quoting some definitions from their paper we first introduce the following convention: when we say 'the cut u', we mean the cut generated by u and given by

$$d^T y + (b - Dy)^T u \leq \xi.$$

Definition 11.3.1 (Dominance) *A cut u dominates a cut $\tilde{u}$ if*

$$d^T y + (b - Dy)^T \tilde{u} \leq d^T y + (b - Dy)^T u, \quad \forall y \in Y,$$

and there exists $\tilde{y} \in Y$ such that

$$d^T \tilde{y} + (b - D\tilde{y})^T \tilde{u} < d^T \tilde{y} + (b - D\tilde{y})^T u.$$

Definition 11.3.2 (Pareto–optimal) *A cut u is Pareto–optimal if no cut dominates it.*

Magnanti and Wong [160] show that a Pareto–optimal cut can be found as follows:

> Solve the subproblem (S(y)) and let its set of optimal solutions be $U(y)$; solve the auxiliary LP problem
>
> $$\max_u \left\{ (b - Dy^{(0)})^T u \ : \ A^T u \leq c, \ u \in U(y) \right\},$$
>
> where $y^{(0)}$ is arbitrary in the relative interior of Y.

The optimal solution of the auxiliary problem is a Pareto–optimal cut (see [160]). Note that the set $U(y)$ might be described using bases, the optimal partition or the optimal value (cf. Chapter 3). The following theorem is a special case of [160, Th. 1]; we give a slightly modified proof.

Theorem 11.3.3 *Let $\overline{y}$ be an interior (Slater) point of Y and let $\overline{u}$ be an optimal solution of $(S(\overline{y}))$. Then $\overline{u}$ is a Pareto–optimal cut.*

Proof: Suppose $\overline{u}$ is not Pareto–optimal, i.e., there exists $\tilde{u} \in \mathcal{B}$ dominating $\overline{u}$. So

$$d^T y + (b - Dy)^T \overline{u} \leq d^T y + (b - Dy)^T \tilde{u} \quad \forall y \in Y. \tag{11.1}$$

Applying (11.1) with $y = \overline{y}$ it follows that $\tilde{u}$ must also be optimal in $(\mathrm{S}(\overline{y}))$, so

$$d^T \overline{y} + (b - D\overline{y})^T \overline{u} = d^T \overline{y} + (b - D\overline{y})^T \tilde{u}. \tag{11.2}$$

Because of dominance there is a $\tilde{y} \in Y$ such that

$$d^T \tilde{y} + (b - D\tilde{y})^T \overline{u} < d^T \tilde{y} + (b - D\tilde{y})^T \tilde{u}. \tag{11.3}$$

Since $\overline{y}$ is an interior point of Y, there is a $\theta > 1$ such that $w = \theta\overline{y} + (1 - \theta)\tilde{y}$, $w \in Y$ (Rockafellar [207, Th. 6.4]). Multiplying (11.2) by θ, (11.3) by $1 - \theta$ and adding gives

$$\begin{aligned} \theta \left(d^T \overline{y} + (b - D\overline{y})^T \overline{u}\right) \quad &+ \quad (1-\theta)\left(d^T \tilde{y} + (b - D\tilde{y})^T \overline{u}\right) \\ &> \theta \left(d^T \overline{y} + (b - D\overline{y})^T \tilde{u}\right) + (1-\theta)\left(d^T \tilde{y} + (b - D\tilde{y})^T \tilde{u}\right), \end{aligned}$$

or equivalently

$$d^T w + (b - Dw)^T \overline{u} > d^T w + (b - Dw)^T \tilde{u},$$

which contradicts (11.1). Hence $\overline{u}$ is Pareto–optimal. □

The following corollary immediately follows from the theorem and is the main observation of this section.

Corollary 11.3.4 *ACCPMs in the Benders decomposition framework generate Pareto–optimal cuts.*

Proof: Theorem 11.3.3 implies that every solution of $(\mathrm{S}(\overline{y}))$ gives a Pareto–optimal cut if $\overline{y}$ is an interior point of Y. ACCPMs only use interior points of Y to feed $(\mathrm{S}(\overline{y}))$, which implies the result. □

A second connection between Pareto–optimal cuts and interior point methods, is via strict complementarity (Chapter 2, Theorem 2.1.2). It is well known (e.g., Güler and Ye [97]) that many interior point methods generate a strictly complementary solution (cf. Theorem 2.2.9). Let $\hat{y} \in Y$ be arbitrary and consider

$$\max_{u,s} \left\{ (b - D\hat{y})^T u \;:\; A^T u + s = c,\; s \geq 0 \right\},$$

$$\min_{x} \left\{ c^T x \;:\; Ax = b - D\hat{y},\; x \geq 0 \right\},$$

with solution $(\hat{x}, \hat{u}, \hat{s})$. We give a condition guaranteeing $\hat{u}$ to be Pareto–optimal, and show that the condition is weakest if $\hat{u}$ is a strictly complementary solution.

Lemma 11.3.5 *Let $\mathcal{B} := \sigma(\hat{x}) = \{i : \hat{x}_i > 0\}$ and $\mathcal{N} := \{1, \ldots, n\} \setminus \mathcal{B}$. If there is an interior point $\tilde{y} \in Y$ such that*

$$b - D\tilde{y} \quad \text{is in the column space of} \quad A_{\mathcal{B}}$$

then $\hat{u}$ is a Pareto-optimal cut.

Proof: Let $\tilde{y}$ satisfy the given condition and let

$$y(\lambda) := (1-\lambda)\hat{y} + \lambda\tilde{y}, \quad \lambda \in (0,1);$$

then $y(\lambda)$ is also an interior point of Y. Let $\tilde{x}_{\mathcal{B}}$ be such that $A_{\mathcal{B}}\tilde{x}_{\mathcal{B}} = b - D\tilde{y}$, and define $\tilde{x}(\lambda) \in \mathbb{R}^n$ by

$$\tilde{x}_{\mathcal{B}}(\lambda) := (1-\lambda)\hat{x}_{\mathcal{B}} + \lambda\tilde{x}_{\mathcal{B}}, \quad \tilde{x}_{\mathcal{N}}(\lambda) := 0.$$

We show the existence of $\hat{\lambda} \in (0,1)$ such that $(\tilde{x}(\hat{\lambda}), \hat{u}, \hat{s})$ solves the pair of LP problems

$$\max_{u,s} \left\{ (b - Dy(\hat{\lambda}))^T u \ : \ A^T u + s = c, \ s \geq 0 \right\}, \tag{11.4}$$

$$\min_{x} \left\{ c^T x \ : \ Ax = b - Dy(\hat{\lambda}), \ x \geq 0 \right\}. \tag{11.5}$$

The proof of the lemma then follows from Theorem 11.3.3. Since $(\hat{u}, \hat{s})$ is feasible in (11.4) and complementarity is satisfied by construction, we only have to check whether $\tilde{x}(\lambda)$ is feasible in (11.5). Note that

$$\begin{aligned} A\tilde{x}(\lambda) &= A_{\mathcal{B}}\tilde{x}_{\mathcal{B}}(\lambda) = A_{\mathcal{B}}((1-\lambda)\hat{x}_{\mathcal{B}} + \lambda\tilde{x}_{\mathcal{B}}) \\ &= (1-\lambda)(b - D\hat{y}) + \lambda(b - D\tilde{y}) = b - Dy(\lambda). \end{aligned}$$

Also, for $i \in \mathcal{B}$, we have $\tilde{x}_i(\lambda) = (1-\lambda)\hat{x}_i + \lambda\tilde{x}_i > 0$ if

$$\lambda < \overline{\lambda} := \min_{i:\tilde{x}_i<0} \left\{ \frac{\hat{x}_i}{\hat{x}_i - \tilde{x}_i} \right\}.$$

If $\tilde{x} \geq 0$ we may take $\hat{\lambda}$ anywhere in $(0,1)$; otherwise we choose $\hat{\lambda}$ arbitrary in the interval $(0, \overline{\lambda})$. This proves the lemma. □

Observe from the proof of the lemma that the fact that $\tilde{y}$ is an interior point of Y is crucial. Furthermore, if the condition in the theorem is satisfied then $\hat{u}$ is optimal in a series of problems since we can take any λ in $(0, \overline{\lambda})$.

Corollary 11.3.6 *The condition in Lemma 11.3.5 is weakest if* $(\hat{x}, \hat{u}, \hat{s})$ *is strictly complementary.*

Proof: For a strictly complementary solution $\hat{x}$ the support $\sigma(\hat{x})$ is maximal and contains the supports of all other optimal solutions. □

From a computational point of view solving the subproblems with interior point methods will not be efficient unless the problem of warm–starting is satisfactorily solved.

11.4 CONCLUDING REMARKS

We laid a connection between a not well–known computational enhancement of Benders decomposition and modern interior cutting plane methods. Specifically it was shown that decomposition methods based on analytic centers generate Pareto–optimal cuts for free. The practical usefulness of Pareto–optimal cuts should be reconsidered in modern implementations of decomposition methods using the simplex method.

This application shows the lesson not to forget old methods, but to put them into new light as (computing) science emerges. As for interior point methods themselves, we have seen this for decomposition techniques. It has also been the case with the simplex method, where for instance the steepest edge pivot rule is born again (Forrest and Goldfarb [57]) and is effectively implemented in CPLEX [35] and OSL [196]; also, implementations of the dual simplex method have attained unexpectedly good results.

A

TECHNICAL RESULTS

This appendix contains some technical lemmas that are used in this book. The first lemma is concerned with an inequality on logarithms.

Lemma A.1 *Let h and w be vectors in $\mathbb{R}^n$ such that $||h|| < 1$, and $w > 0$. Then*

$$\sum_{i=1}^{n} w_i \ln(1+h_i) \geq w^T h + \max(w)\,||h|| + \max(w) \ln(1-||h||).$$

Proof: Using $|h_i| < 1$, $1 \leq i \leq n$, we may write

$$\begin{aligned}
-\sum_{i=1}^{n} w_i \ln(1+h_i) &= -\sum_{i=1}^{n} w_i \left(h_i - \frac{h_i^2}{2} + \frac{h_i^3}{3} - \ldots\right) \\
&\leq -w^T h + \frac{\max(w)||h||^2}{2} - \sum_{i=1}^{n} w_i \left(\frac{h_i^3}{3} - \frac{h_i^4}{4} + \ldots\right) \\
&\leq -w^T h + \frac{\max(w)||h||^2}{2} + \frac{\max(w)||h||^3}{3} + \frac{\max(w)||h||^4}{4} + \ldots \\
&= -w^T h - \max(w)||h|| - \max(w) \ln(1-||h||).
\end{aligned}$$

□

The next two lemmas are used in the analysis of the primal–dual Dikin–affine scaling algorithms.

Lemma A.2 *Let $u \in \mathbb{R}^n$ satisfy $e^T u = n$ and $1 - \beta \leq u_i \leq 1 + \beta$ for all i, where $\beta \in (0,1)$. Then*

$$\|u\|^2 \leq n(1 + \beta^2).$$

Proof: Consider the problem

$$\max_u \left\{ \|u\|^2 \ : \ e^T u = n, \ 1 - \beta \leq u_i \leq 1 + \beta \ \forall i \right\},$$

and let u be a feasible solution for which there exist two distinct indices i and j such that

$$1 - \beta < u_i < 1 + \beta, \quad 1 - \beta < u_j < 1 + \beta, \quad u_i \geq u_j.$$

Let $\hat{u}$ be defined by

$$\hat{u}_i = u_i + \epsilon, \quad \hat{u}_j = u_j - \epsilon, \quad \hat{u}_k = u_k \ \ k \neq i, j.$$

For ϵ positive but sufficiently small $\hat{u}$ is also feasible in the maximization problem. Furthermore,

$$\|\hat{u}\|^2 = \|u\|^2 + 2\epsilon^2 + 2\epsilon(u_i - u_j) > \|u\|^2.$$

This shows that u cannot be optimal. Hence, in any optimal solution of the maximization problem at most one element of u can be not at its bound. This implies for the case n even that $n/2$ elements will have value $1 - \beta$ and the others $1 + \beta$; in case n is odd, one element will have value 1. It is easy to check, that this implies the bound in the lemma. □

Lemma A.3 *Let $u \in \mathbb{R}^n_{++}$ be arbitrary and let $\nu \geq 0$. Define*

$$\psi(\nu) := -\frac{e^T u^{2\nu+2}}{\|u^{2\nu}\|}.$$

The following bounds hold:
(i) If $0 \leq \nu \leq 1$ then $\psi(\nu) \leq -\|u\|^2/\sqrt{n}$;
(ii) If $\nu \geq 1$ then $\psi(\nu) \leq -\omega(u)^{2\nu-2}\|u\|^2/\sqrt{n}$, where $\omega(u) = \min(u)/\max(u)$.

Proof: We first show (i). It is obvious that $\psi(0) = -\|u\|^2/\sqrt{n}$. Hence it suffices to show that the derivative of $\psi(\nu)$ is nonpositive as long as $0 \leq \nu \leq 1$. We differentiate the nominator and denominator separately:

$$\begin{aligned} \frac{\partial(e^T u^{2\nu+2})}{\partial \nu} &= 2\sum_{i=1}^n u_i^{2\nu+2} \ln u_i, \\ \frac{\partial \|u^{2\nu}\|}{\partial \nu} &= \frac{4}{2\|u^{2\nu}\|} \sum_{i=1}^n u_i^{4\nu} \ln u_i = \frac{2\sum_{i=1}^n u_i^{4\nu} \ln u_i}{\|u^{2\nu}\|}. \end{aligned}$$

The sign of $\psi'(\nu)$ is determined by the nominator of the derivative, which is given by

$$-\left(2\sum_{i=1}^{n} u_i^{2\nu+2}\|u^{2\nu}\| \ln u_i - \frac{2\|u^{\nu+1}\|^2}{\|u^{2\nu}\|}\sum_{i=1}^{n} u_i^{4\nu} \ln u_i\right)$$

$$= \frac{2}{\|u^{2\nu}\|}\sum_{i=1}^{n}\left(u_i^{4\nu}\|u^{\nu+1}\|^2 - u_i^{2\nu+2}\|u^{2\nu}\|^2\right) \ln u_i.$$

Note that

$$2\sum_{i=1}^{n}\left(u_i^{4\nu}\|u^{\nu+1}\|^2 - u_i^{2\nu+2}\|u^{2\nu}\|^2\right)\ln u_i = 2\sum_{i=1}^{n}\sum_{j=1}^{n}\left(u_i^{4\nu}u_j^{2\nu+2} - u_i^{2\nu+2}u_j^{4\nu}\right)\ln u_i$$

$$= 2\sum_{i,j=1}^{n}(u_iu_j)^{2\nu+2}(u_i^{2\nu-2} - u_j^{2\nu-2})\ln u_i$$

$$= \sum_{i,j=1}^{n}(u_iu_j)^{2\nu+2}\left((u_i^{2\nu-2} - u_j^{2\nu-2})\ln u_i + (u_j^{2\nu-2} - u_i^{2\nu-2})\ln u_j\right)$$

$$= \sum_{i,j=1}^{n}(u_iu_j)^{2\nu+2}(u_i^{2\nu-2} - u_j^{2\nu-2})\ln \frac{u_i}{u_j}.$$

For $0 \leq \nu \leq 1$ the last expression is nonpositive, hence $\psi(\nu)$ monotonically non–increasing, which proves (i). For (ii) we derive

$$-\psi(\nu) = \frac{e^T u^{2\nu+2}}{\|u^{2\nu}\|\,\|u\|^2}\|u\|^2 \geq \frac{\min(v)^{2\nu-2}e^T u^4}{\max(v)^{2\nu-2}\|u^2\|\,\|u\|^2}\|u\|^2 \geq \frac{\omega(u)^{2\nu-2}}{\sqrt{n}}\|u\|^2.$$

The last inequality follows from

$$\frac{e^T u^4}{\|u^2\|\,\|u\|^2} = \frac{\|u^2\|}{\|u\|^2} \geq \frac{1}{\sqrt{n}},$$

using the Cauchy–Schwarz inequality. This completes the proof. □

The next results are concerned with positive semidefinite matrices.

Lemma A.4 *If* $A, B \in \mathbb{R}^{n\times n}$ *are symmetric positive semidefinite matrices then* $\operatorname{tr}(AB) \geq 0$. *Moreover, if* $\operatorname{tr}(AB) = 0$ *then* $AB = 0$.

Proof: Let orthogonal decompositions of A and B be given by

$$A = Q_A^T \Lambda_A Q_A, \quad B = Q_B^T \Lambda_B Q_B,$$

where Q_A and Q_B are orthogonal $n \times n$ matrices and Λ_A and Λ_B diagonal. Define the symmetric matrices

$$\overline{A} := Q_A^T \Lambda_A^{1/2} Q_A, \quad \overline{B} := Q_B^T \Lambda_B^{1/2} Q_B;$$

then $A = \overline{A}\,\overline{A}$ and $B = \overline{B}\,\overline{B}$. We derive

$$\mathrm{tr}(AB) = \mathrm{tr}(\overline{A}\,\overline{A}\,\overline{B}\,\overline{B}) = \mathrm{tr}(\overline{B}\,\overline{A}\,\overline{A}\,\overline{B}) = \mathrm{tr}((\overline{A}\,\overline{B})^T \overline{A}\,\overline{B}) \geq 0$$

where the inequality follows since $(\overline{A}\,\overline{B})^T(\overline{A}\,\overline{B})$ is positive semidefinite. Equality holds if and only if $(\overline{A}\,\overline{B})^T(\overline{A}\,\overline{B}) = 0$, which implies $\overline{A}\,\overline{B} = 0$. Hence, $AB = \overline{A}(\overline{A}\,\overline{B})\overline{B} = 0$. □

Corollary A.5 *If $A, B \in \mathbb{R}^{n\times n}$ are symmetric positive semidefinite matrices such that* $\mathrm{tr}(AB) = 0$ *and* $0 \leq k \leq n$, *then it holds that if* $\mathrm{rank}\,(B) = k$ *then* $\mathrm{rank}\,(A) \leq n - k$.

Proof: Since $AB = 0$ by Lemma A.5, the result follows immediately. □

Lemma A.6 *Let X be a symmetric $n \times n$ matrix and $y \in \mathbb{R}^n$. Then*

$$\begin{pmatrix} X & y \\ y^T & 1 \end{pmatrix} \succeq 0 \iff X - yy^T \succeq 0.$$

Proof: First we show the if–part. Let $h \in \mathbb{R}^n$ and $h_0 \in \mathbb{R}$ be arbitrary. Then

$$\begin{aligned} \begin{pmatrix} h^T & h_0 \end{pmatrix} \begin{pmatrix} X & y \\ y^T & 1 \end{pmatrix} \begin{pmatrix} h \\ h_0 \end{pmatrix} &= h^T X h + 2y^T h h_0 + h_0^2 \\ &\geq (h^T y)^2 + 2y^T h h_0 + h_0^2 \\ &= (h^T y + h_0)^2 \geq 0. \end{aligned}$$

For the converse, let $h \in \mathbb{R}^n$ be arbitrary. Then

$$\begin{aligned} h^T X h - h^T y y^T h &= h^T X h - 2(h^T y)(h^T y) + (h^T y)^2 \\ &= \begin{pmatrix} h^T & -h^T y \end{pmatrix} \begin{pmatrix} X & y \\ y^T & 1 \end{pmatrix} \begin{pmatrix} h \\ -h^T y \end{pmatrix} \geq 0. \end{aligned}$$

□

REFERENCES

[1] I. Adler, N.K. Karmarkar, M.G.C. Resende, and G. Veiga. An implementation of Karmarkar's algorithm for linear programming. *Mathematical Programming*, 44:297–335, 1989. (Errata in *Mathematical Programming*, 50:415, 1991).

[2] I. Adler and R.D.C. Monteiro. A geometric view of parametric linear programming. *Algorithmica*, 8:161–176, 1992.

[3] N. Agmon, Y. Alhassid, and R.D. Levine. An algorithm for finding the distribution of maximal entropy. *Journal of Comp. Physics*, 30:250–258, 1979.

[4] M. Akgül. A note on shadow prices in linear programming. *J. Opl. Res. Soc.*, 35:425–431, 1984.

[5] F. Alizadeh. *Combinatorial optimization with interior point methods and semi-definite matrices.* PhD thesis, University of Minnesota, Minneapolis, USA, 1991.

[6] E.D. Andersen and Y. Ye. Combining interior-point and pivoting algorithms for linear programming. Technical Report, Department of Management Sciences, University of Iowa, Iowa City, USA, 1994. (To appear in Management Science).

[7] E.D. Andersen and Y. Ye. On a homogeneous algorithm for the monotone complementarity problem. Technical Report, Department of Management Sciences, University of Iowa, Iowa City, USA, 1995.

[8] K.M. Anstreicher. On long step path following and SUMT for linear and quadratic programming. *SIAM Journal on Optimization*, 6:33–46, 1996.

[9] K.M. Anstreicher and R.A. Bosch. A new infinity-norm path following algorithm for linear programming. Working Paper, Department of Management Sciences, University of Iowa, Iowa City, USA, 1993.

[10] D.S. Atkinson and P.M. Vaidya. A scaling technique for finding the weighted analytic center of a polytope. *Mathematical Programming*, 57:163–192, 1992.

[11] D.C. Aucamp and D.I. Steinberg. The computation of shadow prices in linear programming. *J. Opl. Res. Soc.*, 33:557–565, 1982.

[12] O. Bahn, J.-L. Goffin, J.-Ph. Vial, and O. Du Merle. Experimental behavior of an interior point cutting plane algorithm for convex programming: an application to geometric programming. *Discrete Applied Mathematics*, 49:3–23, 1994.

[13] O. Bahn, O. Du Merle, J.-L. Goffin, and J.-Ph. Vial. A cutting plane method from analytic centers for stochastic programming. *Mathematical Programming*, 69:1–44, 1995.

[14] M.L. Balinski and A.W. Tucker. Duality theory of linear programs: a constructive approach with applications. *SIAM Review*, 11:499–581, 1969.

[15] B. Bank, J. Guddat, D. Klatte, B. Kummer, and K. Tammer. *Non-linear parametric optimization*. Birkhauser Verlag, Basel, Switzerland, 1983.

[16] E.R. Barnes. A variation on Karmarkar's algorithm for solving linear programming problems. *Mathematical Programming*, 36:174–182, 1986.

[17] D.A. Bayer and J.C. Lagarias. The nonlinear geometry of linear programming, Part I: Affine and projective scaling trajectories. *Transactions of the American Mathematical Society*, 314:499–526, 1989.

[18] A. Ben-Tal and M.P. Bendsœ. A new method for optimal truss topology design. *SIAM Journal on Optimization*, 3:322–358, 1993.

[19] A. Ben-Tal and G. Roth. A truncated log barrier algorithm for large scale conve programming and minmax problems: implementation and computational results. Research Report #1/94, Optimization Laboratory, Faculty of Industrial Engineering and Managemènt, Technion–Israel Institute of Technology, Haifa, Israel, 1994.

[20] A. Ben-Tal and M. Teboulle. Hidden convexity in some nonconvex quadratically constrained programming. *Mathematical Programming*, 72:51–63, 1996.

[21] A. Ben-Tal, M. Teboulle, and A. Charnes. The role of duality in optimization problems involving entropy functionals with applications to information theory. *Journal of Optimization Theory and Applications*, 58:209–223, 1988.

[22] J.F. Benders. Partitioning procedures for solving mixed variables programming problems. *Numerische Mathematik*, 4:238–252, 1962.

[23] J. Bisschop and R. Entriken. *AIMMS, The modeling system.* Paragon Decision Technology B.V., Haarlem, The Netherlands, 1993.

[24] R. E. Bixby and M. J. Saltzman. Recovering an optimal LP basis from an interior point solution. Technical Report 607, Dept. of Mathematical Sciences, Clemson University, Clemson, USA, 1992.

[25] J.C.G. Boot. *Quadratic programming. Algorithms, anomalies, applications.* Studies in Mathematical and Managerial Economics, Vol. 2. North–Holland Publishing Company, Amsterdam, The Netherlands, 1964.

[26] B. Borchers and J.E. Mitchell. Using an interior point method in a branch and bound algorithm for integer programming. RPI Math. Report No. 195, Department of Mathematical Sciences, Rensselaer Polytechnic Institute, Troy, New York, USA, 1991.

[27] R.A. Bosch and R.V. Torenbeek. A family of algorithms for approximating the smallest eigenvalue of a real matrix with no complex eigenvalues. Technical Report, Department of Mathematics, Oberlin College, Oberlin, Ohio, USA, 1995.

[28] S.E. Boyd, L. El Ghaoui, E. Feron, and V. Balakrishnan. *Linear matrix inequalities in system and control theory.* SIAM Studies in Applied Mathematics, Vol. 15. SIAM, Philadelphia, USA, 1994.

[29] S.F. Bradley, A.C. Hax, and T.L. Magnanti. *Applied mathematical programming.* Addison–Wesley, USA, 1977.

[30] M.G. Breitfeld and D. Shanno. Computational experience with penalty–barrier methods for nonlinear programming. Technical Report RRR17-93, Rutgers Center for Operations Research, Rutgers University, New Brunswick, USA, 1993/4. (To appear in Annals of OR).

[31] A. Brooke, D. Kendrick, and A. Meeraus. *GAMS, Release 2.25, A user's guide.* The Scientific Press, Redwood City, USA, 1992.

[32] T.J. Carpenter, I.J. Lustig, J.M. Mulvey, and D.F. Shanno. High–order predictor–corrector interior point methods with application to quadratic objectives. *SIAM Journal on Optimization*, 3:696–725, 1993.

[33] R.W. Cottle and G.B. Dantzig. Complementary pivot theory of mathematical programming. In G.B. Dantzig and A.F. Veinott, editors, *Mathematics of the Decision Sciences, Part 1*, pages 115–136. American Mathematical Society, Providence, Rhode Island, 1968.

[34] R.W. Cottle, J.S. Pang, and R.E. Stone. *The linear complementarity problem.* Academic Press Inc., San Diego, USA, 1992.

[35] CPLEX user's guide. CPLEX Optimization, Inc., Incline Village, NV, 1993.

[36] J. Cullum, W. Donath, and P. Wolfe. The minimization of certain non-differentiable sums of eigenvalues of symmetric matrices. *Mathematical Programming Study*, 3:35–55, 1975.

[37] G.B. Dantzig. Maximization of a linear function of variables subject to linear inequalities. In T.C. Koopmans, editor, *Activity analysis of production and allocation – Proceedings of a conference*, pages 339–347. John Wiley & Sons, New York, 1951.

[38] G.B. Dantzig. Linear programming under uncertainty. *Management Science*, 1:197–206, 1955.

[39] G.B. Dantzig. *Linear programming and extensions.* Princeton University Press, Princeton, 1963.

[40] G.B. Dantzig, D.R. Fulkerson, and S.M. Johnson. Solution of a large–scale traveling salesman problem. *Operations Research*, 2:393–410, 1954.

[41] G.B. Dantzig and P. Wolfe. Decomposition principle for linear programming. *Operations Research*, 8:101–111, 1960.

[42] I.I. Dikin. Iterative solution of problems of linear and quadratic programming. *Doklady Akademii Nauk SSSR*, 174:747–748, 1967. (Translated in: *Soviet Mathematics Doklady*, 8:674–675, 1967).

[43] I.I. Dikin. On the convergence of an iterative process. *Upravlyaemye Sistemi*, 12:54–60, 1974. (In Russian).

[44] I.I. Dikin. Letter to the editor. *Mathematical Programming*, 41:393–394, 1988.

[45] J. Ding and T.Y. Li. An algorithm based on weighted logarithmic barrier functions for linear complementarity problems. *Arabian Journal for Science and Engineering*, 15:679–685, 1990.

[46] R.L. Dykstra. An algorithm for restricted least squares regression. *Journal of the American Statistical Association*, 78:837–842, 1983.

[47] B.C. Eaves. On quadratic programming. *Management Science*, 17:698–711, 1971.

[48] T. Terlaky (ed.). *Interior point methods in mathematical programming.* Applied Optimization. Kluwer Academic Publishers, Boston MA, 1996.

[49] J. Edmonds and R.M. Karp. Theoretical improvements in algorithmic efficiency for network flow problems. *Journal of the ACM*, 19:248–264, 1972.

[50] J. Elzinga and T.G. Moore. A central cutting plane algorithm for the convex programming problem. *Mathematical Programming*, 8:134–145, 1975.

[51] J.R. Evans and N.R. Baker. Degeneracy and the (mis)interpretation of sensitivity analysis in linear programming. *Decision Sciences*, 13:348–354, 1982.

[52] A.V. Fiacco. *Introduction to sensitivity and stability analysis in nonlinear programming.* Academic Press, New York, USA, 1983.

[53] A.V. Fiacco and G.P. McCormick. *Nonlinear programming: sequential unconstrained minimization techniques.* John Wiley & Sons, New York, 1968. (Reprint: Volume 4 of *SIAM Classics in Applied Mathematics*, SIAM Publications, Philadelphia, USA, 1990).

[54] R. Fletcher. A nonlinear programming problem in statistics (educational testing). *SIAM J. Sci. Stat. Comp.*, 2:257–267, 1981.

[55] O.E. Flippo and B. Jansen. Duality and sensitivity in quadratic optimization over a sphere. Technical Report 92–65, Faculty of Technical Mathematics and Computer Science, Delft University of Technology, Delft, The Netherlands, 1992. (To appear in European Journal of Operational Research).

[56] L.R. Ford and D.R. Fulkerson. *Flows in networks.* Princeton University Press, Princeton, 1962.

[57] J.J. Forrest and D. Goldfarb. Steepest–edge simplex algorithms for linear programming. *Mathematical Programming*, 57:341–374, 1992.

[58] R. Fourer and S. Mehrotra. Solving symmetric indefinite systems in an interior point method. *Mathematical Programming*, 62:15–39, 1993.

[59] R.M. Freund. Theoretical efficiency of a shifted barrier function algorithm for linear programming. *Linear Algebra and Its Applications*, 152:19–41, 1991.

[60] R.M. Freund. Projective transformations for interior–point algorithms, and a superlinearly convergent algorithm for the w–center problem. *Mathematical Programming*, 58:385–414, 1993.

[61] R.M. Freund. Complexity of an algorithm for finding an approximate solution of a semidefinite program with no regularity assumption. Technical Report OR302–94, Operations Research Center, MIT, Boston, USA, 1994.

[62] K.R. Frisch. The logarithmic potential method for convex programming. Unpublished Manuscript, Institute of Economics, University of Oslo, Oslo, Norway, 1955.

[63] T. Fujie and M. Kojima. Semidefinite programming relaxation for nonconvex quadratic programs. Research Report on Information Science Series B: Operations Research B298, Department of Mathematical and Computing Sciences, Tokyo Institute of Technology, Tokyo, Japan, 1995.

[64] T. Gal. *Postoptimal analyses, parametric programming and related topics.* Mac-Graw Hill Inc., New York/Berlin, 1979.

[65] T. Gal. Shadow prices and sensitivity analysis in linear programming under degeneracy, state-of-the-art-survey. *OR Spektrum*, 8:59–71, 1986.

[66] T. Gal. Weakly redundant constraints and their impact on postoptimal analyses in LP. Diskussionsbeitrag 151, FernUniversität Hagen, Hagen, Germany, 1990.

[67] M.R. Garey and D.S. Johnson. *Computers and intractability: a guide to the theory of NP–completeness.* W.H. Freeman and Company, Publishers, San Fransisco, USA, 1979.

[68] J. Gauvin. Quelques precisions sur les prix marginaux en programmation lineaire. *INFOR*, 18:68–73, 1980. (In French).

[69] D.M. Gay. Electronic mail distribution of linear programming test problems. *Mathematical Programming Society COAL News Letter*, pages 10–12, 1985.

[70] A.M. Geoffrion. Elements of large–scale mathematical programming. In A.M. Geoffrion, editor, *Perspectives on optimization*, pages 25–64. Addison–Wesley Publ. Co., 1972.

[71] P.E. Gill, W. Murray, M.A. Saunders, J.A. Tomlin, and M.H. Wright. On projected Newton barrier methods for linear programming and an equivalence to Karmarkar's projective method. *Mathematical Programming*, 36:183–209, 1986.

[72] M.X. Goemans and D.P. Williamson. Improved approximation algorithms for maximum cut and satisfiability problems using semidefinite programming. Working Paper, M.I.T., Cambridge, USA, 1994. (To appear in Journal of Assoc. Comput. Mach.).

[73] J.-L. Goffin, J. Gondzio, R. Sarkissian, and J.-Ph. Vial. Solving nonlinear multicommodity flow problems by the analytic center cutting plane method. Technical Report, HEC, Section of Management Studies, University of Geneva, 1994.

[74] J.-L. Goffin, A. Haurie, and J.-Ph. Vial. Decomposition and nondifferentiable optimization with the projective algorithm. *Management Science*, 38:284–302, 1992.

[75] J.-L. Goffin, A. Haurie, J.-Ph. Vial, and D.L. Zhu. Using central prices in the decomposition of linear programs. *European Journal of Operational Research*, 64:393–409, 1993.

[76] J.-L. Goffin, Z.-Q. Luo, and Y. Ye. Further complexity analysis of a primal–dual column generation algorithm for convex or quasiconvex feasibility problems. Technical report, Faculty of Management, McGill University, Montreal, Canada, 1993.

[77] J.-L. Goffin and J.-Ph. Vial. Cutting planes and column generation techniques with the projective algorithm. *Journal of Optimization Theory and Applications*, 65:409–429, 1990.

[78] J.-L. Goffin and J.-Ph. Vial. On the computation of weighted analytic centers and dual ellipsoids with the projective algorithm. *Mathematical Programming*, 60:81–92, 1993.

[79] D. Goldfarb and D. Xiao. A primal projective interior point method for linear programming. *Mathematical Programming*, 51:17–43, 1991.

[80] A.J. Goldman and A.W. Tucker. Theory of linear programming. In H.W. Kuhn and A.W. Tucker, editors, *Linear Inequalities and Related Systems, Annals of Mathematical Studies, No. 38*, pages 53–97. Princeton University Press, Princeton, New Jersey, 1956.

[81] G.H. Golub and C.F. Van Loan. *Matrix computations.* The John Hopkins University Press, Baltimore, 2nd edition, 1989.

[82] J. Gondzio. Multiple centrality corrections in a primal–dual method for linear programming. Technical report, Section of Management Studies, University of Geneva, Geneva, Switzerland, 1994.

[83] J. Gondzio and T. Terlaky. A computational view of interior–point methods for linear programming. In J. Beasley, editor, *Advances in linear and integer programming*. Oxford University Press, Oxford, UK, 1995.

[84] C.C. Gonzaga. An algorithm for solving linear programming problems in $\mathcal{O}(n^3L)$ operations. In N. Megiddo, editor, *Progress in Mathematical Programming: Interior Point and Related Methods*, pages 1–28. Springer Verlag, New York, 1989.

[85] C.C. Gonzaga. Large steps path–following methods for linear programming, Part I: Barrier function method. *SIAM Journal on Optimization*, 1:268–279, 1991.

[86] C.C. Gonzaga. Search directions for interior linear programming methods. *Algorithmica*, 6:153–181, 1991.

[87] C.C. Gonzaga. Path following methods for linear programming. *SIAM Review*, 34:167–227, 1992.

[88] C.C. Gonzaga. The largest step path following algorithm for monotone linear complementarity problems. Technical Report 94–07, Faculty of Technical Mathematics and Computer Science, Delft University of Technology, Delft, The Netherlands, 1994.

[89] C.C. Gonzaga and R.A. Tapia. On the quadratic convergence of the simplified Mizuno–Todd–Ye algorithm for linear programming. Technical Report 92–41, Dept. of Mathematical Sciences, Rice University, Houston, TX, USA, 1992.

[90] H.J. Greenberg. An analysis of degeneracy. *Naval Research Logistics Quarterly*, 33:635–655, 1986.

[91] H.J. Greenberg. The use of the optimal partition in a linear programming solution for postoptimal analysis. *Operations Research Letters*, 15:179–186, 1994.

[92] M. Grötschel, L.A. Lovász, and A. Schrijver. *Geometric algorithms and combinatorial optimization*. Springer Verlag, Berlin, 1988.

[93] M. Guignard and S. Kim. Lagrangean decomposition: a model yielding stronger Lagrangean bound. *Mathematical Programming*, 39:215–228, 1987.

[94] O. Güler. Path following and potential reduction algorithms for nonlinear monotone complementarity problems. Working Paper, Dept. of Management Science, University of Iowa, Iowa City, USA, 1991.

[95] O. Güler. Existence of interior points and interior paths in nonlinear monotone complementarity problems. *Mathematics of Operations Research*, 18:128–147, 1993.

[96] O. Güler, C. Roos, T. Terlaky, and J.-Ph. Vial. Interior point approach to the theory of linear programming. Cahiers de Recherche 1992.3, Faculté des Sciences Economique et Sociales, Universite de Geneve, Geneve, Switzerland, 1992. (To appear in Management Science).

[97] O. Güler and Y. Ye. Convergence behavior of interior–point algorithms. *Mathematical Programming*, 60:215–228, 1993.

[98] P.T. Harker and J.S. Pang. Finite–dimensional variational inequality and nonlinear complementarity problems: a survey of theory, algorithms and applications. *Mathematical Programming*, 48:161–220, 1990.

[99] C. Helmberg, F. Rendl, R. Vanderbei, and H. Wolkowicz. A primal–dual interior–point method for the max–min eigenvalue problem. *SIAM Journal on Optimization*, 6:342–361, 1996.

[100] D. den Hertog. *Interior point approach to linear, quadratic and convex programming, Algorithms and complexity.* Kluwer Publishers, Dordrecht, The Netherlands, 1994.

[101] D. den Hertog, F. Jarre, C. Roos, and T. Terlaky. A sufficient condition for self–concordance, with application to some classes of structured convex programming. *Mathematical Programming*, 69:75–88, 1995.

[102] D. den Hertog, J. Kaliski, C. Roos, and T. Terlaky. A logarithmic barrier cutting plane method for convex programming. *Annals of Operations Research*, 58:69–98, 1995.

[103] D. den Hertog and C. Roos. A survey of search directions in interior point methods for linear programming. *Mathematical Programming*, 52:481–509, 1991.

[104] D. den Hertog, C. Roos, and T. Terlaky. A polynomial method of weighted centers for convex quadratic programming. *Journal of Information & Optimization Sciences*, 12:187–205, 1991.

[105] D. den Hertog, C. Roos, and T. Terlaky. On the classical logarithmic barrier function method for a class of smooth convex programming problems. *Journal of Optimization Theory and Applications*, 73:1–25, 1992.

[106] D. den Hertog, C. Roos, and T. Terlaky. The linear complementarity problem, sufficient matrices, and the criss–cross method. *Linear Algebra and Its Applications*, 187:1–14, 1993.

[107] H.S. Houthakker. The capacity method of quadratic programming. *Econometrica*, 28:62–87, 1960.

[108] P. Huard. Resolution of mathematical programming with nonlinear constraints by the method of centers. In J. Abadie, editor, *Nonlinear programming*, pages 207–219. North Holland, Amsterdam, 1967.

[109] P.-F. Hung and Y. Ye. An asymptotical $\mathcal{O}(\sqrt{n}L)$–iteration path–following linear programming algorithm that uses wide neighborhoods. Technical Report, Department of Mathematics, University of Iowa, Iowa City, Iowa, USA, 1994. (To appear in SIAM Journal on Optimization).

[110] B. Jansen, J.J. de Jong, C. Roos, and T. Terlaky. Sensitivity analysis in linear programming: just be careful! Report AMER.93.022, Royal/Shell Laboratories Amsterdam, Amsterdam, The Netherlands, 1993.

[111] B. Jansen, C. Roos, and T. Terlaky. An interior point approach to postoptimal and parametric analysis in linear programming. Technical Report 92–21, Faculty of Technical Mathematics and Computer Science, Delft University of Technology, Delft, The Netherlands, 1992.

[112] B. Jansen, C. Roos, and T. Terlaky. A family of polynomial affine scaling algorithms for positive semi–definite linear complementarity problems. Technical Report 93–112, Faculty of Technical Mathematics and Computer Science, Delft University of Technology, Delft, The Netherlands, 1993. (To appear in SIAM Journal on Optimization).

[113] B. Jansen, C. Roos, and T. Terlaky. A polynomial primal–dual Dikin–type algorithm for linear programming. Technical Report 93–36, Faculty of Technical Mathematics and Computer Science, Delft University of Technology, Delft, The Netherlands, 1993. (To appear in Mathematics of Operations Research).

[114] B. Jansen, C. Roos, and T. Terlaky. The theory of linear programming: Skew symmetric self–dual problems and the central path. *Optimization*, 29:225–233, 1994.

[115] B. Jansen, C. Roos, and T. Terlaky. Interior point methods: a decade after Karmarkar. A survey, with special application to the smallest eigenvalue problem. *Statistica Neerlandica*, 50:146–170, 1996.

[116] B. Jansen, C. Roos, T. Terlaky, and J.-Ph. Vial. Primal–dual algorithms for linear programming based on the logarithmic barrier method. *Journal of Optimization Theory and Applications*, 83:1–26, 1994.

[117] B. Jansen, C. Roos, T. Terlaky, and Y. Ye. Improved complexity using higher–order correctors for primal–dual Dikin affine scaling. Technical Report 94–75, Faculty of Technical Mathematics and Computer Science, Delft University of Technology, Delft, The Netherlands, 1994. (To appear in Mathematical Programming).

[118] F. Jarre. *The method of analytic centers for smooth convex programs.* PhD thesis, Institut für Angewandte Mathematik und Statistik, Universität Würzburg, Würzburg, Germany, 1989.

[119] F. Jarre. Interior–point methods for convex programming. *Applied Mathematics & Optimization*, 26:287–311, 1992.

[120] F. Jarre. Interior–point methods via self–concordance or relative Lipschitz condition. *Optimization Methods and Software*, 5:75–104, 1995.

[121] K. Jittorntrum. Solution point differentiability without strict complementarity in nonlinear programming. *Mathematical Programming Study*, 21:127–138, 1984.

[122] J.J. de Jong. A computational study of recent approaches to sensitivity analysis in linear programming. Optimal basis, optimal partition and optimal value approach. Master's thesis, Delft University of Technology, Delft, The Netherlands, 1993.

[123] G. Jongbloed. *Three statistical inverse problems.* PhD thesis, Delft University of Technology, Delft, The Netherlands, 1995.

[124] D.B. Judin and A.S. Nemirovskii. *Problem complexity and method efficiency in optimization.* Wiley–Interscience, Chichester, USA, 1983.

[125] D. Karger, R. Motwani, and M. Sudan. Approximate graph coloring by semidefinite programming. Working Paper, Stanford University, Stanford, USA, 1994.

[126] N.K. Karmarkar. A new polynomial–time algorithm for linear programming. *Combinatorica*, 4:373–395, 1984.

[127] R.M. Karp. Reducibility among combinatorial problems. In R.E. Miller and J.W. Thatcher, editors, *Complexity of computer computations*, pages 85–103. Plenum Press, New York, 1972.

[128] P. Kas and E. Klafszky. On the duality of the mixed entropy programming. *Optimization*, 27:253–258, 1993.

[129] J.E. Kelley. The cutting plane method for solving convex programs. *Journal of the SIAM*, 8:703–712, 1960.

[130] L.G. Khacijan. A polynomial time algorithm in linear programming. *Soviet Mathematics Doklady*, 20:191–194, 1979.

[131] V. Klee and G.J. Minty. How good is the simplex algorithm? In O. Shisha, editor, *Inequalities III*. Academic Press, New York, 1972.

[132] G. Knolmayer. The effects of degeneracy on cost-coefficient ranges and an algorithm to resolve interpretation problems. *Decision Sciences*, 15:14–21, 1984.

[133] M. Kojima, Y. Kurita, and S. Mizuno. Large–step interior point algorithms for linear complementarity problems. *SIAM Journal on Optimization*, 3:398–412, 1993.

[134] M. Kojima, N. Megiddo, and S. Mizuno. A primal–dual infeasible–interior–point algorithm for linear programming. *Mathematical Programming*, 61:263–280, 1993.

[135] M. Kojima, N. Megiddo, and S. Mizuno. Theoretical convergence of large–step–primal–dual interior point algorithms for linear programming. *Mathematical Programming*, 59:1–21, 1993.

[136] M. Kojima, N. Megiddo, T. Noma, and A. Yoshise. *A unified approach to interior point algorithms for linear complementarity problems*, volume 538 of *Lecture Notes in Computer Science*. Springer Verlag, Berlin, Germany, 1991.

[137] M. Kojima, S. Mizuno, and T. Noma. Limiting behavior of trajectories by a continuation method for monotone complementarity problems. *Mathematics of Operations Research*, 15:662–675, 1990.

[138] M. Kojima, S. Mizuno, and A. Yoshise. A polynomial–time algorithm for a class of linear complementarity problems. *Mathematical Programming*, 44:1–26, 1989.

[139] M. Kojima, S. Mizuno, and A. Yoshise. A primal–dual interior point algorithm for linear programming. In N. Megiddo, editor, *Progress in Mathematical Programming: Interior Point and Related Methods*, pages 29–47. Springer Verlag, New York, 1989.

[140] M. Kojima, T. Noma, and A. Yoshise. Global convergence in infeasible–interior–point algorithms. *Mathematical Programming*, 65:43–72, 1994.

[141] J. Kornai and Th. Liptak. Two–level planning. *Econometrica*, 33:141–169, 1965.

[142] K.O. Kortanek, F. Potra, and Y. Ye. On some efficient interior point methods for nonlinear convex programming. *Linear Algebra and Its Applications*, 152:169–189, 1991.

[143] K.O. Kortanek and J. Zhu. A polynomial barrier algorithm for linearly constrained convex programming problems. *Mathematics of Operations Research*, 18:116–127, 1993.

[144] E. Kranich. Interior point methods for mathematical programming: A bibliography. Discussionsbeitrag 171, FernUniversität Hagen, Hagen, Germany, 1991. (Available through NETLIB).

[145] J. Kriens and J.Th. van Lieshout. Notes on the Markowitz portfolio selection method. *Statistica Neerlandica*, 42:181–191, 1988.

[146] H.W. Kuhn and A.W. Tucker. Nonlinear programming. In J. Neyman, editor, *Proceedings of the Second Berkeley Symposium on Mathematical Studies*, pages 481–492. University of California Press, Berkeley, 1951.

[147] M. Laurent and S. Poljak. On a positive semidefinite relaxation of the cut polytope. Rapport de Recherche 27, Departement de Mathematiques et d'Informatique, Ecole Normale Superieure, Paris, France, 1993.

[148] C.E. Lemke. On complementary pivot theory. In G.B. Dantzig and A.F. Veinott, editors, *Mathematics of the Decision Sciences, Part 1*, pages 95–114. American Mathematical Society, Providence, Rhode Island, 1968.

[149] J.K. Lenstra, A.H.G. Rinnooy Kan, and A. Schrijver, editors. *History of mathematical programming. A collection of personal reminiscences.* CWI, North–Holland, The Netherlands, 1991.

[150] P. Ling. A new proof of convergence for the new primal-dual affine scaling interior-point algorithm of Jansen, Roos and Terlaky. Technical Report SYS-C93-09, School of Information Systems, University of East-Anglia, Norwich, Great Britain, 1993.

[151] A. Lisser, R. Sarkissian, and J.-Ph. Vial. Survivability in telecommunication networks. Manuscript, HEC, Section of Management Studies, University of Geneva, 1995.

[152] F.A. Lootsma. Hessian matrices of penalty functions for solving constrained optimization problems. *Philips Research Reports*, 24:322–331, 1969.

[153] L. Lovász. On the Shannon capacity of a graph. *IEEE Trans. on Information Theory*, 25:1–7, 1979.

[154] L. Lovász and A. Schrijver. Cones of matrices and set–functions and 0-1 optimization. *SIAM Journal on Optimization*, 1:166–190, 1991.

[155] I.J. Lustig. Feasibility issues in a primal–dual interior point method for linear programming. *Mathematical Programming*, 49:145–162, 1990/91.

[156] I.J. Lustig, R.E. Marsten, and D.F. Shanno. Computational experience with a primal–dual interior point method for linear programming. *Linear Algebra and Its Applications*, 152:191–222, 1991.

[157] I.J. Lustig, R.E. Marsten, and D.F. Shanno. Interior point methods : Computational state of the art. *ORSA Journal on Computing*, 6:1–15, 1994.

[158] T.L. Magnanti, P. Mireault, and R.T. Wong. Tailoring Benders decomposition for network design. *Mathematical Programming Study*, 26:112–154, 1986.

[159] T.L. Magnanti and J.B. Orlin. Parametric linear programming and anticycling pivoting rules. *Mathematical Programming*, 41:317–325, 1988.

[160] T.L. Magnanti and R.T. Wong. Accelerating Benders decomposition: algorithmic enhancement and model selection criteria. *Operations Research*, 29:464–484, 1981.

[161] T.L. Magnanti and R.T. Wong. Network design and transportation planning: models and algorithms. *Transportation Science*, 18:1–55, 1984.

[162] H.M. Markowitz. The optimization of a quadratic function subject to linear constraints. *Naval Research Logistics Quarterly*, 3:111–133, 1956.

[163] H.M. Markowitz. *Portfolio selection. Efficient diversification of investments.* Cowles foundation for research in economics at Yale University, Monograph 16. John Wiley and Sons, Inc., New York, USA, 1959.

[164] R.E. Marsten, M.J. Saltzman, D.F. Shanno, J.F. Ballintijn, and G.S. Pierce. Implementation of a dual affine interior point algorithm for linear programming. *ORSA Journal on Computing*, 1:287–297, 1989.

[165] MATLAB, User's guide. The MathWorks, Inc., Natick, Mass., USA, 1993.

[166] L. McLinden. The analogue of Moreau's proximation theorem, with applications to the nonlinear complementarity problem. *Pacific Journal of Mathematics*, 88:101–161, 1980.

[167] A.H.M. Meeuwissen. *Dependent random variables in uncertainty analysis.* PhD thesis, Delft University of Technology, Delft, The Netherlands, 1994.

[168] N. Megiddo. Pathways to the optimal set in linear programming. In N. Megiddo, editor, *Progress in Mathematical Programming: Interior Point and Related Methods*, pages 131–158. Springer Verlag, New York, 1989.

[169] N. Megiddo. On finding primal and dual optimal bases. *ORSA Journal on Computing*, 3:63–65, 1991.

[170] S. Mehrotra. On the implementation of a (primal–dual) interior point method. *SIAM Journal on Optimization*, 2:575–601, 1992.

[171] S. Mehrotra and R.D.C. Monteiro. A general parametric analysis approach and its implication to sensitivity analysis in interior point methods. *Mathematical Programming*, 72:65–82, 1996.

[172] S. Mehrotra and Y. Ye. Finding an interior point in the optimal face of linear programs. *Mathematical Programming*, 62:497–515, 1993.

[173] J.E. Mitchell and M.J. Todd. Solving combinatorial optimization problems using Karmarkar's algorithm. *Mathematical Programming*, 56:245–284, 1992.

[174] S. Mizuno. An $\mathcal{O}(n^3L)$ algorithm using a sequence for linear complementarity problems. *Journal of the Operations Research Society of Japan*, 33:66–75, 1990.

[175] S. Mizuno. $\mathcal{O}(n^\rho L)$ iteration $\mathcal{O}(n^3L)$ potential reduction algorithms for linear programming. *Linear Algebra and Its Applications*, 152:155–168, 1991.

[176] S. Mizuno. A new polynomial time method for a linear complementarity problem. *Mathematical Programming*, 56:31–43, 1992.

[177] S. Mizuno. Polynomiality of infeasible–interior–point algorithms for linear programming. *Mathematical Programming*, 67:109–119, 1994.

[178] S. Mizuno and A. Nagasawa. A primal–dual affine scaling potential reduction algorithm for linear programming. Research Memorandum 427, The Institute of Statistical Mathematics, Tokyo, Japan, 1992.

[179] S. Mizuno, M.J. Todd, and Y. Ye. On adaptive step primal–dual interior–point algorithms for linear programming. *Mathematics of Operations Research*, 18:964–981, 1993.

[180] R.D.C. Monteiro and I. Adler. Interior path following primal–dual algorithms: Part I: Linear programming. *Mathematical Programming*, 44:27–41, 1989.

[181] R.D.C. Monteiro and I. Adler. An extension of Karmarkar–type algorithm to a class of convex separable programming problems with global linear rate of convergence. *Mathematics of Operations Research*, 15:408–422, 1990.

[182] R.D.C. Monteiro, I. Adler, and M.G.C. Resende. A polynomial–time primal–dual affine scaling algorithm for linear and convex quadratic programming and its power series extension. *Mathematics of Operations Research*, 15:191–214, 1990.

[183] R.D.C. Monteiro, J.-S. Pang, and T. Wang. A positive algorithm for the nonlinear complementarity problem. *SIAM Journal on Optimization*, 5:129–148, 1995.

[184] R.D.C. Monteiro and S. Wright. Local convergence of interior–point algorithms for degenerate monotone LCP. Preprint MCS-P357-0493, Mathematics and Computer Science Division, Argonne National Laboratory, Argonne, USA, 1993.

[185] R.D.C. Monteiro and F. Zhou. On the existence and convergence of the central path for convex programming and some duality results. Working Paper, School of Industrial and Systems Engineering, Georgia Institute of Technology, Atlanta, USA, 1994.

[186] J.J. Moré. Recent developments in algorithms and software for trust region methods. In A. Bachem et al., editor, *Mathematical Programming. The State of the Art*, pages 258–287. Springer Verlag, Berlin, Germany, 1983.

[187] J.J. Moré and D.C. Sorensen. Newton's method. In G.H. Golub, editor, *Studies in Numerical Analysis*, pages 29–82. The Mathematical Association of America, 1984.

[188] W. Murray. Analytical expressions for the eigenvalues and eigenvectors of the hessian matrices of barrier and penalty functions. *Journal of Optimization Theory and Applications*, 7:189–196, 1971.

[189] W.C. Mylander, R.L. Holmes, and G.P. McCormick. A guide to SUMT-Version 4: the computer program implementing the sequential unconstrained minimization technique for nonlinear programming. Research Paper RAC-P-63, Research Analysis Corporation, McLean, USA, 1971.

[190] J.L. Nazareth. A reformulation of the central path equations and its algorithmic implications. Working Paper, Dept. of Pure and Applied Mathematics, Washington State University, Pullman, WA, USA, 1993.

[191] Y. Nesterov. Long–step strategies in interior point potential reduction methods. Technical Report, University of Geneva, Geneva, Switzerland, 1993.

[192] Y. Nesterov. Complexity estimates of some cutting plane methods based on the analytic barrier. *Mathematical Programming*, 69:149–176, 1995.

[193] Y. Nesterov and A.S. Nemirovskii. *Interior point polynomial algorithms in convex programming.* SIAM Studies in Applied Mathematics, Vol. 13. SIAM, Philadelphia, USA, 1994.

[194] Y. Nesterov and M.J. Todd. Self–scaled barriers and interior–point methods for convex programming. Technical Report 1091, School of OR and IE, Cornell University, Ithaca, New York, USA, 1994. (To appear in Mathematics of Operations Research).

[195] J. Von Neumann and O. Morgenstern. *Theory of games and economic behavior.* Princeton University Press, Princeton, New Jersey, 2nd edition, 1947.

[196] Optimization Subroutine Library, Guide and References. IBM Corporation, Kingston, USA, 1991.

[197] M.W. Padberg. Solution of a nonlinear programming problem arising in the projective method for linear programming. Technical Report, School of Business and Administration, New York University, New York, NY, USA, 1985.

[198] J.S. Pang. Complementarity problems. In R. Horst and P.M. Pardalos, editors, *Handbook of Global Optimization*, pages 271–338. Kluwer Academic Press, The Netherlands, 1995.

[199] J.S. Pang and S.A. Gabriel. NE/SQP: A robust algorithm for the nonlinear complementarity problem. *Mathematical Programming*, 60:295–337, 1993.

[200] S. Poljak, F. Rendl, and H. Wolkowicz. A recipe for semidefinite relaxation for $(0,1)$–quadratic programming. *Global Optimization*, 7:51–73, 1995.

[201] F.A. Potra. An $\mathcal{O}(\sqrt{n}L)$ infeasible–interior–point algorithm for LCP with quadratic convergence. Reports on Computational Mathematics 50, Dept. of Mathematics, The University of Iowa, Iowa City, USA, 1994.

[202] F.A. Potra and Y. Ye. Interior point methods for nonlinear complementarity problems. Reports on Computational Mathematics 15, Dept. of Mathematics, The University of Iowa, Iowa City, USA, 1991. (To appear in Journal of Optimization Theory and Applications).

[203] F.A. Potra and Y. Ye. An interior point algorithm for solving entropy optimization problems with globally linear and locally quadratic convergence rate. *SIAM Journal on Optimization*, 3:843–860, 1993.

[204] F. Pukelsheim. *Optimal design of experiments.* John Wiley and Sons, New York, 1993.

[205] M. Ramana. An exact duality theory for semidefinite programming and its complexity implications. DIMACS Technical Report 95–02, RUTCOR, Rutgers University, New Brunswick, New Jersey, USA, 1995.

[206] J. Renegar. A polynomial–time algorithm, based on Newton's method, for linear programming. *Mathematical Programming*, 40:59–93, 1988.

[207] R.T. Rockafellar. *Convex analysis.* Princeton University Press, Princeton, New Jersey, 1970.

[208] C. Roos and D. den Hertog. A polynomial method of weighted centers for linear programming. Technical Report 89–13, Faculty of Technical Mathematics and Computer Science, Delft University of Technology, Delft, The Netherlands, 1989.

[209] C. Roos and J.-Ph. Vial. Long steps with the logarithmic penalty barrier function in linear programming. In J. Gabszevwicz, J.-F. Richard, and L. Wolsey, editors, *Economic Decision–Making: Games, Economics and Optimization, dedicated to Jacques H. Drèze*, pages 433–441. Elsevier Science Publisher B.V., Amsterdam, The Netherlands, 1990.

[210] C. Roos and J.-Ph. Vial. A polynomial method of approximate centers for linear programming. *Mathematical Programming*, 54:295–305, 1992.

[211] T.J. Van Roy. Cross decomposition for mixed integer progamming. *Mathematical Programming*, 25:46–63, 1983.

[212] D.S. Rubin and H.M. Wagner. Shadow prices: tips and traps for managers and instructors. *Interfaces*, 20:150–157, 1990.

[213] L. Schrage. *User's manual linear, integer, and quadratic programming with LINDO*. Scientific Press, Palo Alto, USA, 2nd edition, 1985.

[214] A. Schrijver. *Theory of linear and integer programming*. John Wiley & Sons, New York, 1986.

[215] D.F. Shanno. Computational experience with logarithmic barrier methods for linear and nonlinear complementarity problems. Technical Report RRR 18-93, Rutgers Center for Operations Research, Rutgers University, New Brunswick, USA, 1993.

[216] R. Sharda. Linear programming software for personal computers: 1995 survey. *OR/MS Today*, pages 49–57, October 1995.

[217] N.Z. Shor. Quadratic optimization problems. *Soviet Journal of Computer and System Sciences*, 25:1–11, 1987.

[218] G. Sonnevend. An "analytic center" for polyhedrons and new classes of global algorithms for linear (smooth, convex) programming. In A. Prekopa, J. Szelezsan, and B. Strazicky, editors, *System Modelling and Optimization: Proceedings of the 12th IFIP-Conference held in Budapest, Hungary, September 1985*, volume 84 of *Lecture Notes in Control and Information Sciences*, pages 866–876. Springer Verlag, Berlin, Germany, 1986.

[219] D.C. Sörensen. Newton's method with a model trust region modification. *SIAM Journal on Numerical Analysis*, 19:409–426, 1982.

[220] R.J. Stern and H. Wolkowicz. Indefinite trust region subproblems and nonsymmetric eigenvalue perturbations. Report CORR 92-38, University of Waterloo, Department of Combinatorics and Optimization, Waterloo, Canada, 1992.

[221] G.W. Stewart. On scaled projections and pseudoinverses. *Linear Algebra and its Applications*, 112:189–193, 1989.

[222] J. Stoer and C. Witzgall. *Convexity and optimization in finite dimensions I.* Springer Verlag, Berlin, 1970.

[223] G. Strang. *Linear algebra and its applications.* Harcourt Brace Jovanovich, Orlando, USA, 1988.

[224] J.F. Sturm and S. Zhang. An $\mathcal{O}(\sqrt{n}L)$ iteration bound primal–dual cone affine scaling algorithm. *Mathematical Programming*, 72:177–194, 1996.

[225] J. Sun, J. Zhu, and G. Zhao. A predictor–corrector algorithm for a class of nonlinear saddle point problems. Technical Report, National University of Singapore, Singapore, 1994.

[226] K. Tanabe. Center flattening transformation and a centered Newton method for linear programming. Technical Report, The Institute of Statistical Mathematics, Minami Azabu, Minato–ku, Tokyo, Japan, 1987.

[227] T. Terlaky. A convergent criss–cross method. *Math. Oper. and Stat. ser. Optimization*, 16:683–690, 1985.

[228] T. Terlaky. On ℓ_p–programming. *European Journal of Operational Research*, 22:70–100, 1985.

[229] T. Terlaky and J.-Ph. Vial. Computing maximum likelihood estimators of convex density functions. Technical Report 95–49, Faculty of Technical Mathematics and Computer Science, Delft University of Technology, Delft, The Netherlands, 1995.

[230] T. Terlaky and S. Zhang. Pivot rules for linear programming: a survey on recent theoretical developments. *Annals of Operations Research*, 46:203–233, 1993.

[231] M.J. Todd and Y. Ye. A centered projective algorithm for linear programming. *Mathematics of Operations Research*, 15:508–529, 1990.

[232] T. Tsuchiya. A new family of polynomial–time interior point algorithms for linear programming. Research Memorandum, The Institute of Statistical Mathematics, Minami–Azabu, Minato–ku, Tokyo, Japan, 1994.

[233] T. Tsuchiya and M. Muramatsu. Global convergence of the long–step affine scaling algorithm for degenerate linear programming problems. *SIAM Journal on Optimization*, 5:525–551, 1995.

[234] A.W. Tucker. Dual systems of homogeneous linear relations. In H.W. Kuhn and A.W. Tucker, editors, *Linear Inequalities and Related Systems, Annals of Mathematical Studies, No. 38*, pages 3–18. Princeton University Press, Princeton, USA, 1956.

[235] H. Väliaho. P_* matrices are just sufficient. *Linear Algebra and Its Applications*, 239:103–108, 1996.

[236] L. Vandenberghe and S.E. Boyd. Semidefinite programming. *SIAM Review*, 38:49–95, 1996.

[237] R.J. Vanderbei. LOQO: an interior point code for quadratic programming. Technical Report, Department of Civil Engineering and Operations Research, Princeton University, Princeton, New Jersey, 1995.

[238] R.J. Vanderbei, M.S. Meketon, and B.A. Freedman. A modification of Karmarkar's linear programming algorithm. *Algorithmica*, 1:395–407, 1986.

[239] S.A. Vavasis. Stable numerical algorithms for equilibrium systems. *SIAM Journal on Matrix Analysis and Applications*, 15, 1994.

[240] J.-Ph. Vial. Computational experience with a primal–dual interior–point method for smooth convex programming. *Optimization Methods and Software*, 3:285–316, 1994.

[241] J. Vöros. An explicit derivation of the efficient portfolio frontier in the case of degeneracy and general singularity. *European Journal of Operational Research*, 32:302–310, 1987.

[242] J.E. Ward and R.E. Wendell. Approaches to sensitivity analysis in linear programming. *Annals of Operations Research*, 27:3–38, 1990.

[243] G.A. Watson. Algorithms for minimum trace factor analysis. *SIAM Journal on Matrix Analysis and Applications*, 13:1039–1053, 1992.

[244] P. Wolfe. The simplex method for quadratic programming. *Econometrica*, 27:382–398, 1959.

[245] M.H. Wright. Some properties of the hessian of the logarithmic barrier function. *Mathematical Programming*, 67:265–295, 1994.

[246] S.J. Wright. A path–following infeasible–interior–point algorithm for linear complementarity problems. *Optimization Methods and Software*, 2:79–106, 1993.

[247] X. Xu, P.-F. Hung, and Y. Ye. A simplified homogeneous and self–dual linear programming algorithm and its implementation. *Annals of Operations Research*, 62:151–172, 1996.

[248] H. Yamashita. A polynomially and quadratically convergent method for linear programming. Working Paper, Mathematical Systems Institute, Inc., Tokyo, Japan, 1986.

[249] H. Yamashita. A globally convergent primal–dual interior point method for constrained optimization. Working Paper, Mathematical Systems Institute, Inc., Tokyo, Japan, 1992.

[250] Y. Ye. A new complexity result on minimization of a quadratic function over a sphere constraint. In C.A. Floudas and P.M. Pardalos, editors, *Recent advances in global optimization*, pages 19–21. Princeton University Press, Princeton, 1992.

[251] Y. Ye. An homogeneous and self–dual algorithm for LCP. Working Paper, Dept. of Management Science, University of Iowa, Iowa City, USA, 1994.

[252] Y. Ye and K.M. Anstreicher. On quadratic and $\mathcal{O}(\sqrt{n}L)$ convergence of predictor–corrector algorithms for LCP. *Mathematical Programming*, 62:537–551, 1993.

[253] Y. Ye and P.M. Pardalos. A class of linear complementarity problems solvable in polynomial time. *Linear Algebra and Its Applications*, 152:3–17, 1991.

[254] Y. Ye, M.J. Todd, and S. Mizuno. An $\mathcal{O}(\sqrt{n}L)$–iteration homogeneous and self–dual linear programming algorithm. *Mathematics of Operations Research*, 19:53–67, 1994.

[255] Y. Zhang. LIPSOL - a MATLAB toolkit for linear programming interior–point solvers. Technical Report, Department of Mathematics and Statistics, University of Maryland at Baltimore County, Baltimore, USA, 1994.

[256] Y. Zhang. On the convergence of a class of infeasible interior–point algorithms for the horizontal complementarity problem. *SIAM Journal on Optimization*, 4:208–227, 1994.

[257] Y. Zhang and D. Zhang. On polynomiality of the Mehrotra–type predictor–corrector interior–point algorithms. *Mathematical Programming*, 68:303–318, 1995.

[258] J. Zhu. A path following algorithm for a class of convex programming problems. *Zeitschrift für Operations Research*, 36:359–377, 1992.

INDEX

Applied Optimization

1. D.-Z. Du and D.F. Hsu (eds.): *Combinatorial Network Theory.* 1996
ISBN 0-7923-3777-8
2. M.J. Panik: *Linear Programming: Mathematics, Theory and Algorithms.* 1996
ISBN 0-7923-3782-4
3. R.B. Kearfott and V. Kreinovich (eds.): *Applications of Interval Computations.* 1996
ISBN 0-7923-3847-2
4. N. Hritonenko and Y. Yatsenko: *Modeling and Optimimization of the Lifetime of Technology.* 1996
ISBN 0-7923-4014-0
5. T. Terlaky (ed.): *Interior Point Methods of Mathematical Programming.* 1996
ISBN 0-7923-4201-1
6. B. Jansen: *Interior Point Techniques in Optimization.* Complementary, Sensitivity and Algorithms. 1997
ISBN 0-7923-4430-8

KLUWER ACADEMIC PUBLISHERS – DORDRECHT / BOSTON / LONDON

MIX
Papier aus verantwortungsvollen Quellen
Paper from responsible sources
FSC® C105338

If you have any concerns about our products,
you can contact us on
ProductSafety@springernature.com

In case Publisher is established outside the EU,
the EU authorized representative is:
Springer Nature Customer Service Center GmbH
Europaplatz 3, 69115 Heidelberg, Germany

Printed by Libri Plureos GmbH
in Hamburg, Germany